ADVANCES IN CHEMICAL ENGINEERING

Volume 18

ADVANCES IN
CHEMICAL ENGINEERING

Editor-in-Chief

JAMES WEI

School of Engineering and Applied Science
Princeton University
Princeton, New Jersey

Editors

JOHN L. ANDERSON

Department of Chemical Engineering
Carnegie Mellon University
Pittsburgh, Pennsylvania

KENNETH B. BISCHOFF

Department of Chemical Engineering
University of Delaware
Newark, Delaware

JOHN H. SEINFELD

Department of Chemical Engineering
California Institute of Technology
Pasadena, California

Volume 18

ACADEMIC PRESS, INC.
Harcourt Brace Jovanovich, Publishers

Boston   San Diego   New York
London   Sydney   Tokyo   Toronto

This book is printed on acid-free paper. ∞

Copyright © 1992 by Academic Press, Inc.

All rights reserved.
No part of this publication may be reproduced or
transmitted in any form or by any means, electronic
or mechanical, including photocopy, recording, or
any information storage and retrieval system, without
permission in writing from the publisher.

ACADEMIC PRESS, INC.
1250 Sixth Avenue, San Diego, CA 92101-4311

United Kingdom Edition published by
ACADEMIC PRESS LIMITED
24-28 Oval Road, London NW1 7DX

Library of Congress Catalog Card Number: 56-6600
International Standard Book Number: 0-12-008518-6

Printed in the United States of America
92 93 94 95 EB 9 8 7 6 5 4 3 2 1

CONTENTS

CONTRIBUTORS vii
PREFACE ix

Microchemical Engineering: The Physics and Chemistry of the Microparticle

E. James Davis

I.	Introduction.	1
II.	Particle Traps	3
III.	Light Scattering	32
IV.	Mass and Heat Transfer	55
V.	Gas/Microparticle Chemical Reactions	81
VI.	Concluding Comments	88
	Acknowledgment	88
	Notation	89
	References	90

Detailed Chemical Kinetic Modeling: Chemical Reaction Engineering of the Future

Selim M. Senkan

I.	Introduction and Overview	95
II.	Computational Quantum Chemistry	101
III.	Thermochemistry	111
IV.	Estimation of Thermochemistry by Conventional Methods .	113
V.	Estimation of Thermochemistry by Quantum Mechanics . . .	126
VI.	Rate Processes	131
VII.	Chemical Processes	132
VIII.	Estimation of Chemical Rate Parameters by Conventional Methods	134
IX.	Estimation of Rate Parameters by Quantum Mechanics . . .	152
X.	Energy-Transfer–Limited Processes	160
XI.	Heterogeneous Reactions	172
XII.	An Example: Detailed Chemical Kinetic Modeling of the Oxidation and Pyrolysis of CH_3Cl/CH_4	175
	Acknowledgments	190
	Note	190
	References	190

Optimization Strategies for Complex Process Models

LORENZ T. BIEGLER

I.	Introduction.	197
II.	Development of Newton-Type Optimization Algorithms	199
III.	Flowsheet Optimization	207
IV.	Differential-Algebraic Optimization.	216
V.	Simultaneous Approaches for Differential/Algebraic Optimization	220
VI.	Summary and Conclusions	249
	References	251

INDEX	257
CONTENTS OF VOLUMES IN THIS SERIAL	261

CONTRIBUTORS

Numbers in parentheses indicate the pages on which the authors' contributions begin.

LORENZ T. BIEGLER, *Department of Chemical Engineering, Carnegie Mellon University, Pittsburgh, Pennsylvania 15213* (197)

E. JAMES DAVIS, *Department of Chemical Engineering, BF-10, University of Washington, Seattle, Washington 98195* (1)

SELIM M. SENKAN, *Department of Chemical Engineering, University of California at Los Angeles, Los Angeles, California 90024* (95)

PREFACE

Volume 18 of *Advances in Chemical Engineering* returns to the dominant twin themes of modern chemical engineering: the microscale of molecules and small particles, and the macroscale of entire engineering systems, of analysis and synthesis.

James Davis is the inventor of the levitation machine, with which a single aerosol particle can be suspended in mid-air in order to study its equilibrium and rate processes without resorting to averaging among many particles. He contributes a very strong chapter on "Microchemical Engineering" that involves chemical reactions, transport processes, thermodynamics and physical processes.

Selim Senkan is noted for his work in environmental engineering, and particularly for his work in the reaction rates of chlorinated hydrocarbons. He writes in "Detailed Chemical Kinetic Mechanisms" on the impact of efficient numerical algorithms and computational quantum mechanics on the prediction of reaction mechanisms and rates.

Lorenz Biegler tackles the ambitious and comprehensive problem of modeling and optimization of complex process models, and features the simultaneous or Newton-type optimization strategies.

Together, these three chapters provide an exciting panorama of vital new frontiers in chemical engineering.

James Wei

MICROCHEMICAL ENGINEERING: THE PHYSICS AND CHEMISTRY OF THE MICROPARTICLE

E. James Davis

Department of Chemical Engineering
University of Washington
Seattle, Washington

I.	Introduction	1
II.	Particle Traps	3
	A. The Electrostatic Balance	3
	B. Electrodynamic Traps	5
	C. Optical Traps	29
III.	Light Scattering	32
	A. Introduction	32
	B. Elastic Scattering	32
	C. Inelastic Scattering	47
IV.	Mass and Heat Transfer	55
	A. Evaporation/Condensation Processes	55
	B. Particle Heating by Electromagnetic Energy Transfer	75
V.	Gas/Microparticle Chemical Reactions	81
	A. Optical Resonance Spectroscopy	81
	B. Raman Spectroscopy	84
VI.	Concluding Comments	88
	Acknowledgment	88
	Notation	89
	References	90

I. Introduction

There has long been an interest in micrometer and submicrometer-size aerocolloidal particles because of their undesirable features as air pollutants, and much of the research on such particles during the past few decades has been stimulated by these objectionable aspects of microparticles. Although pigments such at TiO_2 particles have been produced via aerosol technology for many years, a recent surge of interest in the production of fine particles for ceramics and other applications has led to a great diversity of research on micrometer and nanometer-size particles. Furthermore, new tools have been developed for the measurement of the optical properties of microscopic particles, for the determination of their physical and physicochemical properties, and for the study of the unusual forces on such particles.

A recent example of the interest in small particles was provided by the National Research Council's document entitled *Frontiers in Chemical Engineering* (1988). In the production of optical fibers, a modified chemical vapor deposition process is used to build up layers of glass by deposition of fine particles on the tube wall. One process, shown in Fig. 1, generates SiO_2 and GeO_2 aerosol particles that are driven to the tube wall by thermophoretic forces. The oxides are formed by bubbling oxygen through $SiCl_4$ and $GeCl_4$. The vapors oxidize, condense, and coagulate to produce aerosol particles 0.1–0.3 μm in diameter. The layered deposit of SiO_2 and GeO_2 is annealed by the torch to form a pore-free vitreous layer. This application involves not only the generation of fine particles, but also their controlled transport to a surface via thermophoresis. The thermophoretic force has also been used in clean rooms where electronic chips are produced to inhibit the deposition of small particles on the chip surfaces (Cooper, 1986; Liu and Ahn, 1987). The thermophoretic force, which arises when a microparticle exists in a gas with a large temperature gradient, results from the net momentum transport to the particle due to collisions by gas molecules of different energies on opposite sides of the particle.

Another application of microparticle technology is the production of polymeric microspheres, which are usually produced by emulsion polymerization techniques. But a variety of polymer colloids can be made by aerosol techniques (Partch et al., 1983; Nakamura et al., 1984; Partch et al., 1985). One advantage of the aerosol route is that larger sizes can be attained

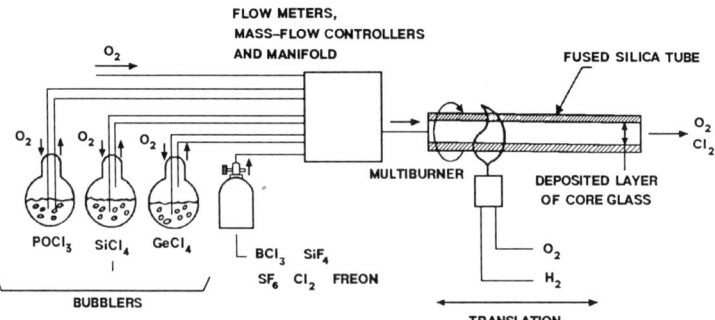

FIG. 1. A modified vapor deposition process for the production of fiber optic rods. (From National Research Council *Frontiers in Chemical Engineering*, 1988).

while limiting the polydispersity of the product. Metal oxides can also be produced via aerosol processes by hydrolysis of their alkoxides (Ingebrethsen and Matijevic, 1984). Ultrafine spheres of titanium, silicon, and aluminum oxides (Okuyama *et al.*, 1986) and mixed titanium/aluminum oxides (Ingebrethsen *et al.*, 1983) have been prepared in this way.

Other applications of microparticles include spray drying, stack gas scrubbing, particle and droplet combustion, catalytic conversion of gases, fog formation, and nucleation. The removal of SO_2 formed in the combustion of high-sulfur coal can be accomplished by adding limestone to coal in a fluidized bed combustor. The formation of CaO leads to the reaction

$$CaO + \tfrac{1}{2}O_2 + SO_2 \to CaSO_4, \qquad (1)$$

which eliminates the SO_2. A problem with this process is that the pores of the CaO particles become plugged with $CaSO_4$ as the reaction proceeds, thereby reducing effective utilization of the CaO.

The production of fine particles that are either desirable (polymer colloids, ceramic precursors, etc.) or undesirable (soot, condensed matter from stack gases, etc.) involves chemical reactions, transport processes, thermodynamics, and physical processes of concern to the chemical engineer. The optimization and control of such processes and the assurance of the quality of the product requires an understanding of the fundamentals of microparticles.

The development of the electrodynamic balance and other particle traps has made it possible to perform precise measurements of the properties of small particles by focusing on the single particle. The variety of processes and phenomena that can be investigated with particle traps is quite extensive and includes gas/liquid and gas/solid chemical reactions, chemical spectroscopies, heat and mass transfer processes, interfacial phenomena, thermodynamic properties, phoretic forces, and other topics of interest to chemical engineers.

It is the purpose of this survey to review the principles of electrical and optical traps, to explore the numerous applications of the devices that have been examined in recent years, and to indicate the use of the instruments for chemical engineering research.

II. Particle Traps

A. THE ELECTROSTATIC BALANCE

The classical electrostatic balance is the Millikan condenser, which was first used to measure the charge on the electron in the famous Millikan oil-drop experiment. In principle, the device can be used to levitate a small charged mass by using the electrical field generated by two flat plates to balance the gravitational force on the particle. If V_{dc} is the dc electrical potential between two large plates separated by distance z_0, the electrical field strength is V_{dc}/z_0, and the electrostatic force on the particle of mass m is the product of the charge, q, on the particle and the electrical field strength. If this force balances gravity, one may write

$$-q\frac{V_{dc}}{z_0} = mg. \qquad (2)$$

If we neglect edge effects at the outer edges of the electrodes, the electrical field is everywhere uniform in the space between the electrodes. As a result, the particle can be balanced at any point in the region between the plates. Because the field is uniform, there is no restoring force to return the particle to its original position if it is perturbed by convection or other forces. Fletcher (1911) overcame the transverse drift of the particle by inserting a smaller disk in the upper electrode. When the particle drifted off center, he applied a higher potential to the small disk to produce a sufficient radial restoring force to move the particle back to the center. Vertical stability of the particle is another matter, and that was not achieved in Millikan's work. That requires some sort of feedback control that senses the vertical position of the particle, then applies a restoring force when it is perturbed from the reference position.

Wyatt and Phillips (1972) commercialized an electrostatic balance with electro-optic feedback control and light-scattering capabilities; the device is shown in Fig. 2. The dc field was generated by a lower electrode and an electrified pin mounted in the upper plate. The particle was illuminated by a laser beam, and light scattered from the particle hit the edge of an adjustable

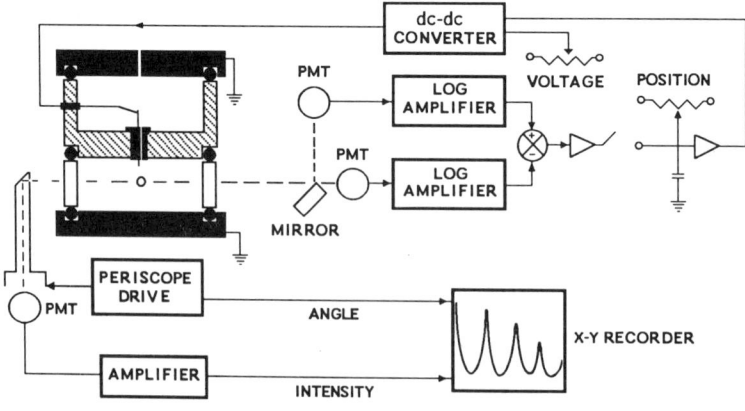

FIG. 2. The light-scattering photometer of Wyatt and Phillips (1972).

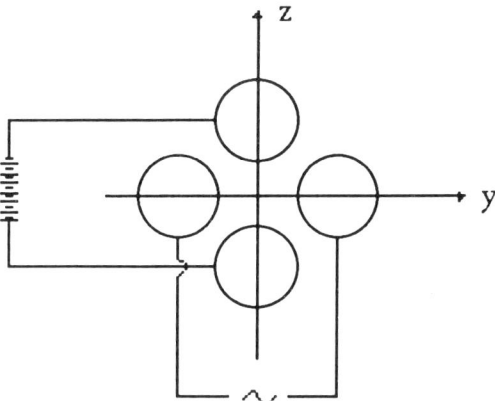

FIG. 3. The electrode configuration of the electric mass filter of Paul and Raether (1955).

mirror to split the beam. Two photomultiplier tubes (PMT) detected the light and controlled the particle position by providing a corrective potential to the pin when the particle moved above or below the midplane of the chamber. A third PMT was moved round the chamber by a stepper motor to record the intensity of the scattered light as a function of the scattering angle in the horizontal plane. This PMT output and the voltage applied to the stepper motor were sent to an $x-y$ plotter to record the angular light-scattering data. A window in the chamber permitted the scattered light to be detected.

The Wyatt and Phillips instrument has the features of primary importance for subsequent electrical levitators—specifically, feedback control for vertical positioning and scattered light detection. However, the restoring forces exerted on the particle were not adequate for a number of applications. Slight convective currents in the chamber would cause the particle to be lost, so it was not possible to provide flow through the device. The electrodynamic balance does not have this difficulty.

B. ELECTRODYNAMIC TRAPS

The electrodynamic balance or quadrupole began with the *electric mass filter* of Paul and Raether (1955), who demonstrated that a charged mass can be stably levitated by means of a quadrupole arrangement of electrodes, one of the simplest configurations being that shown in Fig. 3. An ac potential is applied to the rods in the horizontal plane, and a dc potential is applied to the upper and lower rods as in the Millikan condenser. The ac field has horizontal and vertical components that are 180° out of phase, thereby producing a stable oscillation of the particle, provided that the ac amplitude is not too large. The ac field produces no time-average vertical force on the mass, so the dc field is used to balance the gravitational force. When all time-invariant vertical forces are balanced, the particle can be maintained motionless at the center of the balance.

The electric mass filter is the basis for the electrodynamic trap used for studies of the spectroscopy of atomic ions that earned Paul and Dehmelt the 1989 Nobel Prize in Physics. A wide variety of electrode configurations can be used to trap particles, and a particularly simple design was proposed by Straubel (1956). His dc electrodes were flat plates, and the ac electrode was a simple torus or washer placed at the midplane between the endplates.

Wuerker et al. (1959) introduced bihyperboloidal electrodes, which have been widely used. Figure 4 shows the bihyperboloidal configuration used by Davis and his coworkers (Davis, 1987; Zhang and Davis, 1987; Taflin and Davis, 1987; Taflin et al., 1989; Taflin and Davis, 1990) for a variety of studies involving liquid droplets. The apparatus is equipped with a PID (proportional, integral, and differential) feedback control system that uses a dual photodiode in much the same way that Wyatt and Phillips used a pair of PMTs to position the particle. Integral and derivative control functions were added to permit the device to be operated automatically when the particle mass and/or charge change during an experiment. Gas can be injected into the balance chamber as a laminar jet through a hole in the bottom electrode, and it exits through a larger hole in the top electrode.

A PMT mounted at right angles to the incident laser beam permits continuous recording of the light scattered at that angle. In addition, a 512-element linear photodiode array (not shown) is mounted on the ring electrode

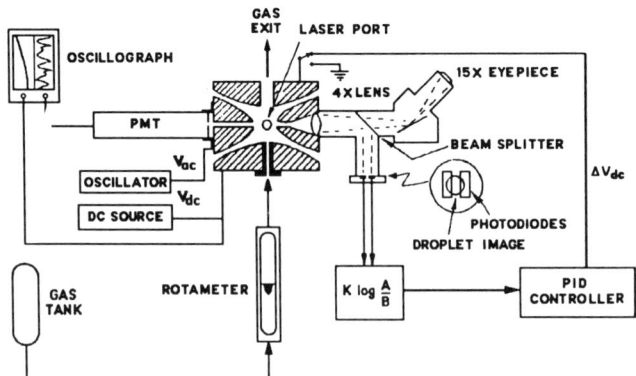

FIG. 4. The electrodynamic balance of Fulton (1985) with the bihyperboloidal configuration of Wuerker et al. (1959).

to obtain angular scattering data in the horizontal plane. The ac frequency can be varied from 20 to 2,000 Hz, and the maximum peak-to-peak ac voltage is 5,000 V. The dc voltage can be varied from 0 to ± 325 V.

The electrodes of the quadrupole of Wuerker et al. are described by the equations

$$2r^2 + z^2 = 2z_0^2 \quad \text{(endcap electrodes)} \tag{3}$$

and

$$2r^2 + z^2 = -2z_0^2 \quad \text{(ring electrode),} \tag{4}$$

where $2z_0$ is the minimum distance between the endcaps in the absence of any holes in them. For Fulton's balance $z_0 = 13$ mm.

The operating parameters and characteristics of the trap are governed by the dynamics of the particle, and it is this stability theory that we examine next. One advantage of the bihyperboloidal configuration is that its theory is well understood (Wuerker et al., 1959; Frickel et al. 1978; and Davis, 1985).

1. *Stability Theory*

The stability of the levitated particle is governed by the vector equation of motion,

$$m\frac{d\mathbf{x}^2}{dt^2} = -m\mathbf{g} + q\mathbf{E}_{ac} + q\mathbf{E}_{dc} - \mathbf{F}_d + \mathbf{F}_p, \tag{5}$$

where $\mathbf{x}$ is the position of the mass m relative to the center of the balance, q is the charge on the particle, $\mathbf{E}_{ac}$ and $\mathbf{E}_{dc}$ are the ac and dc electrical fields, respectively, $\mathbf{F}_d$ is the viscous drag force, and $\mathbf{F}_p$ is any other force exerted on the particle. Other forces might include thermophoretic, photophoretic, or diffusiophoretic forces, or radiation pressure. The thermophoretic force, mentioned above, results from an external temperature gradient, and the analogous diffusiophoretic force arises when there is a concentration gradient in the surrounding gas mixture. The photophoretic force results from nonuniform internal electromagnetic heating of the particle, which is discussed further below. The high intensity of a laser beam can exert a significant radiation pressure on a microparticle because of the momentum of the photons.

For a spherical particle with radius a moving at low Reynolds number, the drag force is Stokesian,

$$\mathbf{F}_d = 6\pi a\mu \frac{d\mathbf{x}}{dt}, \tag{6}$$

where μ is the viscosity of the surrounding fluid. As shown below, Eq. (6) is valid for both solid and liquid spheres because of the lack of internal circulation in the latter.

Let the potential applied to the ring electrode be given by

$$\phi_{ac} = V_b + V_{ac} \cos \omega \tau, \tag{7}$$

where V_b is a dc bias voltage, usually zero; V_{ac} is the amplitude of the ac voltage; t is time; $\omega = 2\pi f$; and f is the frequency of the applied ac potential. A bias voltage can be used to alter the stability characteristics of the system.

Suppose that the particle is negatively charged, that $+V_{dc}$ is the dc potential applied to the upper electrode, and that $-V_{dc}$ is applied to the lower endcap. In the absence of a charged particle, the ac and dc potentials satisfy Laplace's equation,

$$\nabla^2 \phi = 0, \tag{8}$$

and the appropriate boundary conditions. The electrical fields then are given by

$$\mathbf{E} = -\nabla \phi. \tag{9}$$

It is assumed that the field experienced by the charged particle satisfies Eq. (8) and (9). Here ∇ is the vector operator in cylindrical coordinates for an axisymmetric field.

The ac field has the following solutions to first order in r and z:

$$E_{ac,r} = -(V_b + V_{ac} \cos \omega t) \frac{r}{2z_0^2} \tag{10}$$

and

$$E_{ac,z} = (V_b + V_{ac} \cos \omega t) \frac{z}{z_0^2}. \tag{11}$$

Note that the radial and vertical components are out of phase, and that the coefficient multiplying r is only half that multiplying z. Thus, the effect of the ac field is to exert an oscillatory force on the particle with an effective field strength in the vertical direction that is twice the radial field strength. As a result of the larger field strength in the z-direction, the onset of instability is governed by the z-component of the equation of motion, so we need examine only that component.

An approximation for the dc field was developed by Davis (1985), and in the neighborhood of the center of the balance we may write

$$E_{dc,z} = -C_0 \frac{V_{dc}}{z_0}, \tag{12}$$

where C_0 is a geometrical constant. Davis's analytical approximation yielded

$C_0 = 0.8768$ for the configuration described by Eqs. (3) and (4). We note that $C_0 = 1$ for infinite parallel planes. Numerical solutions by Philip et al. (1983) and by Sloane and Elmoursi (1987) yielded $C_0 = 0.80$. The latter investigators performed numerical computations of the balance constant for several electrode shapes.

Using these results for the axial components of the electrical fields, the z-direction equation of motion becomes

$$m\frac{d^2z}{dt^2} = -mg - qC_0\frac{V_{dc}}{z_0} + q\frac{(V_b + V_{ac}\cos\omega t)}{z_0^2}z - 6\pi a\mu\frac{dz}{dt} + F_{p,z}. \quad (13)$$

Equation (13) has a nonoscillatory solution only if its inhomogeneous terms vanish—that is, the dc field must be adjusted to give

$$-mg - qC_0\frac{V_{dc}}{z_0} + F_{p,z} = 0. \quad (14)$$

Thus, in the absence of phoretic and other external forces, the dc voltage must satisfy

$$qC_0\frac{V_{dc}}{z_0} = -mg. \quad (15)$$

This is the basis for the use of the electrodynamic levitator as an analytical balance, for the dc voltage is directly proportional to the mass.

Equation (14) represents a necessary but not sufficient condition for particle stability. We must examine the homogeneous form of Eq. (13) to determine the full stability criteria. To this purpose we introduce the transformations

$$\tau = \frac{\omega t}{2} \quad \text{and} \quad Z = \frac{z}{z_0}\exp(\delta\tau/2), \quad (16)$$

where δ is a nondimensional drag parameter defined by $\delta = 6\pi a\mu/m\omega$.

The equation of motion may be transformed and rearranged to yield Mathieu's equation, which has been studied extensively to establish its stability characteristics (see Abramowitz and Stegun, 1964):

$$\frac{d^2Z}{dt^2} - \beta\left(\frac{\delta^2}{4\beta} + \frac{V_b}{V_{ac}} + \cos 2\tau\right)Z = 0, \quad (17)$$

where β is a nondimensional field strength parameter defined by

$$\beta = \frac{4V_{ac}q}{m\omega^2 z_0^2}. \quad (18)$$

Particle stability is governed by the three parameters, β, δ, and V_b/V_{ac}. Note that since the latter ratio appears as an addition to the ratio $\delta^2/4\beta$ in Eq. (17),

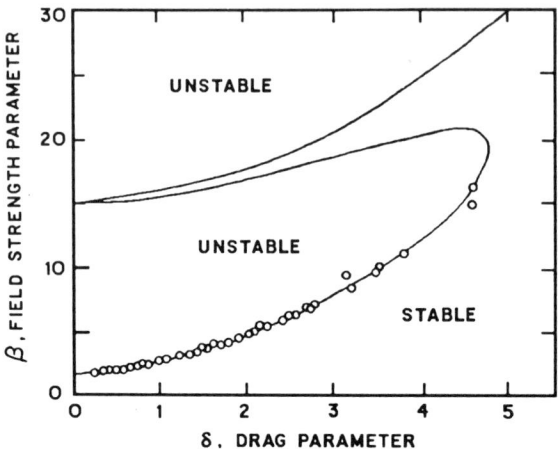

FIG. 5. The first two marginal stability envelopes for $V_b/V_{ac} = 0$ for the bihyperboloidal electrodynamic balance. Also shown as open circles are the experimental stability data of Taflin et al. (1989). Reprinted, in part, with permission from Taflin, D. C., Ward, T. L., and Davis, E. J., Langmuir 5, 376–384, Copyright © 1989 American Chemical Society.

its effect is the same as increasing or decreasing $\delta^2/4\beta$, depending on whether V_b is positive or negative. The solution of the stability equation is presented in Fig. 5 for $V_b = 0$ as a set of marginal stability envelopes that delineate the stable and unstable regions. There exists an infinite set of these envelopes, but only the first two are shown in the figure.

The marginal stability envelopes are shifted when a bias voltage is applied, and recently Iwamoto et al. (1991) prepared a number of stability maps showing the effects of bias voltage. They solved the governing equations numerically.

The critical ac field strength depends on δ, but the minimum critical field strength, which corresponds to $\delta = 0$ is given by $\beta_c = 0.908$. Thus, a particle can be stably levitated provided that $\beta < \beta_c = 0.908$ or, in terms of the ac voltage and frequency, stable trapping occurs provided that

$$V_{ac} < 8.962 \frac{mf^2 z_0^2}{q}. \tag{19}$$

Inequality (19) also represents a rule of thumb for choosing the frequency to be used. For example, if we set $V_{ac} = 1{,}000\,\text{V}$ and take $z_0 = 10\,\text{mm}$, the minimum frequency required to trap a single electron ($q/m = 1.759 \times 10^{11}\,\text{C/kg}$) is 443 MHz. To trap a 1 μm diameter water droplet with 100 units of elementary charge ($q/m = 2.94\,\text{C/kg}$), the minimum frequency is 1.8 kHz based on inequality (19). In fact, micrometer particles can be trapped using even lower frequencies because of the viscous drag, which increases stability. For multimicrometer particles, 60 Hz is usually adequate, which means that a simple transformer can be used for the ac source. In the study of atomic and molecular ions, Dehmelt and his coworkers (Dehmelt and Major, 1962; Richardson *et al.*, 1968; Dehmelt and Walls, 1968) used radiofrequency quadrupole traps.

When the critical voltage is exceeded, the levitated particle suddenly begins to oscillate vigorously and can be lost by hitting an electrode. Stability can be recovered either by reducing V_{ac} or by increasing the frequency. The well-defined onset of instability has been called the *spring point*, and the critical voltage is often referred to as the *spring point voltage*.

The stability characteristics of electrodynamic balances with electrode configurations other than bihyperboloidal are affected by the different electrode geometry. Once the electrical fields are determined for the geometry in question, the stability characteristics can be established quantitatively. Müller (1960) developed expressions for the electrical fields for several configurations, including Straubel's disk and torus system, and Davis *et al.* (1990) analyzed the double-ring which is discussed below.

2. Determination of the Balance Constant, C_0

The geometrical constant is affected by holes drilled in the balance for viewing ports, particle injection, and detectors. In principle, numerical computations can be performed to determine how the electrical fields are perturbed by such additions, but such computations have not been carried out very extensively. A balance can be calibrated by applying the stability theory outlined in the previous section. The procedure, developed by Davis (1985), involves direct observation and measurement of the spring point voltage for a microsphere of known size and density, for example, by injecting a polystyrene latex sphere into the balance chamber. When the size is unknown it is readily established by light-scattering measurements, which can be used to determine the size of a sphere with very great precision.

The particle is first levitated stably in the absence of convective forces or phoretic forces by adjusting the ac and dc voltages and the ac frequency to keep it stationary at the midplane of the balance, as observed through a microscope or video imaging system. The ratio q/m is written in terms of the measured value of V_{dc}, applying Eq. (15), and the result is used to eliminate q/m in Eq. (19). The result is

$$\beta = -\frac{1}{C_0}\frac{V_{ac}}{V_{dc}}\frac{g}{\pi^2 f^2 z_0}. \quad (20)$$

Either V_{ac} or f is varied to reach the spring point, and the critical values of

V_{ac} and f are then recorded. Presumably the size and density are known, so the drag parameter can be written as

$$\delta = \frac{9}{2\pi} \frac{\mu}{a^2 \rho f}, \tag{21}$$

where ρ is the particle density and a its radius. Thus, δ is calculated from the measured critical frequency.

Now assuming a numerical value for C_0 and computing β, the point δ, β should fall on the lowest marginal stability curve of Fig. 5. If not, C_0 must be corrected appropriately. Taflin *et al.* (1989) used this technique to determine the balance constant for a device similar to Fulton's, and their results are presented in Fig. 5 as the open circles. They determined the balance constant to be 0.79 with a standard deviation of 0.020, which is in good agreement with the numerical estimates of Philip *et al.* and Sloane and Elmoursi. The data are seen to follow the marginal stability curve very well.

3. Basic Charge and Mass Measurements

From Eq. (15) it is clear that the charge-to-mass ratio can be determined from the measured dc levitation voltage once C_0 is known. Independent measurement of either charge or mass is required if absolute values of both quantities are needed, but in many applications only the change in one of the quantities is required. For example, the rate of a gas/particle reaction can be followed by recording the levitation voltage. Taflin and Davis (1990) reported such data for the reaction between bromine vapor and 1-octadecene (OCT) to form 1,2-dibromooctadecane (DBO), and Rubel and Gentry (1984a) examined the reaction between a phosphoric acid droplet and ammonia vapor in this manner. These applications are examined below.

In the simplest type of experiment involving a change of mass with no charge loss or gain, the fractional change in mass is easily obtained. Let $V(0)$ be the initial voltage required to levitate mass m_0, and let $V(t)$ be the levitation voltage at any point in time during the experiment when the mass is m. Using Eq. (15), the fractional change of mass is given by

$$\frac{(m - m_0)}{m_0} = \frac{V(t)}{V(0)} - 1. \tag{22}$$

Rubel and Gentry showed that the extent of the reaction, ξ, for the NH_3/H_3PO_4 reaction can be written in terms of the ratio $V(t)/V(0)$ as follows:

$$\xi = \frac{M_P \left[\frac{V(t)}{V(0)} - 1 \right]}{M_A f + M_P(f - 1)}, \tag{23}$$

where the extent of the reaction is defined as the ratio of the number of moles of NH_3 absorbed by the droplet to the initial number of moles of H_3PO_4, and f is the ratio of the mass of water to that of acid. The authors assumed that during the reaction, f remained constant at its initial value at time $t = 0$. Here M_P and M_A are the molecular weights of H_3PO_4 and NH_3, respectively. Samples of their results are presented in Fig. 6 for two droplets of approximately 60 μm diameter. It is seen that with a gas-phase ammonia partial pressure of 11.5 Pa, the data follow a gas-phase diffusion-controlled model for times up to about 25 s, but at the higher concentration of 100 Pa, the results deviate from diffusion-controlled reaction after about 5 s. In both cases gas-phase diffusion does not control the rate at later times, for then simultaneous internal diffusion and chemical reaction govern the rate.

It should be pointed out that in the experiments of Rubel and Gentry, and in many other applications, there is no charge loss encountered during evaporation, condensation, or chemical reaction processes. The absence of charge loss is readily established by independently measuring the droplet size by light-scattering, then calculating the mass from the volume and density of the droplet. Since the charge-to-mass ratio is obtained from the dc voltage, the constancy of the charge can be verified. In the first study of electrodynamic levitation of evaporating droplets, Davis and Ray (1980), working with 1 μm dibutyl sebacate droplets, showed that a droplet with 77 elementary charges retained that charge as evaporation proceeded. Another droplet had 28 elementary charges. The loss of even one charge would have been easily detected in those experiments.

There are important applications when charge-loss is either desired or is unavoidable. One example is that of a radioactive microparticle or a particle exposed to a radioactive gas. Davis et al. (1988) and Ward et al. (1989) showed

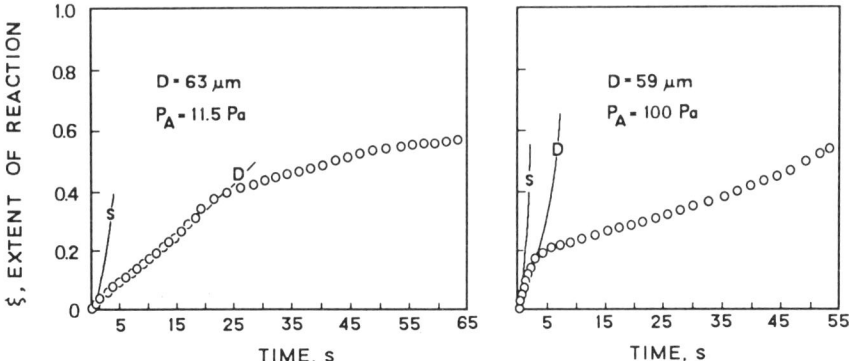

FIG. 6. Chemical reaction rate data for NH_3 reacting with H_3PO_4 solution droplets, from Rubel and Gentry (1984a). The data are compared with theory for surface reaction control (S) and gas-phase diffusion control (D). Reprinted with permission from J. Aerosol Sci. 15, 661–671, Rubel, G. O., and Gentry, J. W., Copyright © 1984, Pergamon Press plc.

that the electrodynamic balance can be used to detect radioactivity by using the charged microparticle as an ion collector. Early in his studies of the charge on oil drops, Millikan (1935) observed that the charge could be altered in the presence of ionizing radiation and by illuminating the particle with UV light.

Recognizing that there is a threshold wavelength for photoionization, Pope (1962a, 1962b) used a Millikan condenser to measure this threshold energy by exposing crystalline particles to UV light from a monochromator that used a 1,000 W Hg–xenon lamp as the source. At a wavelength of approximately 250 nm, photoionization of Cu phthalocyanine crystals occurred, resulting in charge loss. This led Arnold (1979) to develop a technique for determining the absolute charge and mass of a small particle. He used a Millikan chamber with a feedback controller similar to those shown in Fig. 2 and Fig. 4 to levitate a particle, and he illuminated it with a UV source to emit electrons one at a time. Philip et al. (1983) used this *electron-stepping* technique with an electrodynamic balance. A description of the method follows.

Let V_n be the dc voltage required to balance a mass m with a charge q_n equivalent to n electrons ($q_n = ne$, where e is the charge on the electron), and let V_{n-1} be the required voltage when the charge is reduced to $q_{n-1} = (n-1)e$ by emission of one electron. From Eq. (15) we may write

$$q_n = -\frac{mgz_0}{C_0 V_n} \tag{24}$$

and

$$q_{n-1} = -\frac{mgz_0}{C_0 V_{n-1}}. \tag{25}$$

Substracting Eq. (25) from Eq. (24), we have

$$q_n - q_{n-1} = e = \frac{mgz_0}{C_0}\left(\frac{1}{V_{n-1}} - \frac{1}{V_n}\right). \tag{26}$$

Solving for the mass, there results

$$m = \frac{C_0 e V_n V_{n-1}}{gz_0(V_n - V_{n-1})}, \tag{27}$$

where e is considered to be positive here, and $(V_n - V_{n-1}) < 0$. The initial charge, then, is calculated from Eq. (24) once m is determined.

Philip et al. demonstrated that the mass of a polyvinyltoluene microsphere with a nominal diameter of 2.35 μm was 6.979 ± 0.177 pg (1 pg = 10^{-12} g). The microspheres examined had charges equivalent to 39–71 elementary charges. This procedure fails, however, when the number of charges is large, for the voltage change associated with the loss of one electron is then too small to be measured precisely.

Such is the case in the radioactivity experiments of Ward et al. and Davis et al. They introduced radioactive particles with initial charges equivalent to 10^5-10^6 elementary charges into an electrodynamic balance. The particles, solid dotriacontane and low volatility liquid dibutyl phthalate, were labeled with ^{14}C. Emission of beta particles due to radioactive decay led to ionization of the gas in the chamber (helium, nitrogen, or air). Ions of opposite charge were then attracted to the levitated particle, thereby neutralizing it. The rate of charge loss of the microparticle is a measure of the radioactivity of the sample. The rate of charge loss can be determined from the measured levitation voltage using Eq. (15), which indicates that the charge is inversely proportional to the dc voltage. Solving for q and differentiating, there results

$$\frac{dq}{dt} = \frac{mgz_0}{C_0 V_{dc}^2} \frac{dV_{dc}}{dt}. \tag{28}$$

Significant charge loss also occurred when a nonradioactive microdroplet was injected in a ^{14}C-contaminated balance. Figure 7 presents results obtained by Davis et al. for a droplet of dioctyl phthalate in a balance that had been contaminated by prior experiments. The voltage rapidly increased as the droplet charge was neutralized, and within two minutes the charge decreased from about 165,000 elementary charges to about 10,000.

Quantitative interpretation of the charge-loss data to relate it to the level of radioactivity requires analysis of the ion motion in the quadrupole, for the

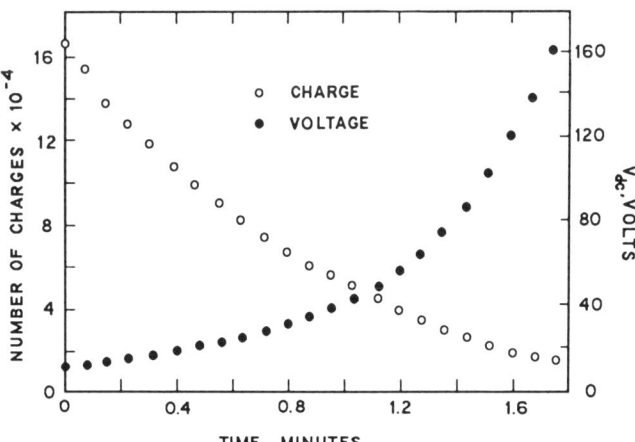

FIG. 7. Droplet charge and levitation voltage for radioactivity detection in a ^{14}C-contaminated electrodynamic balance, from Davis et al. (1988).

electrodes compete with the microparticle in the charge-capture process. Ward and Davis (1990) performed such an analysis for the bihyperboloidal device, and they demonstrated that decay rates of 200 decays per minute are easily measured. They worked with positively charged droplets as well as negatively charged droplets to determine if the mobility of the diffusing ions affected the results. From their work it appears feasible to measure the low level of one decay per minute.

4. *Drag Force Measurements*

When there is gas flow upward through the balance chamber, an aerodynamic drag force is exerted on the suspended mass. If the particle Reynolds number (Re = $2aU_\infty/\nu$) is sufficiently small so that creeping flow may be assumed, Eq. (14) may be written to include the aerodynamic drag, and it becomes

$$-mg - qC_0 \frac{V_{dc}}{z_0} + 6\pi a\mu U_\infty = 0, \quad (29)$$

where U_∞ is the velocity upstream of the particle along the chamber axis. In their studies of the aerodynamic drag and aerodynamic size of microparticles Davis and Periasamy (1985) and Davis *et al.* (1987) introduced gas into an electrodynamic balance chamber as a laminar jet through a hole in the bottom electrode. By varying the gas flow rate they varied the drag force on the particle.

The drag force is determined by two simple measurements. First the particle is levitated with no flow through the chamber. Let V_0 be the voltage required to suspend the particle when $U_\infty = 0$ with no forces other than gravity acting on the mass. Rearranging Eq. (15), we have

$$\frac{V_0}{mg} = -\frac{z_0}{qC_0}. \quad (30)$$

Now let V be the voltage required to maintain the particle at the null point of the chamber when there is a steady upward flow with velocity U_∞. Substituting Eq. (30) in Eq. (29) to eliminate the geometrical constants and the charge, there results

$$F_{\text{drag}} = 6\pi a\mu U_\infty = mg\left(1 - \frac{V}{V_0}\right). \quad (31)$$

Thus, the ratio of the drag force to the particle weight is simply related to the voltage ratio V/V_0. An independent determination of the particle mass, such as by electron-stepping, permits the absolute value of the drag force to be determined.

If the particle is a sphere of known density ρ_1, light-scattering data can be used to obtain the radius, and then the mass is given by

$$m = \frac{4}{3}\pi a^3 \rho_1. \quad (32)$$

In fact, this provides a method for calibrating the laminar jet to determine U_∞ as a function of the metered gas flow rate. Substituting Eq. (32) in Eq. (31), and solving for U_∞, we have

$$U_\infty = \frac{2}{9}\frac{a^2 g \rho_1}{\mu}\left(1 - \frac{V}{V_0}\right). \quad (33)$$

The calibration procedure, then, is to measure voltage V for a given volumetric flow rate and calculate U_∞ from the measured radius and knowledge of the gas viscosity. Note that when the flow rate is set such that no dc voltage is required to levitate the particle—that is, when the aerodynamic drag balances the weight—Eq. (33) can be used to determine the aerodynamic size, provided that the density is known.

Figure 8 shows drag force data obtained by Davis et al. (1987) for an evaporating droplet of α-pinene, a moderately volatile terpene (boiling point 156.2°C). The graph presents the ratio of the drag force to the weight as a function of nitrogen velocity. The droplet size was measured simultaneously by recording the angular scattering intensity, and the initial diameter was 29.8 μm. The solid line in the figure is the F_d/mg ratio predicted on the assumption of a quasi-steady process—that is, at any point in time the drag force is that for a nonevaporating solid sphere. Except for the data point

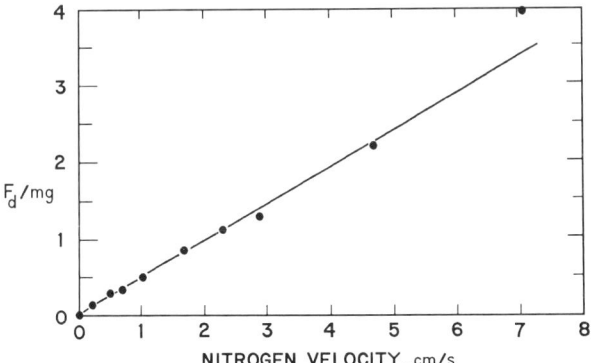

FIG. 8. The ratio of the drag force to the weight of an α-pinene droplet with initial diameter 29.8 μm evaporating in nitrogen at 293 K. The solid line is the prediction based on Stokes' law for the drag force on a sphere, assuming a quasi-steady process.

corresponding to a velocity of 7.2 cm/s, at which the droplet became unstable in the laminar jet, the quasi-steady approximation appears to be valid.

There are three factors that would tend to cause the drag data of Fig. 8 to deviate from Stokes's law. The first is internal circulation; however, based on the equation due to Hadamard (1911) and Rybczynski (1911), one would not expect motion within the droplet. The drag force determined by Hadamard and Rybczynski is

$$F_d = 2\pi a \mu_2 \frac{(2\mu_2 + 3\mu_1)}{(\mu_2 + \mu_1)}, \tag{34}$$

where μ_1 and μ_2 are the viscosities of the liquid and the surrounding gas, respectively. Since $\mu_1 \gg \mu_2$ in the experiments, Eq. (34) reduces to Stokes's result. The second factor is the mass injected into the flow around the sphere due to evaporation. For the very high evaporation rates associated with burning drops, Eisenklam et al. (1966) found the effects on drag to be very significant. Chuchottaworn et al. (1983) performed numerical computations of the effects of injection and suction on flow around a sphere, finding that injection decreases the drag coefficient. The effects are particularly significant for $Re \gg 1$. For the pinene experiments $Re < 0.2$, and apparently the injection rate due to evaporation was not sufficiently large to alter the drag force. Thirdly, distortion of the droplet from a sphere would alter the drag characteristics, but even the slightest distortion from spherical would have a large effect on the light scattering. Comparison of experimental data with Mie theory indicates that no deviation from spherical occurred. Light-scattering measurements and theory are examined below.

The aerodynamic size of a micrometer particle was determined by Sageev et al. (1986) based on terminal velocity measurements. Using an electrodynamic balance, they balanced a particle with dc potential V_0, then changed the voltage to a new value V while simultaneously turning off the ac field. The particle velocity was then determined by measuring the time required for it to move about 1 mm. After that the ac field was activated, and the particle was rebalanced with the dc potential. The particle movement was measured by means of a photomultiplier tube (PMT) triggered to record the light scattered by the particle every millisecond as it moved past a mask placed over a reticle located in the space between the PMT and a light-gathering lens. For a polystyrene latex sphere with a nominal diameter of 9.6 μm, based on the supplier's information, they obtained a diameter of 9.96 μm using the measured fall time and distance. Rearranging Eq. (33), the aerodynamic radius is given by

$$a = \sqrt{\frac{9\mu U_\infty}{2\rho g} \frac{V_0}{(V_0 - V)}}. \tag{35}$$

Using the laminar jet technique to obtain aerodynamic sizes for mixed silicate particles, Davis and Periasamy (1985) measured aerodynamic radii for dozens of particles ranging in diameters from about 1 μm to 40 μm. Many of the particles had melted and had become spheres due to exposure to high temperatures in a solar furnace, and Table I lists a sample of results for spherical particles. The average absolute difference between the two measurements for the silicate particles is 11.3%, but when the two methods are applied to droplets the difference is much less. Scanning electron micrographs of the silicate particles indicated that many of the silicate particles deviated from perfect spherical shape, which would affect the light scattering significantly. The large difference associated with sample number 69C in the table is probably due to deviations from spherical shape.

5. *The Rayleigh Limit of Charge*

Equation (29) offers an alternate method of determining the charge of an evaporating droplet. Using Eq. (32) in Eq. (29), and rearranging the result, one obtains

$$a^2 = \left(\frac{3qC_0}{8\pi g \rho_1 z_0}\right)\frac{V_{dc}}{a} + \frac{9}{2}\frac{\mu_2 U_\infty}{\rho_1 g}. \quad (36)$$

Thus, a plot of a^2 versus V_{dc}/a for an evaporating droplet should yield a

TABLE I

AERODYNAMIC RADII AND LIGHT-SCATTERING RADII MEASURED FOR MIXED SILICATE PARTICLES BY DAVIS AND PERIASAMY (1985)[a]

Sample Number	Aerodynamic Radius, μm	Light-Scattering Radius, μm
57A	3.78	3.32
57E	3.39	3.56
57G	10.0	9.87
60B	1.56	1.79
60C	3.51	3.76
69C	3.17	5.27
69E	1.49	1.48
69F	6.03	6.24
104A	1.43	1.55
104D	3.96	3.76
107C	5.80	6.00
138A	6.10	5.78
138B	3.26	3.90
138C	1.47	1.32

[a] Reprinted with permission from Davis, E. J., and Periasamy, R., *Langmuir* 1, 373, Copyright © 1985 American Chemical Society.

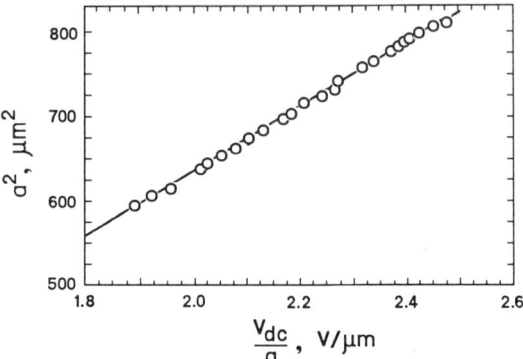

FIG. 9. Droplet radii data for 1-dodecanol from which the charge and convective velocity data are determined (from Bridges, 1990).

straight line with a slope proportional to the surface charge and an intercept proportional to the gas velocity in the chamber. Figure 9 shows such a graph obtained by Bridges (1990) for 1-dodecanol evaporating in air. The radii were obtained by analysis of light-scattering data. The data are seen to lie on a straight line from which extrapolation to $V_{dc}/a = 0$ yields the desired intercept and measurement of the slope permits one to determine the droplet surface charge.

There is a limit to the charge that the surface of a droplet can sustain, for the electrical stress generated by the surface charges can balance the surface tension forces. At that point the droplet becomes unstable and breaks up. Lord Rayleigh (1882) analyzed the criterion of instability for a conducting droplet using spherical harmonics to describe the modes of oscillation. The natural frequency of the nth mode of oscillation of a droplet was found to be given by

$$f_n = \frac{1}{2\pi} \sqrt{\frac{n(n-1)}{\rho_1 a^3}\left[(n+2)\sigma_{lg} - \frac{q^2}{16\pi^2\varepsilon_0 a^3}\right]}. \tag{37}$$

Here σ_{lg} is the surface tension, and ε_0 is the permittivity of free space. The mode $n = 0$ corresponds to the equilibrium sphere, and $n = 1$ is a purely translational mode. The first unstable mode is $n = 2$. The critical charge q_c for this mode is given by setting Eq. (37) to zero, which yields the Rayleigh limit of charge,

$$q_c = 8\pi\sqrt{\varepsilon_0 \sigma_{lg} a^3}. \tag{38}$$

A number of investigators attempted to verify the Rayleigh limit using Millikan condensers and quadrupoles (Doyle et al., 1964; Abbas and Latham, 1967; Berg et al., 1970), but these studies were not sufficiently precise because of the relatively crude methods of determining the droplet size. Using an electrodynamic balance of the Straubel type, Ataman and Hanson (1969) and Schweizer and Hanson (1971) reported confirmation of the Rayleigh limit to within $\pm 4\%$. There is some question of their method of calibrating the balance. Rhim et al. (1987a, 1987b) used a hybrid electrostatic–acoustic technique to levitate charged droplets of order 1 mm, and they found that the drops burst prior to reaching the Rayleigh limit.

Taflin et al. (1989) reviewed the experimental studies of the Rayleigh limit and made an extensive study of droplet explosions using electrodynamic levitation for a variety of hydrocarbons. They followed the explosions via optical resonance measurements, which are discussed further below. Such measurements clearly show the droplet fission and permit the size change and charge change to be determined with good precision. The dc voltage trace and a comparison between experimental and theoretical resonance spectra are provided in Fig. 10 for a droplet of dodecanol evaporating in nitrogen at room temperature. The voltage trace tracks the decrease in droplet mass as evaporation proceeds at constant surface charge up to the point of droplet explosion. There is then a sharp increase in the dc levitation voltage as charge and mass are lost. This occurred when the light-scattering size

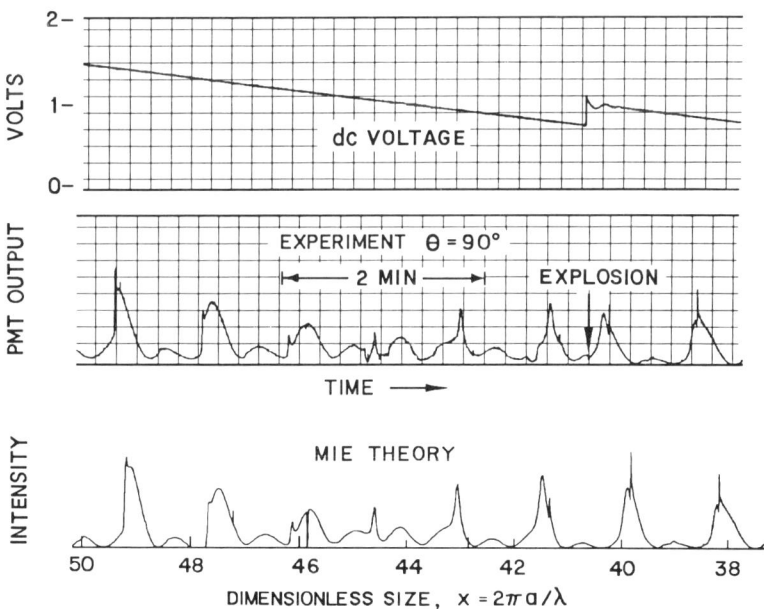

FIG. 10. The dc voltage trace and resonance spectra for an evaporating droplet of dodecanol, showing the effect of explosion on the resonance spectrum.

($x = 2\pi a/\lambda$, where λ is the wavelength of light) was approximately 40.5 ($a = 4.08\ \mu$m). Prior to and subsequent to the explosion, the experimental and theoretical resonance spectra are in good agreement, but there is a discontinuity in size, which occurs at $x = 40.5$. Note that the distance between the resonance peaks at $x = 39.8$ and $x = 41.7$ in the theoretical and experimental spectra do not match owing to the change in droplet size associated with the explosion.

Table II summarizes the droplet explosion data of Taflin et al. The table presents the charge q_- and radius a_- prior to explosion, and the fractional changes in the charge $\Delta q/q_-$ and mass $\Delta m/m_-$. The ratio q_-/q_c of the charge prior to explosion to that calculated using Eq. (38) for the Rayleigh limit is also listed. For the data in Table II, the average charge loss was approximately 15% and the average mass loss was about 1.5%. These results suggest that the explosion resulted from a localized increase in charge, which produced a protuberance that was subsequently expelled from the surface, carrying away relatively little mass but a significant amount of surface charge. Bridges extended this work in an attempt to measure the surface tension via droplet fission measurements using an additional ac field of higher frequency than the

TABLE II
CHARGE LOSS AND MASS LOSS DATA ASSOCIATED WITH THE EXPLOSION OF LEVITATED DROPLETS OF SEVERAL HYDROCARBON COMPOUNDS FROM TAFLIN et al. (1989)[a]

Compound[b]	a_-, μm	$q_- \times 10^{13}$, C	$\Delta m_-/m_-$	$\Delta q/q_-$	q_-/q_c
BDD	21.74	−9.47	—	0.12	0.715
DBO	19.237	−10.03	0.0155	0.14	0.863
	17.490	−8.63	0.0189	0.18	0.856
	15.366	−7.10	0.016	0.14	0.854
	13.861	−6.089	0.0223	0.18	0.856
DBP	9.96	3.24	—	—	0.747
DOD	17.78	8.13	0.020	0.13	0.849
	16.13	7.10	—	0.17	0.859
HXD	16.583	6.15	0.016	0.18	0.734
	14.560	5.06	0.015	0.18	0.734
	32.58	17.11	—	0.14	0.742
HPD	18.072	7.51	0.023	0.13	0.783
	16.363	6.51	0.021	0.10	0.787
	17.047	−7.09	0.0098	0.095	0.805
	15.963	−6.41	0.0150	0.14	0.805
	14.536	−5.51	0.0138	0.14	0.796

[b]BDD = bromododecane, DBO = dibromooctane, DBP = dibutyl phthalate, DOD = dodecanol, HXD = hexadecane, HPD = heptadecane.
[a]Reprinted, in part, with permission from Taflin, D. C., Ward, T. L., and Davis, E. J., *Langmuir* 5, 376–384, Copyright © 1989 American Chemical Society.

basic levitation frequency to examine the effects of an external frequency which could be in resonance with surface waves. He found no significant effect of the additional field.

Richardson et al. (1989) performed similar measurements for droplets of sulfuric acid and dioctyl phthalate (DOP) in a quadrupole. Sulfuric acid droplets exploded prior to the Rayleigh limit (at $84 \pm 20\%$ of the Rayleigh limit), and the DOP droplets fissioned approximately at the Rayleigh limit $\pm 6\%$.

The discrepancies in the droplet fission results of various investigators appear to be due to the external field that is applied to trap the droplet in an electrodynamic balance. Recently, Yang and Carleson (1990) analyzed the linear oscillations of a drop in uniform alternating electric fields, finding that the imposed electric field distrubs the interface by interacting with the induced electric charge. This interaction can lead to instability at lower charges than Rayleigh's theory predicts. Rayleigh's classical theory does not take into account the external electric field, and as a result the droplet fission experiments performed in electrodynamic balances need not conform to Rayleigh's prediction.

If the effects of the external electric field are taken into account by the analysis of Yang and Carleson, it should be possible to use droplet explosion measurements to determine the surface tension.

6. Phoretic and Radiometric Force Measurement

Numerous forces can be exerted on a microparticle, and direct measurement of the force-to-weight ratio follows that of the drag force. The examples of phoretic and radiometric forces should suffice to make the point.

When a temperature gradient is imposed on the gas surrounding an aerocolloidal particle, the particle can experience a *thermophoretic force* resulting from imbalanced momentum exchange between the gas molecules and the particle surface. The thermophoretic force acts in the direction opposite the imposed temperature gradient. The force is not usually significant in the continuum regime, but at reduced gas pressures the effect can be quite large compared with the gravitational force on the particle. In the near-continuum regime or slip-flow regime, the thermophoretic force is given by (Jacobsen and Brock, 1965)

$$\mathbf{F}_{th} = -12\pi a^2 \mu_2 \nabla T_\infty C_{tc} \mathrm{Kn} \frac{(\kappa_2/\kappa_1 + C_t \mathrm{Kn})}{(1+3C_m \mathrm{Kn})(1+2\kappa_2/\kappa_1 + 2C_t \mathrm{Kn})}, \quad (39)$$

where C_{tc}, C_t, and C_m are the thermal creep, temperature jump, and velocity slip coefficients, respectively; Kn is the Knudsen number (the ratio of the mean free path of the gas phase molecules to the particle radius); κ_1 and κ_2

are the thermal conductivities of the interior and exterior phases, respectively; and μ_2 is the surrounding fluid (gas) viscosity.

For thermal creep Maxwell (see Kennard, 1938) proposed the equation

$$C_{tc} = \frac{3}{4} \frac{v_2}{T_2 \Lambda}, \qquad (40)$$

where v_2 is the kinematic viscosity of the gas, T_2 is its temperature, and Λ is the mean free path of the gas molecules. Many other expressions have been proposed for the thermal creep coefficient, and there is considerable uncertainty about the values of the temperature jump and velocity slip coefficients.

Brock (1962) compared Eq. (39) with data available in the literature at the time, reporting $C_m = 1.23$ and $C_t = 1.94$ based on data of Rosenblatt and LaMer (1946) for tricresyl phosphate aerosol droplets in air, and $C_m = 1.42$ and $C_t = 2.57$ from Schmitt's (1959) data for silicon oil droplets in air. Jacobsen and Brock measured the thermal force on spherical NaCl particles in argon using a modified Millikan condenser. Their results, which are in the Knudsen number range $\text{Kn} \leqslant 0.2$, indicate that it is very important to include the temperature jump and slip velocity boundary conditions, but Eq. (39) was found to be in poorer agreement with their data than with an equation based on flux equations obtained from the solution of the Boltzmann equation. Jacobsen and Brock concluded that the Maxwell approximation for C_{tc} is not generally adequate when large temperature gradients are involved.

Temperature variations along the surface of a microparticle, which lead to a phoretic force, can be produced by nonuniform heat generation within the particle as well as by external temperature gradients. When a particle is illuminated by electromagnetic radiation, anisotropic heat generation can occur, which leads to a nonuniform surface temperature. The phoretic force resulting from anisotropic internal heat generation is called the *photophoretic force*. The simplest case is that of strong absorption of the incident beam energy by a small sphere of radius a illuminated on one side only by a beam of intensity I_0. If Q_{abs} is the energy absorption efficiency of the sphere, and if all of the energy is absorbed in thin layer near the surface, the heat source can be approximated as a surface heat flux q_s given by

$$q_s = \frac{1}{4} H\left(\frac{\pi}{2} - \theta\right) Q_{abs} I_0 \cos\theta, \qquad (41)$$

where H is the Heaviside function defined by

$$H\left(\frac{\pi}{2} - \theta\right) = \begin{cases} 0 \text{ for } \theta < \frac{\pi}{2} \\ 1 \text{ for } \theta > \frac{\pi}{2} \end{cases}. \qquad (42)$$

For a dielectric sphere the absorption efficiency can be computed from knowledge of the size and complex refractive index ($N = n + ik$) of the object by means of Mie theory. The complex refractive index varies with wavelength, and the imaginary component, which relates to the energy absorption, is particularly sensitive to wavelength. As a result, the heat source associated with electromagnetic energy absorption is sensitive to the wavelength of the incident radiation.

In the case of one-sided illumination of a strongly absorbing sphere, the photophoretic force for the slip-flow regime is given by (Reed, 1977)

$$F_{ph} = -\frac{2\pi a^2 \mu_2}{\kappa_1} \frac{C_{tc} \operatorname{Kn} Q_{abs} I_0}{(1 + 3C_m \operatorname{Kn})(1 + 2\kappa_2/\kappa_1 + 2C_t \operatorname{Kn})}. \tag{43}$$

In the problem formulation it was assumed that the electromagnetic radiation propagates in the negative z-direction, so the minus sign indicates that the photophoretic force acts in the direction of propagation of the radiation. In this event the photophoresis is said to be positive. As we shall discuss later, it is possible for the force to be in the opposite direction because of greater heat generation on the side of the sphere opposite the illumination. In that event, negative photophoresis is said to occur. Understanding the highly anisotropic electromagnetic heating that occurs in a dielectric sphere requires some discussion of light-scattering theory, which is provided later.

Since the photophoretic force depends on the electromagnetic absorption efficiency Q_{abs}, which is sensitive to wavelength, photophoretic force measurements can be used as a tool to study absorption spectroscopy. This was first recognized by Pope *et al.* (1979), who showed that the spectrum of the photophoretic force on a 10 μm diameter perylene crystallite agrees with the optical spectrum. This was accomplished by suspending a perylene particle in a Millikan chamber with electro-optic feedback control and measuring the photophoretic force as a function of the wavelength of the laser illumination. Improvements on the technique and additional data were obtained by Arnold and Amani (1980), and Arnold *et al.* (1980) provided further details of their *photophoretic spectrometer*. A photophoretic spectrum of a crystallite of cadmium sulfide reported by Arnold and Amani is presented in Fig. 11.

Measurements of the photophoretic force on crystalline ammonium sulfate particles were made by Lin and Campillo (1985) using an electrodynamic balance. The measurement procedure is identical to that for any such force, that is, the levitation voltage is measured in the absence of the photophoretic force and then when the force is exerted. A force balance yields an equation for the photophoretic force similar to Eq. (31):

$$\frac{F_{ph}}{mg} = -\frac{(V - V_0)}{V_0} = -\frac{\Delta V}{V_0}, \tag{44}$$

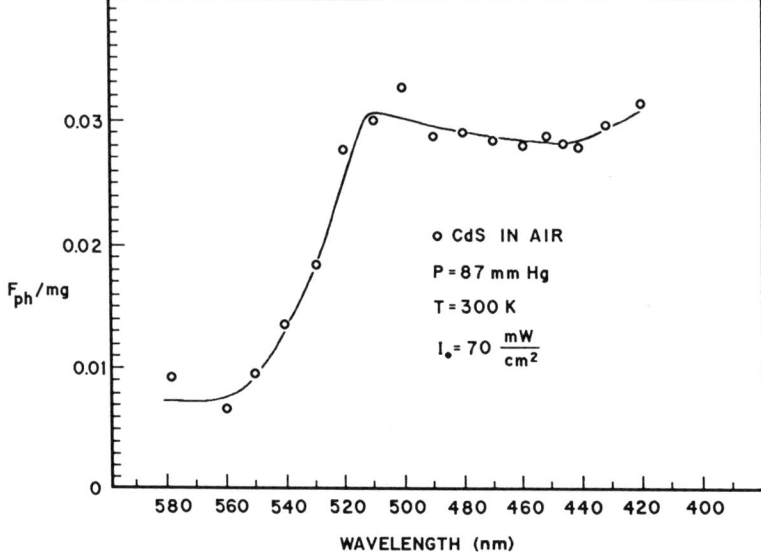

FIG. 11. A photophoretic spectrum of a CdS crystallite obtained by Arnold and Amani (1980) with a modified Millikan condenser. Reprinted with permission from Arnold, S., and Amani, Y., *Optics Letters* 5, 242–244, Copyright © 1980, The Optical Society of America.

where V_0 is the voltage required to balance the gravitational force alone, and V is the voltage required to balance the photophoretic and gravitational forces. Figure 12 presents the data of Lin and Campillo for two different microparticles of ammonium sulfate, one with a dimension of 8.4 μm, and the other, 11.7 μm. The chamber pressure was varied to examine the effects of Knudsen number on the photophoretic force. Note that for both particles the photophoretic force changed direction at lower pressures. Furthermore, for the smaller particle the photophoretic force was found to be four times the gravitational force at a pressure of 760 torr, which indicates that the photophoretic force is not an insignificant one here.

A force that is as large as the gravitational force can be used to suspend a particle against gravity, provided that it can be controlled and directed upward to balance gravity. One such force is the radiation pressure force or radiometric force. Ashkin and Dziedzic (1977), whose work is discussed in the next section, were the first to use the radiation pressure to levitate a microsphere stably. It was demonstrated by Allen et al. (1991) that the radiometric force can be measured with the electrodynamic balance, and they used the technique to determine the absolute intensity of the laser beam illuminating a suspended particle. This was accomplished in the apparatus displayed in Fig. 13. The laser illuminated the microparticle from below, and

MICROCHEMICAL ENGINEERING 27

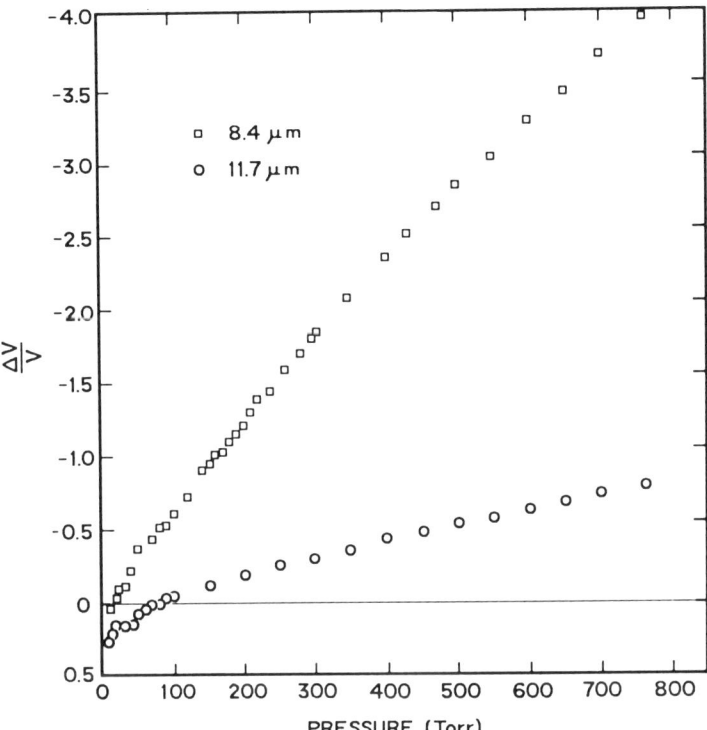

FIG. 12. Photophoretic force data of Lin and Campillo (1985) for crystalline ammonium sulfate particles levitated in an electrodynamic balance. Reprinted with permission from Lin, H.-B., and Campillo, A. J., *Applied Optics* **24**, 244, Copyright © 1985, The Optical Society of America.

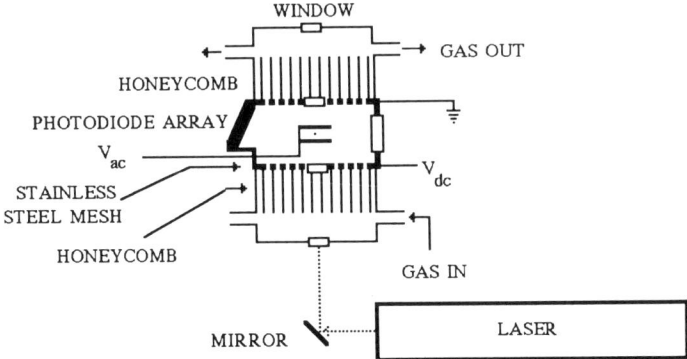

FIG. 13. The double-ring electrodynamic balance used by Allen *et al.* (1991) to measure the radiation pressure force on a microparticle.

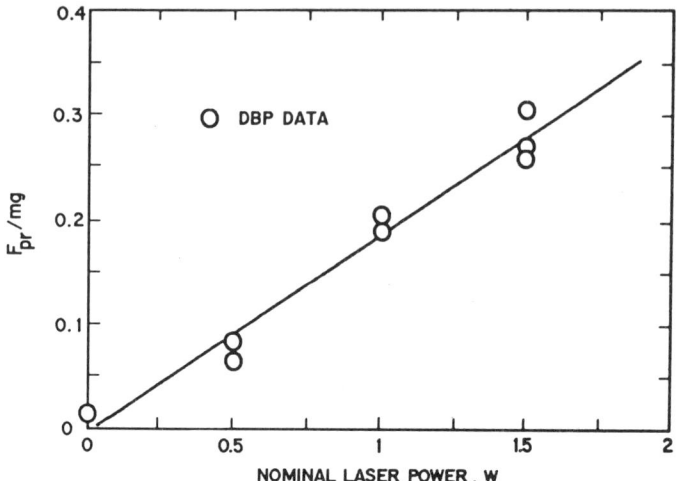

FIG. 14. The ratio of the radiometric force to the gravitational force as a function of laser power for a levitated microdroplet of dibutyl phthalate, from Buehler (1991).

the dc field was used to balance the gravitational and radiometric forces. The dc electrodes were stainless steel mesh, which permitted gas flow through the chamber, and the ac electrode consisted of two parallel rings mounted above and below the midplane of the device. A 512-element photodiode array was mounted as shown to obtain the intensity of the scattered light as a function of angle. The droplet size was determined from such light-scattering data. A large window in the chamber was used together with appropriate optics (not shown) to collect the Raman-shifted light using a dual monochromator to obtain Raman spectra.

The radiometric force was measured in the same way that Lin and Campillo determined the photophoretic force. The levitation voltage V_0 required to balance the gravitational force was first measured using very low laser power, then the laser power was increased and the voltage recorded. Equation (44) may be applied to calculate the ratio of the radiometric, F_{pr}, and gravitational forces from the measured voltages. Typical results are plotted in Fig. 14 as F_{pr}/mg versus nominal laser power. The latter was indicated by a wattmeter on the laser power supply. The droplet used for the measurements was dibutylphthlate, a relatively nonvolatile and nonabsorbing chemical, and the droplet diameter was approximately 40 μm. For such a large droplet there was insufficient laser power to balance the gravitational force radiometrically, but for smaller particles optical levitation can be achieved with modest laser power. The scatter in the data of Fig. 14 is a result of morphological resonances, which will be examined below.

C. Optical Traps

Ashkin and Dziedzic (1977) used the radiation pressure force of a laser beam to levitate microdroplets with the apparatus presented in Fig. 15. A polarized and electro-optically modulated laser beam illuminated the particle from below. The vertical position of the particle was detected using the lens and split photodiode system shown. When the particle moved up or down a difference signal was generated; then a voltage proportional to the difference and its derivative were added, and the summed signal used to control an electro-optic modulator to alter the laser beam intensity. Derivative control serves to damp particle oscillations, while the proportional control maintains the particle at the null point.

For a gaussian laser beam, lateral or transverse positioning of the particle is not a problem, for the net transverse force exerted on a microsphere centers it. This is illustrated in Fig. 16 using ray optics. When the particle moves off the centerline of the beam, a restoring force is exerted on the particle, which may be understood by considering parallel rays *a* and *b* in the figure. Assume that the beam is weakly focused, that is, the rays are nearly parallel, and that ray *a* is of greater intensity than ray *b*. Bending of the rays due to refraction produces components of the beam momentum in the axial and transverse directions.

The result is a net axial force in the direction of the electromagnetic energy propagation, and a net transverse force towards the centerline of the beam. Thus, the particle motion is opposite the gradient of the beam intensity.

When a strongly divergent beam is used to illuminate the particle, it is possible to generate a *negative light pressure* which can be used to trap a particle. This phenomenon was discovered by Ashkin *et al.* (1986), and it is illustrated in Fig. 17. In this case rays A and A, which we assume are of equal

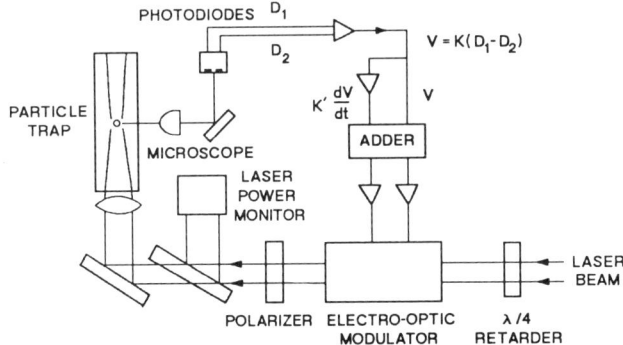

Fig. 15. The radiation pressure levitator of Ashkin and Dziedzic (1977).

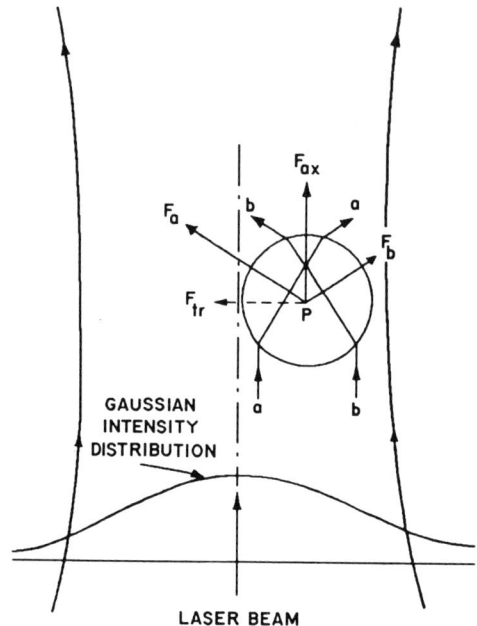

FIG. 16. The forces on a microsphere in a gaussian laser beam.

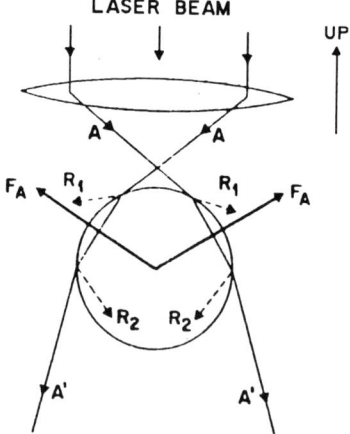

FIG. 17. The optical gradient trap of Ashkin et al. (1986). Reprinted with permission from Ashkin, A., Dziedzic, J. M., Bjorkholm, J. E., and Chu, S., *Optics Letters* **11**, 288–290, Copyright © 1986, The Optical Society of America.

intensity, have a large angle of incidence and undergo the refractions (A') and reflections (beams R_1 and R_2) indicated in the figure. This results in forces F_A whose transverse components are balanced and whose axial components act in the upward direction, a rather surprising result.

Ashkin et al. analyzed the forces acting on a microsphere in their *single-beam gradient force trap*. For a Rayleigh scatterer—that is, for a sphere small compared with the wavelength λ of light—they established criteria for axial stability. First, the gradient force F_{grad} in the direction of the intensity gradient must be greater than the scattering force F_{sca} exerted in the direction of the incident beam. They obtained the criterion

$$\frac{F_{\text{grad}}}{F_{\text{sca}}} = \frac{3\sqrt{3}}{64\pi^5}\left(\frac{m^2+2}{m^2-1}\right)\frac{N_b^2 \lambda^5}{a^3 w_0^2} \geq 1, \tag{45}$$

where N_b is the refractive index of the sphere of radius a, m is the ratio of the refractive index of the scatterer to that of the surrounding medium, and w_0 is the spot size of the focused gaussian beam.

Equation (45) represents a necessary condition for trapping, but it is also necessary to have sufficient power to generate forces larger than destabilizing forces, such as Brownian motion. As a result, the additional requirement for trapping is

$$\exp\left(\frac{N_b \alpha_p E^2}{2k_B T}\right) \gg 1, \tag{46}$$

where α_p is the polarizability of the particle, E is the incident electrical field strength, and k_B is Boltzmann's constant.

Ashkin and his coworkers used their single-beam optical trap or *optical tweezers* to suspend colloidal, aerocolloidal, and biocolloidal particles, including colloidal silica and latex particles. A particularly interesting application was explored by Ashkin and Dziedzic (1987), who trapped viruses and bacteria with the technique. A significant advantage of the device is that it is not necessary that the particle be charged, and hence it can be used with particles in aqueous media as well as in gases. To the author's knowledge, quantitative measurements of force with the trap have not been performed.

Several investigators have used light pressure to levitate microparticles since Ashkin and Dziedzic first demonstrated stable levitation. Thurn and Kiefer (1984a, 1984b, 1985), Preston et al. (1985) and Schweiger (1990b) used the method to obtain Raman spectra of microparticles as discussed later. The relatively high intensities needed to balance the gravitational force are also needed to obtain Raman spectra, so optical levitation serves two purposes when applied to such spectroscopic measurements.

III. Light Scattering

A. INTRODUCTION

A major advantage in the study of microspherical systems is the power of light-scattering measurements together with theoretical interpretation of such data to yield highly precise determination of size, refractive index, molecular bond information, and composition. One example is enough to excite the interested researcher. By illuminating a microdroplet with a laser beam and recording the scattered light intensity at right angle (or any angle) to the beam as the object changes size and/or refractive index, the size and refractive index can be determined to one part in 10^5. Such measurements make it possible to determine evaporation or condensation rates with great precision.

When a small particle is illuminated by a laser beam or some other source of electromagnetic radiation, most of the light is scattered *elastically*, that is, the wavelength of the scattered light is the same as that of the incident light. The electromagnetic radiation can also produce molecular transitions that lead to a shift in wavelength of the emitted radiation, *inelastic scattering*. If certain electronic transitions occur, the inelastic scattering phenomenon is termed *fluorescence*, and if vibrational modes are involved, there is a Raman shift in the wavelength. Elastic and inelastic scattering processes are both of great interest for the study of small particles, and some details of these processes should be examined to provide the basis for understanding the application of scattering phenomena to microchemical engineering.

Many chemical engineering researchers are not familiar with the principles of light scattering, so in the following sections the theories of elastic and inelastic scattering are outlined to make this survey reasonably self-contained. For the reader desirous of more details of elastic scattering, the monographs of Kerker (1969), Born and Wolf (1980), van de Hulst (1981), and Bohren and Huffman (1983) are recommended. We shall use the notation of Bohren and Huffman here. The Raman effect has been presented by many authors, and the works of Anderson (1971) and Koningstein (1972) provide the essential information.

B. ELASTIC SCATTERING

Lord Rayleigh (1871) recognized that the blue of the sky is the result of light scattering. It is obvious from the numerous pictures sent back from outer space that the blackness observed by astronauts is quite different from our earthly observations of a blue sky. Rayleigh developed the theory of light scattering for the case in which the scatterer is very small compared with the wavelength of light, and for unpolarized incident light with irradiance I_i he obtained

$$I_s = \frac{8\pi^4 N a^6}{\lambda^4 r^2} \left|\frac{m^2 - 1}{m^2 + 2}\right|^2 (1 + \cos^2\theta) I_i, \quad (47)$$

where I_s is the irradiance of the scattered light, N is the refractive index of the sphere of radius a, r is the distance from the center of the sphere, and θ is the scattering angle measured from the direction of propagation of the electromagnetic wave. Note that the scattered intensity is inversely proportional to the wavelength to the fourth power. Thus, the shorter-wavelength blue light can be expected to be scattered more strongly than the longer wavelengths. This result is observed as the blue of the sky.

The more general theory of light scattering by an object with a size of the order of the wavelength of light is generally credited to Mie (1908), and the solution of Debye (1909) is often cited. Kerker (1982) pointed out that earlier Lorenz (1890) analyzed the scattering of electromagnetic waves by an isotropic sphere, and he rightfully suggested that we speak of Lorenz–Mie theory rather than what is usually termed simply *Mie theory*.

1. Governing Equations

The electromagnetic fields $\mathbf{E}(\mathbf{x}, t)$ and $\mathbf{H}(\mathbf{x}, t)$ associated with scattering from a microsphere satisfy Maxwell's equations. For a homogeneous, isotropic linear material the time-harmonic electrical field $\mathbf{E}$ and the magnetic field $\mathbf{H}$ satisfy vector wave equations, which in SI units are (Bohren and Huffman, 1983)

$$\nabla^2 \mathbf{E} + k^2 \mathbf{E} = 0, \quad (48)$$

and

$$\nabla^2 \mathbf{H} + k^2 \mathbf{H} = 0, \quad (49)$$

where $k^2 = \omega^2 \varepsilon \mu, \varepsilon = \varepsilon_0(1 + \chi) + i\sigma/\omega$, ω is the angular frequency, ε_0 is the permittivity of free space, σ is the conductivity of the medium, μ is its permeability, and χ is the electric susceptibility, which measures how easily the material can be polarized by an electric field. Herein we shall consider that μ for the dielectric medium equals that *in vacuo*.

If there exist no free charges within the domain of interest, the time-harmonic electromagnetic field must also satisfy zero divergence conditions

$$\nabla \cdot \mathbf{E} = 0, \text{ and } \nabla \cdot \mathbf{H} = 0. \quad (50)$$

The electric vector and the magnetic vector are also related by

$$\nabla \times \mathbf{E} = i\omega\mu\mathbf{H}, \quad (51)$$

and

$$\mathbf{V} \times \mathbf{H} = -i\omega\varepsilon\mathbf{E}. \qquad (52)$$

Since **H** can be determined from **E** using Eq. (52), we need only write expressions for **E** in the following discussion.

The solution of the vector wave equation can be written in terms of the *generating function* ψ, which is a solution of the scalar wave equation

$$\nabla^2\psi + k^2\psi = 0. \qquad (53)$$

Let $\mathbf{M} = \mathbf{V} \times (\mathbf{r}\psi)$ and $\mathbf{N} = (\mathbf{V} \times \mathbf{M})/k$ with ψ given by Eq. (53), where **r** is the radius vector of the spherical coordinate system. Then both **M** and **N** satisfy the vector wave equation and both have zero divergence. In addition, **M** and **N** satisfy

$$\mathbf{V} \times \mathbf{N} = k\mathbf{M}. \qquad (54)$$

Thus, **M** and **N** have all of the properties required of the electromagnetic field. Furthermore, ψ satisfies the scalar wave equation in spherical coordinates,

$$\frac{1}{r^2}\frac{\partial}{\partial r}\left(r^2\frac{\partial\psi}{\partial r}\right) + \frac{1}{r^2\sin\theta}\frac{\partial}{\partial\theta}\left(\sin\theta\frac{\partial\psi}{\partial\theta}\right) + \frac{1}{r^2\sin\theta}\frac{\partial^2\psi}{\partial\phi^2} + k^2\psi = 0. \qquad (55)$$

2. *Spherical Harmonic Analysis*

Equation (55) can be solved by the conventional product method to yield the following even and odd solutions

$$\psi_{emn} = \cos m\phi P_n^m(\cos\theta)z_n(kr), \qquad (56)$$

$$\psi_{omn} = \sin m\phi P_n^m(\cos\theta)z_n(kr), \qquad (57)$$

where m and n are integers given by $m = 0, 1, 2, \ldots$ and $n = m, m+1, m+2, \ldots$.

The functions $P_n^m(\cos\theta)$ are associated Legendre functions of the first kind of degree n and order m, and $z_n(kr)$ denotes any of four spherical Bessel functions. The choice of the spherical Bessel function depends on the domain of interest, that is, on whether we are looking for the solution inside the sphere ($r < a$) or outside the sphere ($r > a$). For the internal field we choose $z_n(kr) = j_n(kr)$, where $j_n(kr)$ is the spherical Bessel function of the first kind of order n. The solution for the external field can be written in terms of spherical Bessel functions $j_n(kr)$ and $y_n(kr)$, where the latter is the spherical Bessel function of the second kind, but it is more convenient to introduce the spherical Hankel function $h_n^{(l)}(kr)$ to determine ψ for the outer field.

The vector spherical harmonics generated by the even and odd generating functions defined by Eq. (56) and (57) become

$$\mathbf{M}_{emn} = \nabla \times (\mathbf{r}\psi_{emn}), \qquad \mathbf{M}_{omn} = \nabla \times (\mathbf{r}\psi_{omn}),$$

$$\mathbf{N}_{emn} = \frac{\nabla \times \mathbf{M}_{emn}}{k}, \qquad \text{and } \mathbf{N}_{omn} = \frac{\nabla \times \mathbf{M}_{omn}}{k}.$$

The solutions of the inner and outer fields can now be written as expansions in these spherical harmonic functions or vector eigenfunctions, once the incident irradiation and the boundary conditions are specified.

Let us consider the incident beam to be a plane wave polarized with the electric vector in the x-direction as indicated in Figure 18. In polar coordinates the incident electric vector $\mathbf{E}_i$ may be written

$$\mathbf{E}_i = E_0 e^{ikr\cos\theta}(\sin\theta\cos\phi\,\mathbf{e}_r + \cos\theta\sin\phi\,\mathbf{e}_\theta - \sin\phi\,\mathbf{e}_\phi), \tag{58}$$

where $\mathbf{e}_r$, $\mathbf{e}_\theta$, and $\mathbf{e}_\phi$ are unit vectors in the r, θ, and ϕ directions, respectively. Writing $\mathbf{E}_i$ in terms of spherical harmonic functions, one obtains

$$\mathbf{E}_i = E_0 \sum_{n=1}^{\infty} i^n \frac{2n+1}{n(n+1)} (\mathbf{M}_{oln}^{(1)} - i\mathbf{N}_{eln}^{(1)}), \tag{59}$$

where the superscript (1) indicates that the r-dependent eigenfunction used is $j_n(kr)$.

The incident, scattered, and internal electric vectors are related through boundary conditions. The tangential components of the electric vectors are continuous across the surface of the sphere, that is,

$$(\mathbf{E}_i + \mathbf{E}_s - \mathbf{E}_1) \times \mathbf{e}_r = \mathbf{0}. \tag{60}$$

Here subscripts s and 1 refer to the scattered field and internal field, respectively.

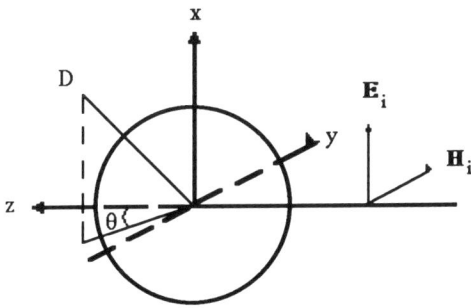

FIG. 18. The coordinate system, showing a detector at D.

For a charged surface the tangential components of the magnetic vectors must satisfy

$$(\mathbf{H}_i + \mathbf{H}_s - \mathbf{H}_1) \times \mathbf{e}_r = \mathbf{K}, \quad (61)$$

where $\mathbf{K}$ is the surface current density of free charges at the surface, which is usually zero. In the case of the electrodynamically levitated sphere there exists a surface charge, and little is known about the surface current density. Bohren and Hunt (1977) addressed the problem of scattering by a charged sphere, concluding that for a metallic sphere with a/λ small the surface charge does not affect the scattering appreciably. Experimental comparisons of theory for an uncharged sphere with data for charged spheres also indicate that the charge densities usually encountered are not sufficient to affect the scattering. Thus, it appears that one can take $\mathbf{K} = \mathbf{0}$ in the boundary condition on $\mathbf{H}$ without introducing measurable error.

Let us write the expansions of the internal electric vector $\mathbf{E}_1$ and the scattered field electric vector $\mathbf{E}_s$ in the forms

$$\mathbf{E}_1 = E_0 \sum_{n=1}^{\infty} i^n \frac{2n+1}{n(n+1)} (c_n \mathbf{M}_{o1m}^{(1)} - i d_n \mathbf{N}_{e1n}^{(1)}), \quad (62)$$

and

$$\mathbf{E}_s = E_0 \sum_{n=1}^{\infty} i^n \frac{2n+1}{n(n+1)} (i a_n \mathbf{N}_{e1n}^{(3)} - b_n \mathbf{M}_{o1n}^{(3)}). \quad (63)$$

The superscript (3) signifies that the r-dependent eigenfunction is the spherical Hankel function $h_n^{(1)}(kr)$. The vectors $\mathbf{M}_n$ and $\mathbf{N}_n$ are called the *normal modes* of the sphere.

It remains to apply the boundary conditions to obtain the scattering coefficients, a_n and b_n, and the interior field coefficients, c_n and d_n. The results, for $\mu = \mu_1$, are

$$a_n = \frac{m^2 j_n(mx)[x j_n(x)]' - j_n(x)[m x j_n(mx)]'}{m^2 j_n(mx)[x h_n^{(1)}(x)]' - h_n^{(1)}(x)[m x j_n(mx)]'}. \quad (64)$$

The primes indicate differentiation with respect to the argument of the relevant function, either x or mx.

$$b_n = \frac{j_n(mx)[x j_n(x)]' - j_n(x)[m x j_n(mx)]'}{j_n(mx)[x h_n^{(1)}(x)]' - h_n^{(1)}(x)[m x j_n(mx)]'}, \quad (65)$$

$$c_n = \frac{j_n(x)[x h_n^{(1)}(x)]' - h_n^{(1)}(x)[x j_n(x)]'}{j_n(mx)[x h_n^{(1)}(x)]' - h_n^{(1)}(x)[m x j_n(mx)]'}, \quad (66)$$

$$d_n = \frac{m j_n(x)[x j_n^{(1)}(x)]' - m h_n^{(1)}(x)[x j_n(x)]'}{m^2 j_n(mx)[x h_n^{(1)}(x)]' - h_n^{(1)}(x)[m x j_n(mx)]'}. \quad (67)$$

In these expressions x is the *size parameter* and m is the *relative refractive index*, defined by

$$x = ka = \frac{2\pi Na}{\lambda} \quad \text{and} \quad m = \frac{k_1}{k} = \frac{N_1}{N}.$$

There are two particularly important aspects of these results for the study of microparticles. The first is the angular dependence of the energy of the scattered light, which is implicit in Eq. (63), and the second is that the denominators of the scattering coefficients, which are complex numbers, can become small. If a detector is used to measure the scattered light as it is moved round the sphere in the *plane of observation*, the output of that detector plotted as a function of the scattering angle is often called the *phase function*. The frequencies for which the scattering coefficients a_n and b_n have poles are the *natural* frequencies of the sphere. There is an infinite set of poles corresponding to complex values of x or of mx, but not all of these resonances can be observed. When the imaginary part of such an x is small, the resonance is sharply defined, but the resulting peak in the scattered intensity might be too narrow to be observed. When the imaginary part is large, the peak is broad and other resonances can dominate. Many well-defined peaks can be observed with a detector as either m or x change. A graph of the detector output as a function of one of these parameters is often called the *morphological* or *structural resonance spectrum*. Phase functions and resonance spectra are extensively used to measure the properties of microparticles.

The angular dependence of a detector signal can be made explicit by writing out the vectors $\mathbf{M}_{oln}$ and $\mathbf{N}_{eln}$, that is,

$$\mathbf{M}_{o1n} = \cos\phi\, \pi_n(\cos\theta) z_n(\rho) \mathbf{e}_\theta - \sin\phi\, \tau_n(\cos\theta) z_n(\rho) \mathbf{e}_\phi, \tag{68}$$

and

$$\mathbf{N}_{e1n} = n(n+1)\cos\phi \sin\theta\, \pi_n(\cos\theta) \frac{z_n(\rho)}{\rho} \mathbf{e}_r$$

$$+ \cos\phi\, \tau_n(\cos\theta) \frac{[\rho z_n(\rho)]'}{\rho} \mathbf{e}_\theta - \sin\phi\, \pi_n(\cos\theta) \frac{[\rho z_n(\rho)]'}{\rho} \mathbf{e}_\phi. \tag{69}$$

Here $\rho = kr$, and the functions π_n and τ_n are defined by

$$\pi_n(\cos\theta) = \frac{P_n^1(\cos\theta)}{\sin\theta} \quad \text{and} \quad \tau_n(\cos\theta) = \frac{d}{d\theta}[P_n^1(\cos\theta)].$$

Note that the normal mode $\mathbf{M}_{o1n}$, which appears in the expansion of the electric vector for the scattered light, has no radial component, so it is called a *transverse electric mode*. Similarly, $\mathbf{M}_{e1n}$, which appears in the expansion of the magnetic vector, has no radial component, so it is called a *transverse magnetic mode*.

These results can be put in a more useful and simpler form if kr is sufficiently large to permit asymptotic forms of the spherical Bessel functions and spherical Hankel functions to be applied. In this case the transverse components of the scattered electric vector are

$$E_{s\theta} \cong -E_0 \frac{e^{ikr}}{ikr} \cos \phi S_2(\cos \theta), \qquad (70)$$

and

$$E_{s\phi} \cong E_0 \frac{e^{ikr}}{ikr} \sin \phi S_1(\cos \theta), \qquad (71)$$

where

$$S_1 = \sum_n \frac{2n+1}{n(n+1)} [a_n \pi_n(\cos \theta) + b_n \tau_n(\cos \theta)], \qquad (72)$$

and

$$S_2 = \sum_n \frac{2n+1}{n(n+1)} [a_n \tau_n(\cos \theta) + b_n \pi_n(\cos \theta)]. \qquad (73)$$

The summation is over a suitably large number of terms. Wiscombe (1980) developed criteria for the number of terms required, and for $8 < x < 4{,}100$ he recommended that the number of terms be $n = x + 4.05x^{1/3} + 2$.

The scattering functions S_1 and S_2 are particularly useful for interpreting experimental data. For incident light polarized *parallel* to the scattering plane, typically chosen to be the horizontal plane for which $\phi = 90°$, the ratio of the scattered irradiance to the incident irradiance is given by

$$i_\parallel = \frac{1}{k^2 r^2} |S_2|^2, \qquad (74)$$

and for incident light polarized *perpendicular* to the scattering plane the ratio is

$$i_\perp = \frac{1}{k^2 r^2} |S_1|^2. \qquad (75)$$

For unpolarized light the irradiance ratio is the average of the ratios for parallel and perpendicular polarization

$$i_u = \frac{1}{2k^2 r^2} (|S_1|^2 + |S_2|^2). \qquad (76)$$

The factor $1/k^2r^2$ is often omitted from these expressions, since it is only the angular dependence that is usually of interest.

When the size parameter x is sufficiently small, that is, when the particle is small compared with the wavelength of light, only the leading term in the normal mode expansion for the spherical harmonic functions is needed. In this case Eq. (76) reduces to Rayleigh's result, Eq. (47), for the ratio of the scattered irradiance to the incident irradiance.

Although the irradiance of the scattered light is of major interest to the experimentalist, there are several other aspects of light scattering that should be included in this review. These include *cross-sections*, radiation pressure, and electromagnetic energy absorption. The latter two phenomena can be related to cross-sections.

3. Cross-Sections, Efficiencies, and Radiation Pressure

Suppose that we place a detector in a light beam with its surface normal to the direction of electromagnetic energy propagation to record the power of the beam. If we now place a microparticle between the light source and the detector, the power measured by the detector will decrease. We say that there has been *extinction* of the incident beam. The extinction could be the result of scattering by the particle or absorption of electromagnetic radiation by the particle. The magnitude and direction of the rate of transfer of electromagnetic energy is given by the *Poynting vector* $\mathbf{S}$ defined by

$$\mathbf{S} = \mathbf{E} \times \mathbf{H}. \tag{77}$$

If we construct an imaginary sphere around the particle, the net rate at which electromagnetic energy crosses surface A of the sphere inwardly is the rate of absorption of that energy, given by

$$W_{abs} = -\int_A \mathbf{S} \cdot \mathbf{e}_r dA. \tag{78}$$

For a nonabsorbing medium surrounding the particle, the rate of extinction W_{ext} is the sum of the net rate of absorption by the particle and the rate at which energy is scattered across surface A, W_{sca}, that is,

$$W_{ext} = W_{abs} + W_{sca}. \tag{79}$$

The scattered energy is obtained from the scattered fields, Eq. (63), and the associated magnetic vector.

One can now define the *extinction cross-section* by dividing W_{ext} by the incident irradiance,

$$C_{ext} = \frac{W_{ext}}{I_i}. \tag{80}$$

Similarly, absorption and scattering cross-sections are defined to give

$$C_{ext} = C_{abs} + C_{sca}. \tag{81}$$

Efficiencies or *efficiency factors* Q are defined by dividing the cross-section by the cross-sectional area of the particle projected onto the plane perpendicular to the incident beam. For a spherical particle of radius a one writes

$$Q_{ext} = \frac{C_{ext}}{\pi a^2}, \quad Q_{sca} = \frac{C_{abs}}{\pi a^2}, \quad Q_{abs} = \frac{C_{abs}}{\pi a^2}.$$

From Eq. (81) there results

$$Q_{ext} = Q_{abs} + Q_{sca}. \tag{82}$$

In terms of the scattering coefficients the scattering and extinction cross-sections are given by

$$C_{sca} = \frac{2\pi}{k^2} \sum_{n=1}^{\infty} (2n + 1)(|a_n|^2 + |b_n|^2), \tag{83}$$

and

$$C_{ext} = \frac{2\pi}{k^2} \sum_{n=1}^{\infty} (2n + 1)\text{Re}\{a_n + b_n\}. \tag{84}$$

The incident beam has momentum as well as energy associated with it, and absorption and scattering of the beam lead to a change in the component of momentum in the direction of electromagnetic energy propagation. This results in the radiation pressure discussed above, which was used by Ashkin and Dziedzic to levitate microparticles. Debye (1909) first analyzed the radiation pressure on a microsphere, and to express it in terms of the scattering coefficients one introduces the *asymmetry parameter*, which involves the average cosine of the scattering angle. The asymmetry parameter may be written in the form of an efficiency factor,

$$Q_{sca}\langle\cos\theta\rangle = \frac{4}{x^2}\left[\sum_n \frac{n(n + 2)}{n + 1} \text{Re}\{a_n a_{n+1}^* + b_n b_{n+1}^*\} + \sum_n \frac{2n + 1}{n(n + 1)} \text{Re}\{a_n b_n^*\}\right], \tag{85}$$

where the asterisk denotes the complex conjugate of the scattering coefficient.

The efficiency for the radiation pressure Q_{pr} becomes

$$Q_{pr} = Q_{ext} - Q_{sca}\langle\cos\theta\rangle, \tag{86}$$

and the force on the microsphere produced by the radiation pressure is

$$F_{pr} = \frac{\eta a^2 I_i Q_{pr}}{C} \tag{87}$$

where C is the velocity of light.

We note that since Q_{pr} involves the scattering coefficients, the radiation pressure force has resonance or near-resonance behavior. This first was observed and analyzed by Ashkin and Dziedzic (1977) in their study of microparticle levitation by radiation pressure. They made additional measurements (Ashkin and Dziedzic, 1981) of the laser power required to levitate a microdroplet, and Fig. 19 presents their data for a silicone droplet. The morphological resonance spectrum for the 180° backscattered light shows well-defined peaks at wavelengths corresponding to frequencies close to natural frequencies of the sphere. The laser power shows the same resonance structures in reverse, that is, when the scattered intensity is high the laser power required to levitate the droplet is low.

4. *Phase Functions and Morphological Resonances*

Chylek et al. (1983) showed that, by comparing experimental resonance spectra with spectra computed using Mie theory, the size and refractive index of a microsphere can be determined to about one part in 10^5. Numerous investigators have used resonance spectra to determine the optical properties of microspheres since Ashkin and Dziedzic observed resonances. A recent example is the droplet evaporation study of Tang and Munkelwitz (1991), who measured the vapor pressures of the low-volatility species dioctyl phthalate (DOP), glycerol, oleic acid, and methanesulfonic acid (MSA). This

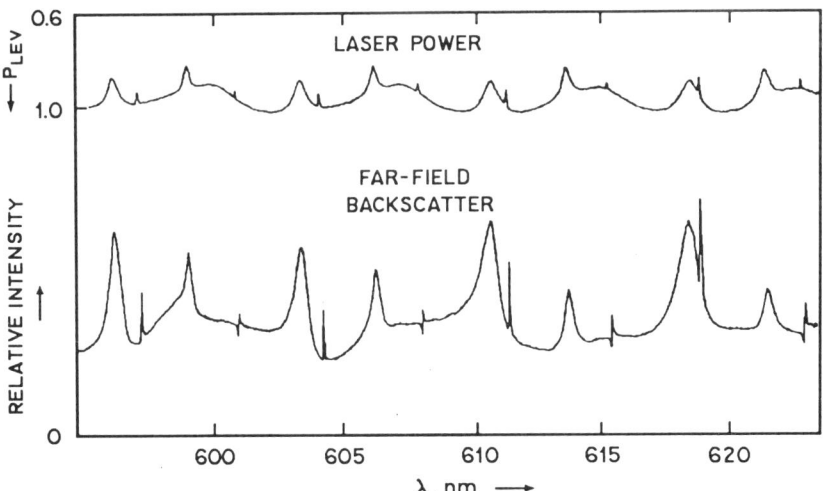

FIG. 19. The levitation power and backscatter intensity recorded by Ashkin and Dziedzic (1981) as a function of laser wavelength in nanometers for a silicone oil droplet with a diameter of 11.4 μm and $N = 1.47$. Reprinted with permission from Ashkin, A., and Dziedzic, J. M., *Applied Optics* **20**, 1803, Copyright © 1981, The Optical Society of America.

was accomplished by recording resonance spectra of evaporating electrodynamically levitated droplets. Samples of their resonance data and calculated spectra for MSA are presented in Fig. 20.

There is very good correspondence between the theoretical and experimental spectra shown in Fig. 20, and several sharp resonance peaks exist, which makes it possible to determine the size with the precision claimed by Chylek et al.

Phase functions can also be used to measure the size and refractive index of a microsphere, and they have been used by colloid scientists for many years to determine particle size. Ray et al. (1991a) showed that careful measurements of the phase function for an electrodynamically levitated microdroplet yield a fine structure that is nearly as sensitive to the optical parameters as are resonances. This is demonstrated in Fig. 21, which presents experimental and theoretical phase functions obtained by Ray and his coworkers for a droplet of dioctylphthalate. The experimental phase function is compared with two

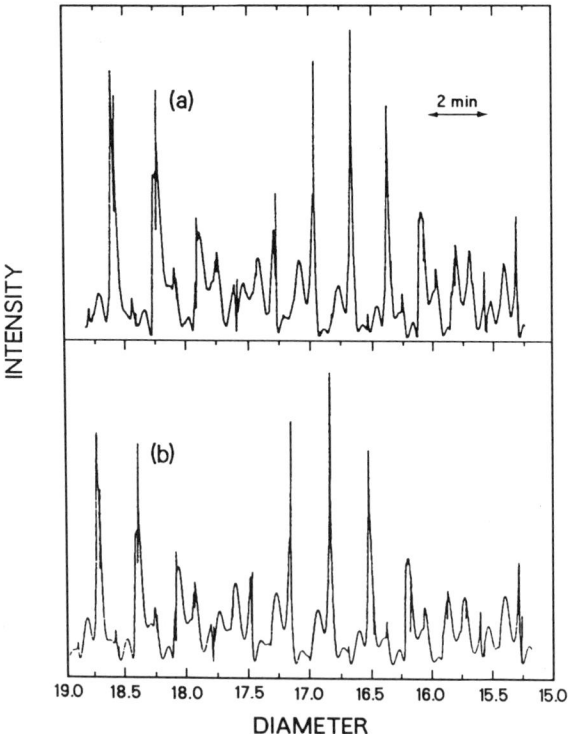

FIG. 20. (a) Experimental and (b) theoretical spectra for an evaporating droplet of methanesulfonic acid ($N = 1.430$) obtained at $\theta = 90.1°$ by Tang and Munkelwitz (1991).

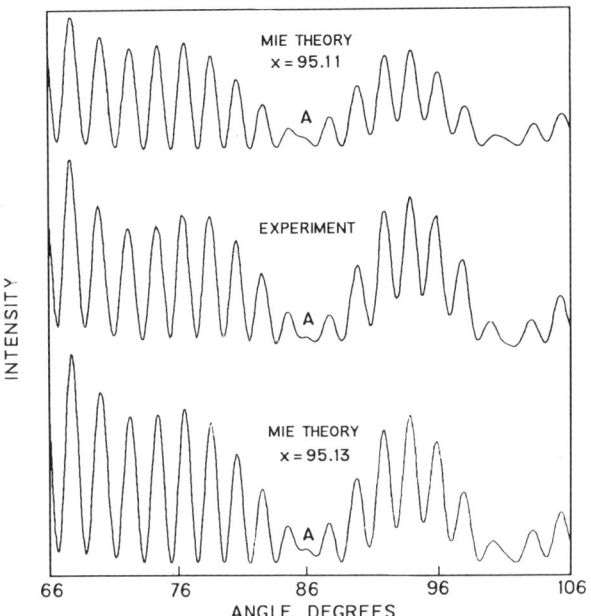

FIG. 21. Experimental and theoretical ($N = 1.4860$) phase functions for a dioctyl phthalate droplet. Reprinted with permission from Ray, A. K., Souyri, A., Davis, E. J., and Allen, T. M., *Applied Optics* **30**, Copyright © 1991, The Optical Society of America.

computed angular distributions to show the sensitivity of the phase function to the size parameter. For $x = 95.13$ the experimental and theoretical results are nearly identical, and for $x = 95.11$ there is generally good agreement between theory and experiment except for some small discrepancies, one of which is the region labeled A in the figure. These comparisons suggest that the phase function can be used to determine the size to about two parts in 10^4 for a single component droplet. The refractive index can also be determined from the data by varying both optical parameters in the computations required to match data and theory.

Volatile and/or multicomponent droplets introduce a number of complications into the interpretation of light-scattering data, a major complication being rapid size change. Furthermore, with multicomponent droplets the refractive index is likely to change as distillative evaporation of the more volatile components occurs, and the homogeneity of the sphere comes into question. If immiscible or partially miscible components are involved, the droplet can be layered. Light-scattering theory for concentric spheres was first analyzed by Aden and Kerker (1951), and Bohren and Huffman (1983) presented the results of such theory in their monograph. More recently, Hightower *et al.* (1988) made measurements of scattering by layered microspheres and Ray et al. (1991b), as discussed below, theoretically and experimentally examined the effect of a low-volatility coating on the evaporation rate of the core droplet.

Aqueous salt solutions are particularly volatile in a dry gas, and they become supersaturated as evaporation proceeds, for in the absence of solid boundaries heterogeneous nucleation does not occur. Homogeneous nucleation of crystals ultimately occurs to complicate the scattering process. Highly supersaturated solutions can be examined using droplet levitation, and studies related to concentrated electrolyte solutions are surveyed later.

If measurements are to be made using a volatile droplet, it is necessary to obtain the phase function over a scan period small compared with the timescale of the evaporation process. To achieve high scan rates Fulton (1985) mounted a 512-element linear photodiode array on the ring electrode of a quadrupole. The array covered an arc of about 38° in the forward scattering region $33° \leqslant \theta \leqslant 71°$. The 512 pixels were read in 40 ms and recorded using a microcomputer. Details of the device have been published by Davis (1987), and Fig. 22 presents representative data obtained for an evaporating droplet of hexadecane in nitrogen at room temperature. A vertically polarized He–Ne laser ($\lambda_i = 632.8$ nm) was used for the light source.

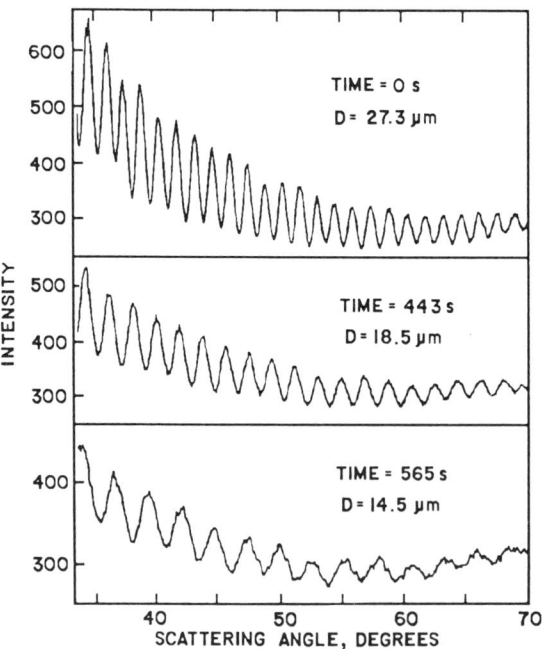

FIG. 22. Phase functions for a levitated droplet of hexadecane evaporating in nitrogen. The data were obtained with the photodiode array of Fulton (1985).

Note that the number of diffraction peaks decreases with time as the droplet diameter decreases, and the number density of peaks is very nearly proportional to the droplet size. The intensity of the scattered light also decreases with size. The resolution of the photodiode array is not adequate to resolve the fine structure that is seen in Fig. 21, but comparison of the phase functions shown in Fig. 22 with Mie theory indicates that the size can be determined to within 1% without taking into account the fine structure. In this case, however, the results are not very sensitive to refractive index. Some information is lost as the price of rapid data acquisition.

5. *Electromagnetic Energy Absorption*

Before we leave the theoretical aspects of light scattering, there is one more issue that should be addressed, and that is microparticle heating by electromagnetic energy absorption. Experiments at elevated temperatures can be performed either by introducing hot gas into the levitation chamber or by electromagnetically heating the particle in a cold gas. Both techniques involve experimental difficulties. The latter method has received more attention than the former because a wide range of heating rates can be achieved, and the levitation chamber does not have to be designed to operate at elevated temperatures.

Maloney and his colleagues at the Morgantown Energy Technology Center performed a number of studies involving microparticle heating by an infrared laser beam, particularly the explosive boiling of coal/water slurry droplets (Maloney and Spann, 1988). The MIT combusion group of Sarofim and Longwell has also used electromagnetic heating for a variety of studies with carbonaceous microparticles, and they developed techniques for the measurement and control of particle temperature (Spjut et al., 1985). Laser heating of a microdroplet is also the basis of the method developed by Arnold et al. (1984) for obtaining an infrared spectrum of a single microdroplet. These studies are examined in more detail later, but first we shall examine the theoretical aspects of electromagnetic energy absorption.

The rate of heat generation per unit volume, the heat source function, of a sphere is given by

$$Q(r, \theta, \phi, t) = \frac{1}{2}\sigma \mathbf{E}_1 \cdot \mathbf{E}_1^*, \qquad (88)$$

where $\mathbf{E}_1$ is the internal electric vector given by Eq. (62), and $\mathbf{E}_1^*$ is its complex conjugate. The electrical conductivity σ of the medium is given by

$$\sigma = \frac{4\pi}{\lambda_i \mu_1 c} \operatorname{Re}(N_1) \operatorname{Im}(N_1). \qquad (89)$$

Note that the heat source function is proportional to the imaginary component of the refractive index, which depends on the wavelength of the electromagnetic radiation.

The heat source function for a small sphere can be highly anisotropic, leading to a nonuniform temperature distribution within the medium. The magnitude and direction of the photophoretic force are governed by the internal heat source. The heat source function computed by Allen et al. (1991) for a weakly absorbing 41.2 µm diameter droplet of 1-octadecene ($N = 1.4521 + i1.30 \times 10^{-7}$) is displayed in Fig. 23. Most of the heat is generated in the forward region (forward in the sense of the direction of propagation of the electromagnetic radiation), but a large spike is seen near the illuminated surface along the axis. Examination of the heat source leads one to expect the illuminated surface to be cooler than the opposite side of the sphere, and solution of the heat conduction equation shows that to be the case (Allen et al., 1991).

Sitarski (1987) computed heat source functions for layered spheres consisting of a core of coal surrounded by a layer of water. One of these source functions is shown in Fig. 24 as a slice through the equatorial plane of the composite sphere. The anisotropy is quite significant with a source spike at the front of the sphere. There is relatively little heat generated in the core, for the electrical field is strongest near the surface. That is quite typical of a strongly absorbing sphere, and Allen and her coworkers showed similar results for a carbonaceous microsphere illuminated by visible and by infrared sources.

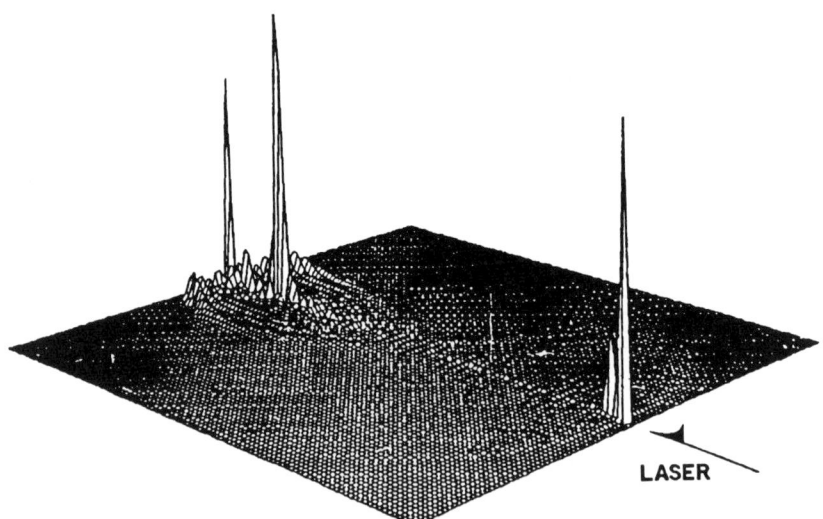

FIG. 23. The heat source function $Q(r, \theta, 0)$ for a droplet of 1-octadecene, computed by Allen et al. (1991) for $I_i = 1$, $x = 265$, and $N = 1.4521 + i1.30 \times 10^{-7}$.

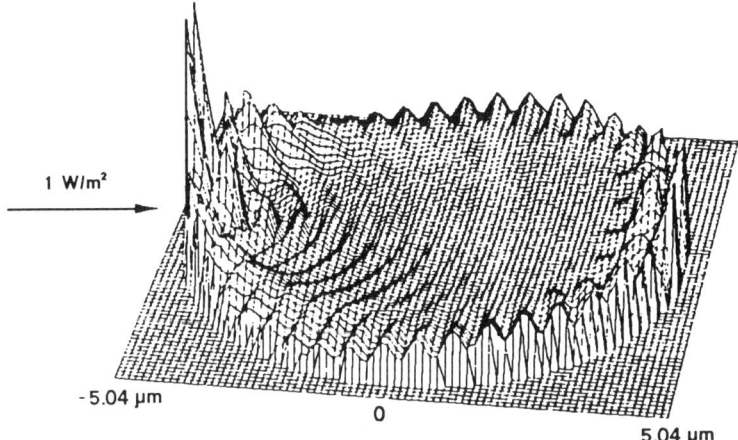

FIG. 24. The heat source function for concentric spheres with an outer diameter of 10.08 μm. The core, consisting of coal, has a diameter of 9.24 μm, and the outer layer is water. The incident wavelength is 1,450 nm. (From Sitarski, 1987.)

We have seen that the electrical field associated with electromagnetic radiation plays an important role in elastic scattering and in microparticle heating. It plays a no less important role in the inelastic scattering processes of fluorescence and Raman spectroscopy, which we examine next.

C. INELASTIC SCATTERING

The interaction of electromagnetic radiation with matter produces the elastic scattering discussed earlier, but the internal electrical fields generated also provide the molecular excitation that leads to scattered light having a wavelength and frequency different from the incident light. As a result, the frequency shift in the scattered light provides information about the chemical bonds in the medium. Fluorescence, which can involve scattered irradiation nearly as intense as elastic scattering, results from transitions in the electronic energy levels of the molecule. The Raman effect is associated with molecular vibrational transitions induced by the incident electrical field.

1. *Photoluminescence*

The absorption of a photon of electromagnetic radiation corresponding to a transition from a ground electronic state to a more energetic electronic state, followed by a release of electromagnetic energy as the molecule returns to a lower electronic state, is called *photoluminescence*. If the emission of energy is from an excited singlet state, the process is termed *fluorescence*, and if the release of electromagnetic energy is a delayed release involving transition to a triplet state, the process is called *phosphorescence*. There can also be *delayed fluorescence* resulting from transitions from singlet to triplet and then from triplet to singlet intermediate states.

The fluorescence process usually involves absorption of a photon in which the transition occurs between the ground state and an excited singlet state. During the time in the excited state, vibrational energy is dissipated, and the electron reaches the lowest vibrational level of the excited electronic state. If the excess energy is not dissipated by molecular collisions, the electron returns to the ground electronic state with a corresponding emission of energy. The energy change ΔE is related to the frequency v and wavelength λ of the emitted light by

$$\Delta E = hv = h\frac{c}{\lambda}, \qquad (90)$$

where h is Planck's constant ($h = 6.626176 \times 10^{-34}$ J-s). Since the energy absorbed is greater than the energy emitted, the wavelength of the emitted radiation is greater than the wavelength of that absorbed. In 1852 Stokes described this dual process of absorption and emission, and he named the process fluorescence after the mineral fluorspar, which he observed to exhibit blue-white fluorescence. As a result of Stokes' investigation, the shift to longer wavelengths is called the *Stokes shift*.

Molecular structure and the microscopic environment determine the photoluminescence characteristics of a molecule. Aromatic molecules having a rigid planar structure fluoresce intensely, but very few aliphatic and saturated cyclic organic compounds fluoresce or phosphoresce. The effect of molecular rigidity is to decrease vibrational amplitudes, which reduces the efficiency of transitions that compete with fluorescence. Rigid molecules such as fluorescein, naphthalene, anthracene, and similar compounds exhibit strong fluorescence, as do dyes such as rhodamines and xanthenes. There is an interesting and useful class of compounds that do not fluoresce in solvents having low viscosity, but fluoresce strongly in highly viscous media such as glycerol. Oster and Nishijima (1956) showed that substances possessing substituted phenyl groups that are capable of internal rotation have this property. They found that diphenylmethane and triphenylmethane dyes, substituted aminostilbene derivatives, and substituted benzophenones show appreciable enhancement of fluorescence as the viscosity of the solvent increases. The cationic dye auramine O (tetramethyldiaminodiphenyl-ketomine hydrochloride), was found to be particularly useful for following viscosity changes.

Oster and Yang (1968) surveyed the subject of photopolymerization of vinyl monomers, and pointed out that one can use auramine O and similar dyes to follow polymerization reactions by recording the increasing fluorescence intensity as the viscosity increases due to polymerization. Ward et al. (1987) applied this method to follow the photochemically initiated polymerization of acrylamide monomer and monomer–glycerol solution microparticles. Earlier, Partch et al. (1983) and Nakamura et al. (1984) had shown that highly monodisperse polymeric microspheres can be produced via aerosol processing in a laminar flow reactor. They produced spheres of styrene polymer and divinylbenzene and ethylvinylbenzene copolymer having diameters in the range 10–20 µm.

To explore the kinetics of such aerosol processes, Ward and his coworkers levitated single microparticles of acrylamide monomer in the bihyperboloidal electrodynamic balance shown in Fig. 25. A monochromator and a PMT detector were mounted at right angles to the laser beam to record the fluorescence intensity. In the absence of UV light, the particle merely evaporated, but when a levitated particle was illuminated by a mercury vapor lamp, polymerization occurred, and a microsphere of polyacrylamide was formed. They added a small amount of auramine O to an aqueous solution of the monomer, and injected a droplet of the mixture into the electrodynamic trap. The water quickly evaporated, leaving auramine-doped acrylamide monomer. They also explored the encapsulation of glycerol by using acrylamide/glycerol solution droplets.

The photochemical reaction was initiated with a UV source, and an argon-ion laser was used as the fluorescence excitation source ($\lambda = 457.9$ nm). A

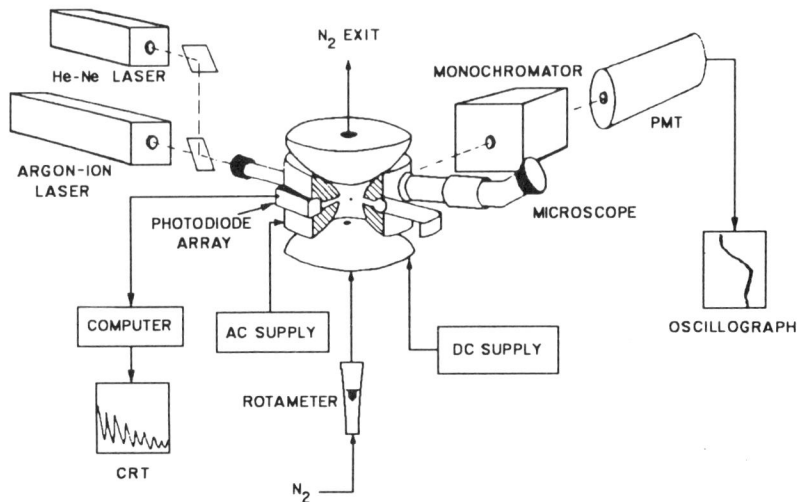

FIG. 25. The apparatus used by Ward et al. to measure polymerization rates of microparticles.

vertically polarized He–Ne laser ($\lambda = 632.8$ nm) was used to size the particle by recording phase functions for the angle range $31° \leq \theta \leq 71°$ in the horizontal plane using a photodiode array mounted on the ring electrode. For an excitation wavelength of 457.9 nm, the maximum intensity of the fluorescence spectrum for auramine O in acrylamide was found to occur at $\lambda = 505$ nm, so the monochromator was set to record the intensity at that wavelength. Figure 26 presents the output of the monochromator/PMT detector as a function of time during polymerization.

The data show that just after the UV source was turned on, the fluorescence signal increased until a maximum was reached, the point at which polymerization was complete. Quenching of the fluorescence by oxygen then occurred, and in later experiments the quenching was eliminated by sweeping the levitation chamber with nitrogen during the polymerization. The product was a microsphere with a size that could be varied by varying the amount of the initial monomeric mass. This suggests that aerosol technology can be used to produce polymeric microspheres with diameters larger than those usually produced via emulsion polymerization.

When one measures fluorescence spectra of a microsphere, a complication occurs that is not encountered in the study of bulk samples. That complication is the effect of elastic resonances on the inelastic scattering. Chew et al. (1976) predicted that the high internal electric field associated with morphological resonances leads to enhancement of fluorescence and Raman emissions, and Benner and his coworkers (1980) were among the first to observe morphological resonances associated with fluorescing molecules embedded in a microsphere.

Interpretation of fluorescence spectra for microspheres should take into account the effects of morphological resonances on the spectra. It is likely

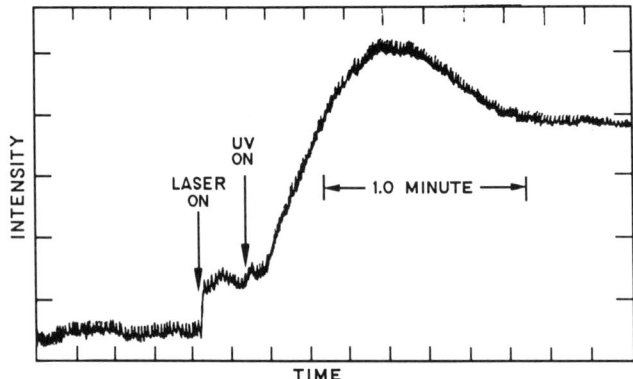

FIG. 26. The output of the monochromator/PMT detector during the polymerization of a microparticle of acrylamide doped with auramine O, from Ward et al. (1987).

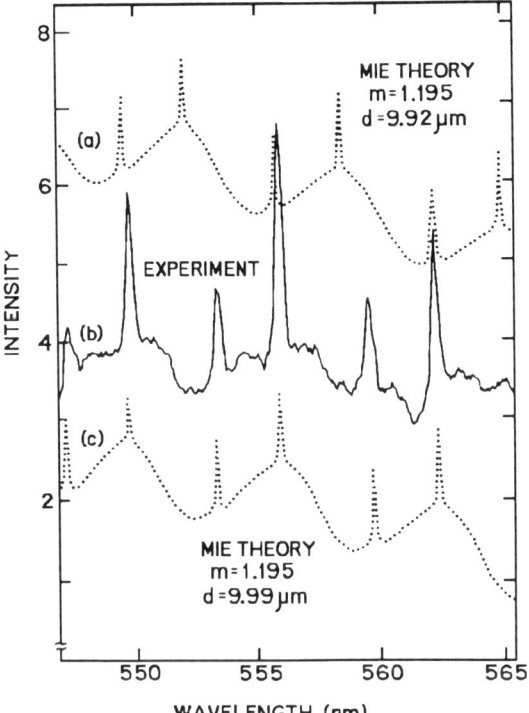

FIG. 27. Fluorescence date of Benner et al. (1980) for polystyrene microspheres that show fluorescence enhancement compared with elastic scattering theory.

that much of the noise in the data shown in Fig. 26 is due to resonances arising because of the change in size and/or refractive index during the polymerization.

Benner et al. suspended in water relatively monodisperse polystyrene microspheres, which contained a fluorescent dye, and they recorded fluorescence spectra. The spheres had a nominal diameter of 9.92 μm (with a 2% standard deviation). Figure 27 presents one of their experimental fluorescence spectra (b) together with elastic scattering spectra computed for a diameter of 9.92 μm (a) and for 9.99 μm (c). The structure of the fluorescence spectrum is not in agreement with the Mie theory computation for a 9.92 μm diameter sphere, but it is quite consistent with the result predicted for a 9.99 μm diameter sphere.

2. *The Raman Effect*

The distinctive feature of the Raman effect compared with fluorescence or infrared spectroscopies is that the photon is not actually absorbed in the process. Instead, the incident electric field induces a dipole moment in a polarizable molecule. Some of the more significant features of the effect can be illustrated using simple classical theory. Let α be the polarizability, which, in general, is a second-order tensor with elements that depend on the configuration of the molecule. The dipole moment **M** induced by the incident light wave may be written as

$$\mathbf{M} = \alpha \mathbf{E}_i \cos(2\pi v_i t), \tag{91}$$

where $\mathbf{E}_i$ is the incident electric vector with frequency v_i.

From classical electromagnetic theory (Jackson, 1975) for an oscillating dipole, the power radiated from the dipole oscillator is given by

$$I = \frac{16\pi^4 v^4}{3c^3} |\mathbf{M}|^2. \tag{92}$$

For a given normal mode of frequency v_m we may write the polarizability as the sum of the polarizability in the equilibrium position α_0 and the induced polarizability due to molecular vibrations,

$$\alpha = \alpha_0 + \alpha_m \cos(2\pi v_m t), \tag{93}$$

where α_m is the maximum change of the polarizability, and v_m is the frequency of the mth vibrational mode.

Using Eq. (93) in Eq. (91), the dipole moment becomes

$$\mathbf{M} = [\alpha_0 + \alpha_m \cos(2\pi v_m t)] \mathbf{E}_i \cos(2\pi v_i t). \tag{94}$$

Expanding this result and applying trigonometric identities, one obtains

$$\mathbf{M} = \alpha_0 \mathbf{E}_i \cos(2\pi v_i t) + \frac{1}{2} \alpha_m \mathbf{E}_i \{\cos[2\pi(v_i + v_m)t] + \cos[2\pi(v_i - v_m)t]\}. \tag{95}$$

The first term in Eq. (95) represents the elastic scattering or *Rayleigh scattering*, the second is associated with the *anti-Stokes* Raman line, and the third represents *Stokes* Raman scattering. The anti-Stokes signal represents a shift to higher frequencies and shorter wavelengths, while the Stokes shift is to lower frequencies and longer wavelengths. The classical treatment of the Raman effect yields many of the primary features of the Raman-scattered light, but the magnitudes of the irradiances of the Stokes and anti-Stokes scattering are not correctly predicted. A quantum mechanical treatment of the Raman effect yields the ratio of the anti-Stokes and Stokes intensities as

$$\frac{I_a}{I_s} = \left(\frac{v_i + v_m}{v_i - v_m}\right) \exp\left(-\frac{hv}{k_B T}\right), \tag{96}$$

where I_a is the intensity of the anti-Stokes radiation, and I_s is the corresponding intensity of the Stokes radiation. We note that if the intensities of the anti-Stokes and Stokes Raman lines are measured, the temperature can be determined using Eq. (96).

At room temperature the anti-Stokes irradiance is much smaller than the irradiance of the Stokes-shifted light, and both are generally much smaller than the irradiance of the elastically scattered light. The molecular scattering cross-section for elastically scattered light is typically 10^{-27} cm^2, which is much larger than the molecular cross-section of 10^{-30} cm^2 for Raman scattering. The small scattering cross-section for Raman scattering makes microparticle Raman spectroscopy experimentally challenging, because the Raman signal is very weak. Furthermore, an even greater challenge arises because the irradiance of the Raman scattering depends on the incident electrical field, and for a microsphere, that introduces complexity related to the existence of morphological resonances. The Raman signal is enhanced by such resonances compared with bulk Raman signals because of the high internal electrical field associated with a resonance. The vexing question is whether or not Raman measurements can be used for quantitative as well as qualitative analysis of microparticles. Resonances make the interpretation of data more difficult. This issue will be addressed further in the section on the measurement of the chemical composition of microdroplets.

Chew, McNulty, and Kerker were the first to recognize theoretically that the relatively large internal electrical field associated with a resonance could be expected to enhance inelastic scattering. They applied a classical treatment in their analysis of inelastic scattering from a molecule embedded in a sphere. In their initial work they limited the analysis to a sphere of the order of the wavelength of light, but Kerker and Druger (1979) extended the theory to larger dielectric spheres.

Thurn and Kiefer (1984a,b) were the first to record experimentally the effects of morphological resonances on Raman scattering. Figure 28 presents Raman data for an optically levitated water/glycerol microdroplet. The wavenumber range shown corresponds to the O–H stretching region of the spectrum. Also presented in the figure is the spectrum for bulk material and a spectrum calculated using an *ad hoc* superposition of Mie theory and the spectrum for bulk material. The correspondence between the resonance peaks computed using Mie theory and the droplet data is quite good, suggesting that the peaks observed in the Raman spectrum are, indeed, the result of morphological resonances associated with the elastic scattering.

Since 1984 several laboratories have obtained Raman spectra of droplets and crystallites. Chang and his colleagues at Yale (Snow *et al.*, 1985; Qian *et al.*, 1985, 1986; Qian and Chang, 1986a, 1986b) and Schweiger (1987, 1990a) used droplet streams generated by a pulsed droplet generator. Schweiger (1990b, 1990c) also used Ashkin and Dziedzic's optical levitation technique to

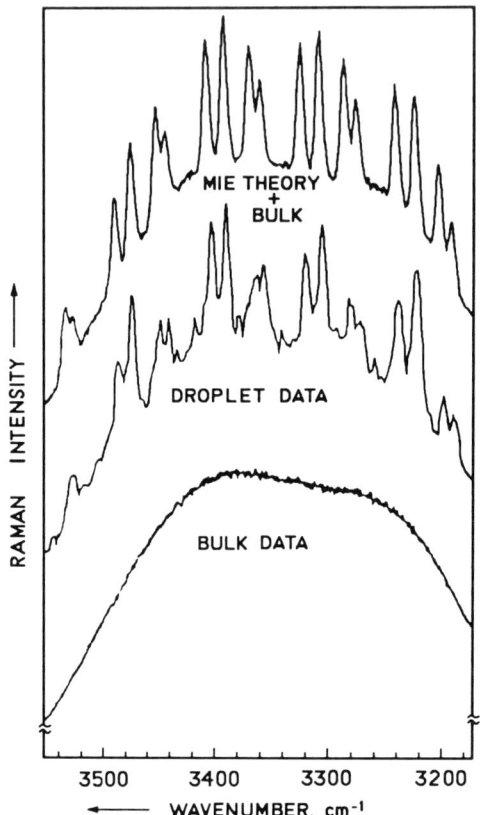

FIG. 28. Raman scattering data for bulk material and a droplet of glycerol/water compared with a calculated spectrum. Reprinted with permission from Thurn, R., and Kiefer, W., *Applied Optics* **24**, 1515–1519, Copyright © 1985, The Optical Society of America.

explore resonances in the Raman spectra of droplets, as did Lettieri and Preston (1985) earlier, and Schweiger reviewed the recent theoretical and experimental work on microparticle Raman spectroscopy.

Fung and Tang used electrodynamic levitation of single droplets to obtain Raman spectra for aqueous solution droplets of sodium nitrate and ammonium sulfate (1988a) and of supersaturated ammonium bisulfate droplets (1988b, 1989a). They obtained additional Raman spectra for crystalline ammonium sulfate particles and potassium nitrate particles, as well as mixed sulfate and nitrate particles (1989b). Recently (1991) they reported relative Raman scattering cross-section measurements of suspended crystals of sodium nitrate, ammonium nitrate, sodium sulfate, and ammonium sulfate.

They gave the cross-section relative to that of sodium nitrate. The experiments were all performed using a bihyperboloidal electrodynamic levitator.

IV. Mass and Heat Transfer

A. EVAPORATION/CONDENSATION PROCESSES

Evaporation and condensation rates for microdroplets are readily measured using electrodynamic trapping and laser light scattering. Davis and Chorbajian (1974) used an electrostatic balance of the type shown in Fig. 2 to measure the evaporation rate of dibutyl phthalate in the diffusion-controlled regime, and Chang and Davis (1976) studied Knudsen aerosol evaporation in the same apparatus. Davis and Ray (1977) showed that gas-phase diffusion coefficients can be determined by means of diffusion-controlled evaporation experiments, and Ray et al. (1979) applied the method to determined ultralow vapor pressures. Davis et al. (1980) reviewed the principles and techniques, and since that time numerous investigators have used microdroplet evaporation measurements for vapor pressure determination and for studying a variety of other phenomena.

1. *Pure Component Evaporation in a Stagnant Medium*

For a pure component droplet evaporating into a stagnant gaseous medium in the continuum regime, the quasi-steady rate of change of droplet radius a with time is given by an equation attributed to Maxwell (1890),

$$\frac{da}{dt} = -\frac{D_{ij}}{\rho_1 a}(c_a - c_\infty), \qquad (97)$$

where D_{ij} is the diffusivity for vapor i in carrier gas j, ρ_1 is the droplet density, c_a is the concentration of vapor at the surface, and c_∞ is the vapor concentration far from the interface. If the gas/vapor mixture is ideal, we may write the concentrations as

$$c_a = \frac{M_i p_i^0}{RT_a} \quad \text{and} \quad c_\infty = \frac{M_i p_{i,\infty}}{RT_\infty}, \qquad (98)$$

where M_i is the molecular weight of the vapor, p_i^0 is the vapor pressure at the droplet surface temperature T_a and p_∞ is the partial pressure of vapor far from the interface where the temperature is T_∞.

If the vapor is removed by flowing a purge gas through the balance chamber during the evaporation process, and/or if the chamber volume is sufficiently large, $p_\infty \approx 0$. In the following discussion we shall assume that p_∞ is negligibly small, so Eq. (97) reduces to

$$\frac{da}{dt} = -\frac{D_{ij}}{\rho_1 a} \frac{M_i p_i^0}{RT_a}. \tag{99}$$

Continuum theory applies when the mean free path of the vapor Λ_i is small compared with the droplet radius, that is, when the Knudsen number Kn is small (Kn = $\Lambda_i/a \ll 1$). From the kinetic theory of gases (Jeans, 1954), the mean free path of the vapor in a binary system is given by

$$\Lambda_i = \frac{1}{\pi(n_i \sigma_{ii}^2 \sqrt{2} + n_j \sigma_{ij}^2 \sqrt{1 + m_i/m_j})}, \tag{100}$$

where n is the number density of the ith or jth species, m is the molecular mass, and σ_{ii} and σ_{ij} are collision diameters for like- and unlike-molecule collisions, respectively. For low-volatility species, $n_i \ll n_j$. The collision diameter σ_{ij} may be written in terms of the like-molecule collision diameters σ_{ii} and σ_{jj} by their mean value (Hirschfelder et al., 1967):

$$\sigma_{ij} = \frac{\sigma_{ii} + \sigma_{jj}}{2}. \tag{101}$$

Values of σ_{jj} for common gases have been tabulated by Hirschfelder et al. (1967) and by Bird et al. (1960).

Rapid evaporation introduces complications, for the heat and mass transfer processes are then coupled. The heat of vaporization must be supplied by conduction heat transfer from the gas and liquid phases, chiefly from the gas phase. Furthermore, convective flow associated with vapor transport from the surface, *Stefan flow*, occurs, and thermal diffusion and the thermal energy of the diffusing species must be taken into account. Wagner (1982) reviewed the theory and principles involved, and a higher-order quasi-steady-state analysis leads to the following energy balance between the net heat transferred from the gas phase and the latent heat transferred by the diffusing species:

$$\kappa_j \phi_H (T_\infty - T_a) = \lambda_{\text{vap}} \frac{D_{ij} \phi_M p_i^0 M_i}{RT_a}, \tag{102}$$

where κ_j is the thermal conductivity of the gas, λ_{vap} is the latent heat of vaporization, T_a is the surface temperature of the droplet, and ϕ_H and ϕ_M are heat and mass transfer correction factors, respectively, which take into account Stefan flow and sensible and latent heat transport by the diffusing vapor. For a droplet not near its boiling point, ϕ_H and ϕ_M are nearly unity, and in the remaining discussion we shall assume that $\phi_H = \phi_M = 1$. In this case the surface temperature satisfies the transcendental equation

$$T_a = T_\infty - \lambda_{vap} \frac{D_{ij} p_i^0(T_a) M_i}{\kappa_j R T_a}, \tag{103}$$

which may be solved for T_a based on knowledge of the vapor pressure as a function of temperature.

When the vapor pressure is sufficiently low, the surface temperature is closely approximated by the bulk gas temperature T_∞. In this event, Eq. (99) may be integrated to yield

$$a^2 = a_0^2 - \frac{2 D_{ij} M_i p_i^0(T_\infty)}{\rho_1 R T_\infty}(t - t_0), \tag{104}$$

where a_0 is the droplet radius at time t_0. Thus, a graph of a^2 versus time should yield a straight line with slope S_{ij} given by

$$S_{ij} = -\frac{2 D_{ij} M_i p_i^0(T_\infty)}{\rho_1 R T_\infty}. \tag{105}$$

Taflin et al. (1988) reported evaporation data for several low-vapor-pressure species in the form of a graph of a^2 versus time. The droplet size was determined from phase function measurements as well as from the resonance spectra. Figure 10 displays a typical resonance spectrum and the corresponding spectrum computed using Eq. (75) and Eq. (72) with $m = 1.4395$. In this case the incident beam was polarized perpendicular to the scattering plane. The resonance spectrum was obtained with a PMT mounted at $\theta = 90°$. The figure is for a droplet of dodecanol evaporating in air at a temperature of 295 K. There is excellent agreement between the two spectra in both the fine structure and overall topography, except at the discontinuity associated with the explosion discussed above.

The results obtained by Taflin et al. for several different chemicals are presented in Fig. 29 as a^2 versus time. The data are seen to fall on straight lines, as predicted by Eq. (104).

Equation (105) is the basis for the determination of gas-phase diffusion coefficients and ultra low vapor pressures using the methods proposed by Davis and Ray (1977), Ravindran et al. (1979), and Ray et al. (1979). Additional information can be gained by writing the Chapman–Enskog first approximation for the gas-phase diffusivity (Chapman and Cowling, 1970),

$$D_{ij} = \frac{3}{8 P \sigma_{ij}^2 \Omega_{ij}(T^*)} \sqrt{\frac{(k_B T)^3 (m_i + m_j)}{2\pi m_i m_j}}. \tag{106}$$

Here P is the total pressure of the system, and Ω_{ij} is the *collision integral*, which is a function of the reduced temperature $T^* = k_B T / \varepsilon_{ij}$. The molecular

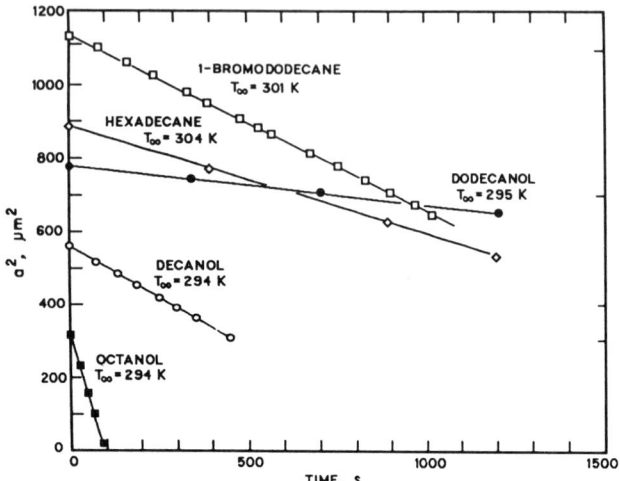

FIG. 29. Evaporation rate data for several low-volatility species in N_2 measured by Taflin et al. (1988) by levitating microdroplets in an electrodynamic balance. From "Measurement of Droplet Interfacial Phenomena by Light-Scattering Techniques," by Daniel C. Tafflin, S. H. Zhang, Theresa Allen and E. James Davis, AIChE Journal, 34, No. 8, pp. 1310–1320, reproduced by permission of the American Institute of Chemical Engineers © 1988 AIChE.

interaction force constant ε_{ij} enters as a result of the use of the Lennard–Jones 6–12 molecular interaction potential in the derivation of Eq. (106),

$$\Phi_{ij} = 4\varepsilon_{ij}\left[\left(\frac{\sigma_{ij}}{r}\right)^{12} - \left(\frac{\sigma_{ij}}{r}\right)^{6}\right], \quad (107)$$

where r is the intermolecular distance.

The force constant for unlike molecules may be written as the geometric mean of the force constants for like-molecule interactions,

$$\varepsilon_{ij} = \sqrt{\varepsilon_{ii}\varepsilon_{jj}}. \quad (108)$$

The collision integral has been evaluated and tabulated by Chapman and Cowling and Hirschfelder, Curtiss, and Bird. Bird, Stewart, and Lightfoot listed the values published by Hirschfelder et al. (1949), and they also tabulated values of ε_{jj} for common gases. Numerous correlations have been proposed for estimating ε_{jj} and σ_{jj}, and Ravindran et al. (1979) reviewed and compared the various methods.

For a pure component droplet of unknown vapor pressure, Eq. (104) together with Eqs. (101), (106), and (108) constitute a set of relationships involving three unknown parameters, p_i^0, ε_{ii}, and σ_{ii}. Evaporation measurements in three different carrier gases at the same temperature provide the data needed to determine the three parameters. Data obtained by Davis and

Ray (1978) for dibutyl sebacate (DBS, $i = 1$) evaporating in He ($j = 2$), N_2 ($j = 3$), and CO_2 ($j = 4$) at atmospheric pressure and 293 K are presented in Fig. 30.

The data were obtained in an electrostatic balance, and phase functions were recorded to determine the droplet size as a function of time. The slopes of the three data sets are: $S_{12} = -7.54 \times 10^{-4} (\mu m)^2/s$, $S_{13} = -1.70 \times 10^{-4} (\mu m)^2/s$, and $S_{14} = -1.10 \times 10^{-4} (\mu m)^2/s$. From these and additional data obtained at different temperatures, Ravindran et al. reported $\sigma_{DBS} = 9.97 \pm 0.26 \text{ Å}$, $\varepsilon_{DBS}/k_B = 688 \pm 72 \text{ K}$, and $p^0_{DBS}(293 \text{ K}) = 207 \, \mu\text{Pa}$. They also reported Lennard–Jones parameters and vapor pressures measured in this way for dibutyl phthalate and dioctyl phthalate.

Recently, Tang and Munkelwitz (1991) applied a variation on this theme by determining very low vapor pressures of levitated droplets from data obtained at atmospheric pressure and in an evacuated chamber. In both cases they used data from the literature obtained at temperatures much higher than room temperature together with their new data to fit the constants in vapor pressure equations of the form

$$\ln p^0_i = A + \frac{B}{T} + C \ln T + DT^k, \qquad (109)$$

where A, B, C, D, and k are constants to be determined.

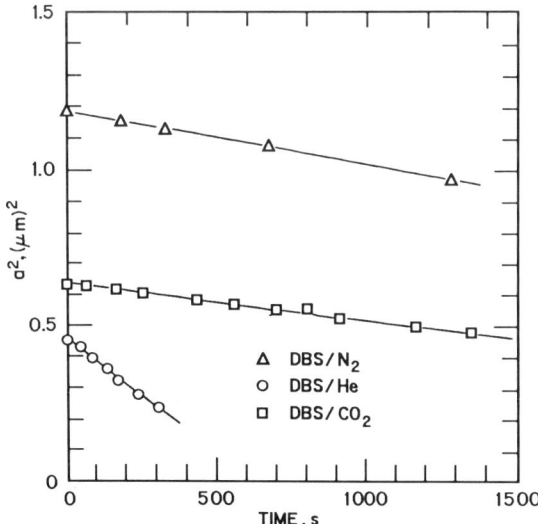

FIG. 30. Evaporation rate data obtained by Davis and Ray (1978) for dibutyl sebacate evaporating in helium, nitrogen, and carbon dioxide at 293 K. Reprinted with permission from J. Aerosol Sci. 9, 411–412, Davis, E. J., and Ray, A. K., Copyright © 1978, Pergamon Press plc.

For the data in the continuum regime they wrote the approximation

$$\frac{D_{ij}(T)}{D_{ij}(T_0)} = \left(\frac{T}{T_0}\right)^2, \quad (110)$$

which is equivalent to assuming that the collision integral in Eq. (106) is proportional to $1/\sqrt{T}$. Thus, the slope S_{ij} defined by Eq. (105) may be written

$$S_{ij} = -p_i^0(T)T\frac{2D_{ij}(T_0)M_i}{\rho_1 R T_0^2}, \quad (111)$$

and using this result, we have

$$\ln p_i^0 = \ln\frac{-S_{ij}(T)}{T} + \ln\beta. \quad (112)$$

Here β is the group for constants defined by

$$\beta = \frac{\rho_1 R T_0^2}{2M_i D_{ij}(T_0)}, \quad (113)$$

which Tang and Munkelwitz (1991) treated as an adjustable parameter in fitting data to Eq. (109).

For data obtained in the free-molecule regime the procedure is somewhat different. In this case da/dt is given by

$$\frac{da}{dt} = -e_v\sqrt{\frac{m_i}{2\pi k_B T}}\frac{p_i^0}{\rho_1}, \quad (114)$$

where e_v is the evaporation coefficient, which incorporates deviations from the kinetic theory of effusion. We note that in the application of the theory for the free-molecule regime, no assumption about the diffusion coefficient is involved.

Letting $S' = -2da/dt$, Eq. (114) may be written in the logarithmic form

$$\ln p_i^0(T) = \ln[S'(T)\sqrt{T}] + \ln\beta', \quad (115)$$

where β' is defined by

$$\beta' = \frac{\rho_1}{e_v}\sqrt{\frac{\pi R}{2M_i}}. \quad (116)$$

Tang and Munkelwitz assumed $e_v = 1$ and used their data obtained under vacuum conditions together with data from the literature to fit the constants of Eq. (109). They reported results for glycerol and methanesulfonic acid (MSA) based on measurements in the continuum regime, and for dioctyl phthalate (DOP) and oleic acid based on experiments involving the free-molecule regime. Their results for DOP are in good agreement with the vapor pressures reported by Ray et al. (1979), who performed their measurements in the continuum regime. This indicates the validity of assuming an evaporation coefficient of unity for DOP, and shows that experiments performed in the two limits of large and small Knudsen number yield the same vapor pressure. Furthermore, these results indicate that extremely low vapor pressures can be determined with good accuracy by microdroplet evaporation methods.

If the vapor pressure is determined by measurements in the free-molecule regime, then evaporation rate measurements in the continuum regime permit the diffusion coefficient to be determined from a single experiment in the carrier gas of interest and at the temperature of interest.

The results obtained by Tang and Munkelwitz for the continuum regime do not appear to be consistent with other data. Their results for glycerol are presented in Fig. 31 and compared with data from the literature and with Eq. (109). The solid line in the figure is Eq. (109) with the constants reported by Tang and Munkelwitz. Since the data of Weast and Selby (1988) and Tang and Munkelwitz were used to fit the constants in the equation, those data agree with the correlation. But, using estimates of the gas phase diffusion

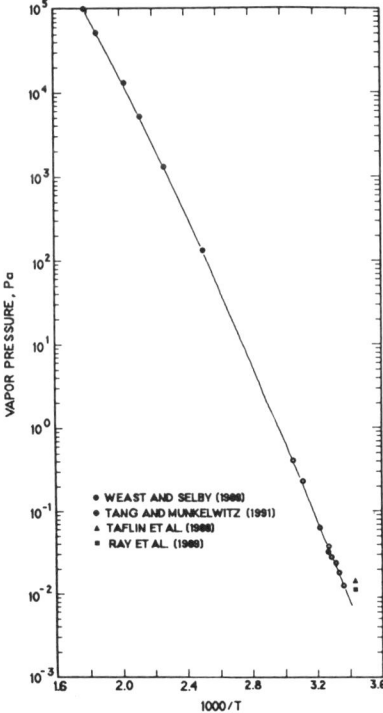

FIG. 31. Vapor pressure data for glycerol. The solid line represents Eq. (109) with the constants reported by Tang and Munkelwitz (1991).

coefficient for glycerol and the evaporation rate data of Taflin et al. (1988) and Ray et al. (1991b) for 293 K, one calculates the points lying well above the correlation line in the lower right corner of the figure. Depending upon how one estimates the diffusion coefficient, the rate data of Ray and his coworkers yield a vapor pressure in the range 0.0115–0.0132 Pa, and the results of Taflin et al. give 0.0142–0.0164 Pa. Taflin and his coworkers did not have precise temperature control on their balance chamber, so there is a greater uncertainty about the temperature in those experiments.

Recently, Ray et al. (1991b) reported the vapor pressure of glycerol at 298 K to be 0.0223 Pa. Equation (109) yields a value of 0.0138 Pa at that temperature. It appears that the correlation of Tang and Munkelwitz underpredicts the vapor pressure near room temperature because of the procedure used to determine the constants in the equation. The approximation written as Eq. (110) is not sufficiently accurate to permit its application over a wide range of temperatures. As shown by Fuller et al. (1966), data for numerous low–molecular-weight binary systems agree with the approximation that D_{ij} is proportional to $T^{1.75}$. An exponent between 1.75 and 2 would be a better choice and would lead to larger values of the vapor pressure at the lower temperatures.

Richardson, Hightower and Pigg (1986) used a quadrupole mounted in a vacuum chamber to measure the vapor pressure of sulfuric acid as a function of temperature in the range $263 \leqslant T \leqslant 303$ K. They operated well into the free-molecule regime and assumed an evaporation coefficient of unity. Richardson et al. correlated their data by means of the equation

$$\ln p_i^0 = (20.70 \pm 1.74) - \frac{(9360 \pm 499)}{T} \quad \text{(in units of Torr)}. \qquad (117)$$

The results are in reasonably good agreement with other data in the literature, except for the data of Gmitro and Vermeulen (1964), which are at least two orders of magnitude larger than the values reported by more recent investigators. The vapor pressures reported by Richardson and his coworkers are not truly sulfuric acid vapor pressures, for under the high vacuum conditions of the experiments H_2SO_4 decomposes to H_2O and SO_3, and the liquid attains its azeotropic composition of 98.479 mass % H_2SO_4 and 1.521 mass % H_2O.

Most of the studies of levitated droplets have involved low–vapor-pressure materials, but Taflin and his coworkers reported data for water droplets evaporating in dry nitrogen. The rapid evaporation of a water droplet requires that the experiment be automated, and this was accomplished by injecting the droplet by means of a 3,000 V dc electrical pulse applied to a flat-tipped hypodermic needle. The pulse triggered the data collection system so that phase functions and the resonance spectrum were obtained during the less than three-second duration of an experiment. From the phase function data, the size was determined as a function of time, and Fig. 32 is a graph of a^2 versus time for two different drops in nitrogen at two different temperatures, 283.2 K and 293.2 K. Also plotted on the figure are the predicted evaporation rates based on evaporation at the ambient temperature, using Eq. (104), and based on surface temperature lowering due to evaporation, using Eq. (103) to determine the surface temperature.

If the surface temperature is not corrected for evaporative cooling, the evaporation rate is greatly overpredicted, as indicated by the difference between the experimental data points and the dotted lines of Fig. 32. When Eq. (103) is used to calculate the surface temperature, theory and experiment are in reasonably good agreement, although there is some evidence that the evaporation rate decreases near the end of the experiment. This is likely due to an increasing concentration of trace contaminants originally in the droplet.

2. Multicomponent Droplet Evaporation

If more than one component is present in the droplet, the evaporation rate depends on the miscibility of the components. For totally miscible components, distillation of the more volatile component(s) occurs, and the droplet composition changes as evaporation proceeds. If the components are partially miscible or insoluble, evaporation may be greatly retarded by the formation of a surface layer. In the case of evaporation of a surfactant

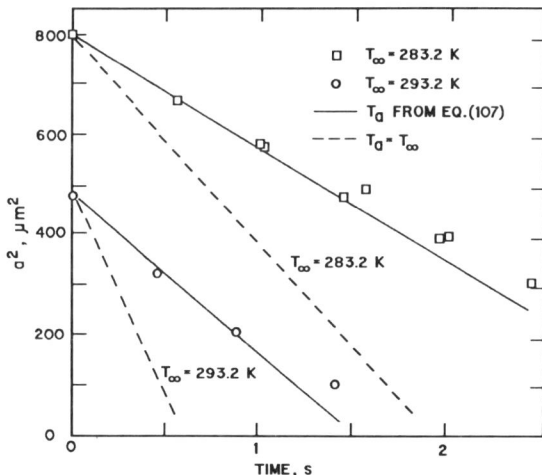

FIG. 32. A comparison between experimental and predicted evaporation rates for water droplets evaporting in dry nitrogen. The dashed line is the theoretical prediction based on evaporation at the surrounding gas temperature.

solution, which is encountered in spray drying processes in the manufacture of detergent powders, the effect of an insoluble monolayer is quite dramatic. Taflin et al. (1988) injected a droplet consisting of a dilute aqueous solution of sodium dodecyl sulfate (SDS) into an electrodynamic balance and quickly trapped the droplet. They recorded phase functions and the resonance spectrum during the evaporation process, and Fig. 33 is a typical resonance spectrum for an SDS solution droplet. The initial concentration of SDS was 10% of the critical micelle concentration (CMC). The resonance spectrum shows a very high frequency of resonance peaks for the first few seconds of the experiment, followed by a longer period in which the resonances are well defined. Near the end of the experiment the droplet became crystalline, and the light scattering became irregular and uninterpretable. Crystalline materials rotate in the quadrupole trap and appear to twinkle when observed through the microscope. The resulting resonance spectrum is chaotic.

Using phase function data and the resonance spectrum, Taflin and his associates determined the droplet size for various times during the evaporation process, and Fig. 34 presents size data as a graph of a^2 versus time for a representative SDS experiment. The effect of the surfactant is quite dramatic, for the figure shows that for about eight seconds rapid evaporation occurred. Suddenly the evaporation rate was reduced by a factor of 230. The initial slope of the a^2 versus time is $-155\,\mu m^2/s$, and the slope at later times is $-0.67\,\mu m^2/s$. The rate-controlling process was initially diffusion-controlled evaporation of water vapor into the surrounding nitrogen, and the formation of an insoluble surface layer resulted in diffusion through the surface layer as the rate-controlling mechanism.

Rubel and Gentry (1984b) showed similar effects in their investigation of phosphoric acid droplet evaporation in the presence of hexadecanol, which was introduced into the gas phase. The hexadecanol concentration was varied to alter the surface coverage. They reported their results as an

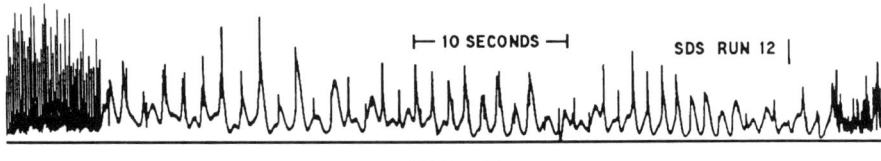

FIG. 33. A resonance spectrum for a droplet of sodium dodecyl sulfate in water evaporating in dry nitrogen at 293 K. From "Measurement of Droplet Interfacial Phenomena by Light-Scattering Techniques," by Daniel C. Tafflin, S. H. Zhang, Theresa Allen and E. James Davis, AIChE Journal, 34, No. 8, pp. 1310–1320, reproduced by permission of the American Institute of Chemical Engineers © 1988 AIChE.

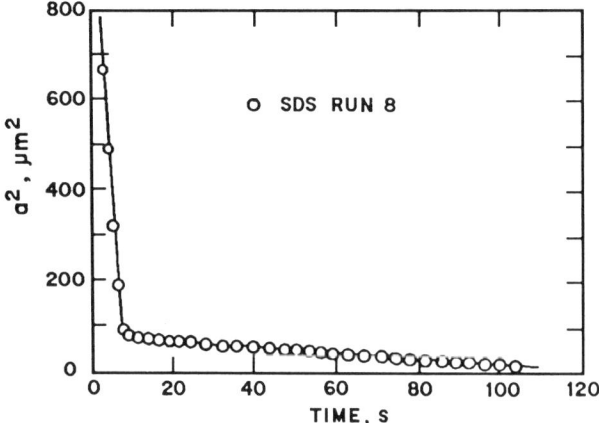

FIG. 34. A graph of a^2 versus time for a droplet of SDS in water evaporating in dry nitrogen at 293 K. From "Measurement of Droplet Interfacial Phenomena by Light-Scattering Techniques," by Daniel C. Tafflin, S. H. Zhang, Theresa Allen and E. James Davis, *AIChE Journal*, **34**, No. 8, pp. 1310–1320, reproduced by permission of the American Institute of Chemical Engineers © 1988 AIChE.

accommodation coefficient (evaporation coefficient). Accommodation coefficients as low as 2.0×10^{-5} were reported.

Partially miscible systems exhibit a greater complexity of behavior, for the evaporation kinetics depends upon whether the volatile component is in the core or in the outer layer of a layered sphere. The evaporation rate also depends upon the thickness of the outer layer(s). Ray et al. (1991b) explored the evaporation characteristics of single binary droplets of glycerol and dioctyl phthalate (DOP). A droplet was generated from a homogeneous solution of DOP and glycerol in the highly volatile cosolvent ethyl alcohol and trapped in a two-ring electrodynamic balance mounted in a temperature-controlled chamber. Their apparatus is presented in Fig. 35.

The two-ring balance was wired such that the ac potential was applied to both rings, and dc potentials of opposite polarity were superposed on the ac potential. Both the frequency and amplitude of the ac potential could be varied to operate in the stable regime of the balance. The two-ring device was introduced by Weiss-Wrana (1983) and is a simple and effective version of the electrodynamic balance. Davis et al. (1990) analyzed the electrical fields generated by this geometry and the stability characteristics of the two-ring configuration.

Ray et al. (1991b) wrote conservation equations for the two species in the droplet and solved the governing equations to yield the evaporation rate in terms of the square of the droplet radius. If the outer material is relatively nonvolatile, the core material must diffuse through the coating of constant

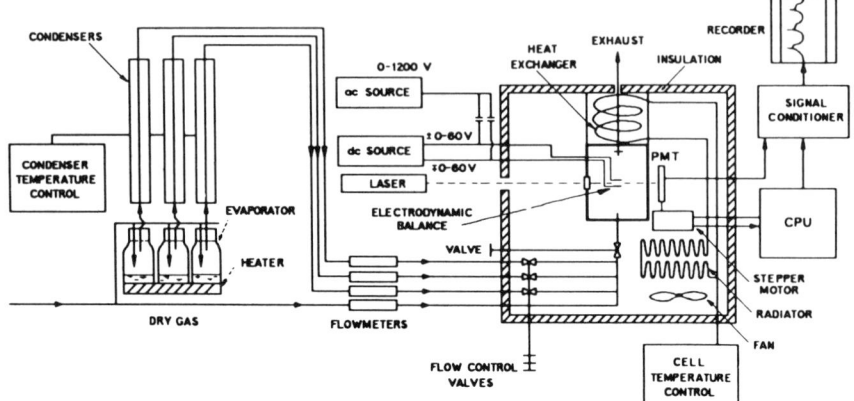

FIG. 35. The apparatus used by Ray et al. for the investigation of glycerol spheres coated with dioctyl phthalate. Reprinted with permission from Ray, A. K., Devakottai, B., Souyri, A., and Huckaby, J. L., Langmuir 7, 525–531, Copyright © 1991, American Chemical Society.

mass before evaporating at the gas–liquid interface. The droplet size is governed by diffusion through the coating and gas-phase diffusion of the volatile species. For the case in which the core material A has very low solubility in the coating material, Ray and his coworkers obtained the following equation for the outer radius as a function of time:

$$a^2 + \frac{\alpha\beta}{(1-\alpha\beta)}(a^3 - a_0^3 + a_{c0}^3)^{2/3} = a_0^2 + \frac{\alpha\beta}{(1-\alpha\beta)}a_{c0}^2 - 2\frac{D_G p_A^0 M_A \gamma_A x_{Am}}{(1-\alpha\beta)\rho_A RT} t.$$

(118)

Here x_{Am} is the mole fraction of A in the outer layer, γ_A is its activity coefficient, and a_0 and a_{c0} are the outer and inner initial radii, respectively. The parameters α and β are defined by

$$\alpha = \gamma_{Am}\left[1 - \frac{2A^{*2}x_{Am}}{B^*\left\{1 + \frac{A^*x_{Am}}{B^*(1-x_{Am})}\right\}^3 (1-x_{Am})^2}\right],$$

(119)

and

$$\beta = \frac{D_G p_A^0}{D_L RTC_L}.$$

(120)

Here D_G and D_L are the diffusivities of A in the gas phase and the coating liquid, respectively, and C_L is the total molar concentration of the shell phase.

The activity coefficients were assumed to satisfy van Laar equations of the form

$$\ln \gamma_A = \frac{A^*}{\left[1 + \dfrac{A^* x_A}{B^*(1 - x_A)}\right]^2} \tag{121}$$

and

$$\ln \gamma_B = \frac{B^*}{\left[1 + \dfrac{B^*(1 - x_A)}{A^* x_A}\right]^2}, \tag{122}$$

where A^* and B^* are empirical constants.

The outer and core radii were determined from optical resonance measurements using Mie theory solutions (Aden and Kerker, 1951; Bohren and Huffman, 1983) for concentric spheres to interpret the resonance spectra. Figure 36 presents some of the data of Ray et al. for a pure component glycerol droplet and for a coated droplet having an initial coating thickness given by $y_0 = 0.321$. Here y is a reduced thickness defined by $y = (a - a_c)/a$.

The data of Fig. 36 show the somewhat surprising result that the coated

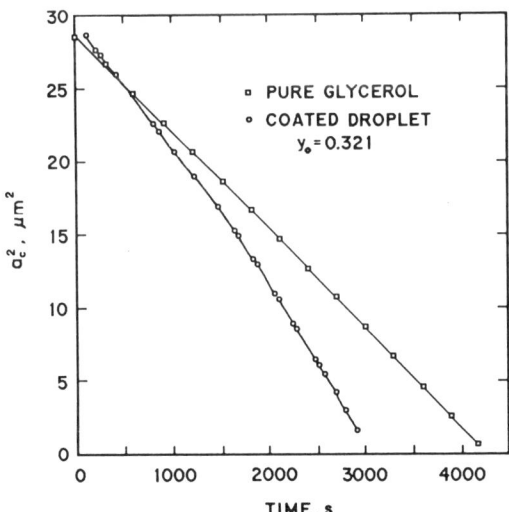

Fig. 36. Evaporation rate data from Ray et al. for a pure glycerol droplet and a glycerol core coated with dioctyl phthalate. Reprinted with permission from Ray, A. K., Devakottai, B., Souyri, A., and Huckaby, J. L., *Langmuir* 7, 525–531, Copyright © 1991 American Chemical Society.

droplet evaporates more rapidly than the droplet of pure glycerol. In view of the fact that the glycerol must diffuse through the shell of nonvolatile DOP, one might anticipate a lower evaporation rate for the binary system. That this is not the case is due to the increased surface area of the coated droplet compared with the surface area of a pure glycerol droplet containing the same mass of glycerol. An activity coefficient much larger than unity would also lead to more rapid evaporation of the binary drop.

The issue of activity coefficient measurement for binary droplets was addressed by Allen et al. (1990). For low–vapor-pressure species, their diffusional fluxes in the gas phase are independent because of their low gas phase concentrations, and for a quasi-steady process each flux may be written

$$J_i = \frac{\phi_i z_i \gamma_i}{a}, \tag{123}$$

where z_i is the mole fraction of species i, γ_i is its activity coefficient, and ϕ_i is a pure component evaporation parameter defined by

$$\phi_i = \frac{D_{ij} p_i^0}{RT}. \tag{124}$$

A total material balance on a binary droplet yields

$$\frac{d}{dt}\left(\frac{4\pi a^3}{3 V_m}\right) = -4\pi a^2 (J_1 + J_2). \tag{125}$$

Allen and her colleagues wrote the molar volume V_m for the solution as the approximation $V_m = z_1 V_1 + (1 - z_1) V_2$, where V_1 and V_2 are component molar volumes. With this approximation and using Eq. (124) and (125), one obtains expressions for da^2/dt and dz_1/dt,

$$\frac{da^2}{dt} = -2(\phi_1 \gamma_1 V_1 - \phi_2 \gamma_2 V_2) z_1 - 2\phi_2 \gamma_2 V_2, \tag{126}$$

and

$$\frac{dz}{dt} = \frac{3(\phi_1 \gamma_1 - \phi_2 \gamma_2)}{a^2}[(V_1 - V_2)z_1^3 + (2V_2 - V_1)z_1^2 - V_2 z_2]. \tag{127}$$

If the activity coefficients are known (unity for ideal solution behavior), this coupled set of first-order differential equations can be solved numerically to obtain the radius and composition as functions of time.

Equations (126) and (127) can be used to calculate activity coefficients from evaporation data, for a^2, z_1, da^2/dt, and dz_1/dt are measurable quantities. The resonance spectrum of an evaporating droplet is highly sensitive to both size and refractive index, and the refractive index of a binary system is a unique function of its composition. Solving for γ_1 and γ_2 in terms of $a^2, z_1, da^2/dt$, and dz_1/dt, one obtains

$$\gamma_1 = -\frac{\dfrac{1}{2}\dfrac{da^2}{dt} + \dfrac{V_2 a^2}{3z_1 V_m}\dfrac{dz_1}{dt}}{\phi_1 V_m}, \tag{128}$$

and

$$\gamma_2 = \frac{-\dfrac{1}{2}\dfrac{da^2}{dt} - \dfrac{V_1 a^2}{3(1-z_1)V_m}\dfrac{dz_1}{dt}}{\phi_2 V_m}. \tag{129}$$

Allen et al. examined binary pairs selected from the set 1,8-dibromooctane (DBO), 1-bromododecane (BDD), heptadecane (HPD), and hexadecane (HXD). They first measured the evaporation rates of pure component droplets to obtain the parameter ϕ for each chemical, and then measured evaporation rates of four binary pairs. Two of the pairs were expected to have ideal solution behavior, and the other two were thought to be nonideal. Figure 37 presents results for a nearly ideal pair (BDD/HXD), and Fig. 38 shows the results for a highly nonideal pair (DBO/HXD). The solid lines in the figures correspond to the solution of Eq. (126) and (127) under the assumption that the activity coefficients are unity. For the system BDD/HXD, the size and composition are in quite good agreement with the computed results, so we can conclude that the system is very nearly ideal. The data for DBO/HXD show significant deviations from ideal solution behavior, for both the composition and size change more rapidly with time than predicted for an ideal solution.

Note that the composition data of Fig. 37a are fairly scattered later in the

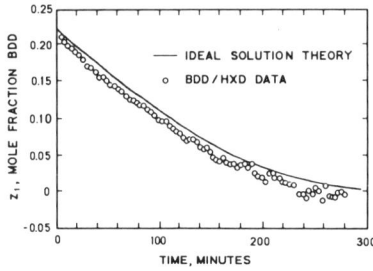
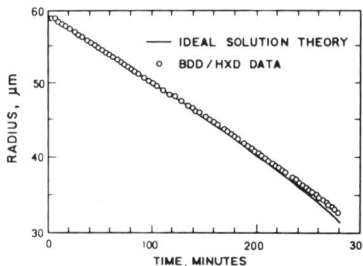

FIG. 37. (a) Composition data and theory for a BDD/HXD droplet evaporating in nitrogen at 295 K, from Allen et al. (1990). (b) Size data and theory for a BDD/HXD droplet evaporating in nitrogen at 295 K. Reprinted with permission from Allen, T. M., Taflin, D. C., and Davis, E. J., Ind. Eng. Chem. Res. 29, 682–690, Copyright © 1990, American Chemical Society.

experiment. This is due to the increasing concentration of nonvolatile impurities that build up as the droplet decreases in size. A factor of 10 in size change corresponds to a factor of 1,000 in the concentration of components that do not transfer across the droplet/gas interface. As a result, data obtained near the end of an experiment should be discarded because of contamination.

Using the data of Fig. 38 and data obtained by attempting to duplicate the run, Allen and her coworkers determined the activity coefficients presented in Fig. 39. The two sets of data are in quite good agreement except at lower mole fractions of DBO, which correspond to the later phases of a run when contamination became significant. Since the activity coefficients for each of the two species were determined from the data, the consistency of the results can be tested by applying the Gibbs–Duhem equation,

$$\frac{d \ln \gamma_2}{d \ln \gamma_1} = -\frac{z_1}{z_2}. \qquad (130)$$

To examine the thermodynamic consistency of the results, let

$$\ln \gamma_1 = \frac{K z_2^2}{RT}, \qquad (131)$$

and

$$\ln \gamma_2 = \frac{K z_1^2}{RT}, \qquad (132)$$

where K is obtained by fitting one set of activity coefficients. Note that Eqs. (131) and (132) satisfy the Gibbs–Duhem equation. Allen *et al.* obtained K via point-by-point fitting of $\ln \gamma_1$ versus z_2^2. Once K has been determined from

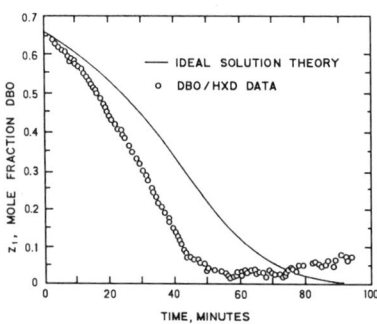 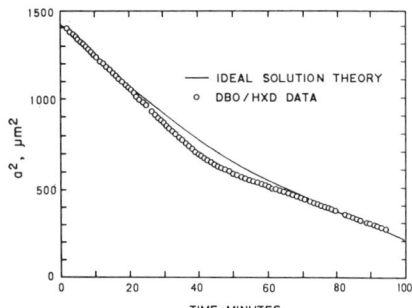

FIG. 38. (a) Composition data and theory for a DBO/HXD droplet evaporating in nitrogen at 295 K, from Allen *et al.* (1990). (b) Size data and th theory for a binary droplet of DBO and HXD evaporating in nitrogen at 295 K. Reprinted with permission from Allen, T. M., Taflin, D. C., and Davis, E. J., *Ind. Eng. Chem. Res.* **29**, 682–690, Copyright © 1990, American Chemical Society.

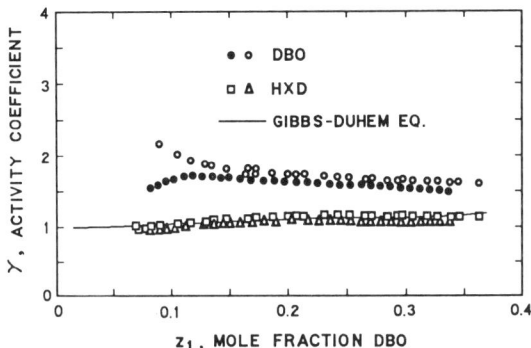

FIG. 39. Activity coefficients calculated from the data of Fig. 37 and from an additional set of data obtained from a duplicate run compared with the Gibbs-Duhem equation. Reprinted with permission from Allen, T. M., Taflin, D. C., and Davis, E. J., *Ind. Eng. Chem. Res.* **29**, 682–690, Copyright © 1990, American Chemical Society.

the results for γ_1, the values of γ_2 calculated using Eq. (132) should agree with the measured values. The solid line in Fig. 39 represents the activity coefficients predicted for HXD using Eq. (132), with K based on the activity coefficients of DBO. The predicted values of γ_{HXD} for the two different runs fall on the same line within the accuracy of the graph, so only one predicted curve is displayed.

This application of multicomponent droplet evaporation to the determination of activity coefficients is only one of several such applications to the study of solution thermodynamics via microdroplet experiments. Tang and Munkelwitz (1984) evaporated solution droplets of $NaCl-H_2O$ and $(NH_4)_2SO_4-H_2O$ in an electrodynamic balance to measure the concentrations at which nucleation commenced. Richardson and his colleagues (Kurtz and Richardson, 1984; Richardson and Spann, 1984; Spann and Richardson, 1985; Richardson and Hightower, 1987; and Hightower and Richardson, 1988) studied nucleation and deliquescence by electrodynamic levitation of single droplets for a number of aqueous systems, including $(NH_4)_2SO_4-H_2O$, $LiI-H_2O$, $NH_4NO_3-H_2O$, mixed ammonium acid sulfate solutions, and $(NH_4)_2SO_4-NH_4NO_3-H_2O$. Droplet levitation is a particularly effective way to achieve the high supersaturations needed for homogeneous nucleation, for this *containerless processing* method eliminates the effects of solid surfaces.

Cohen *et al.* (1987c) used electrodynamic levitation to study the nucleation of several common inorganic salts and mixtures of them. The salts were NaCl, NaBr, KCl, KBr, NH_4Cl, Na_2SO_4, $(NH_4)_2SO_4$, $MnCl_2$, and $FeCl_3$. They also measured water activities for a number of single-electrolyte (1987a) and mixed-electrolyte solution droplets (1987b).

3. Convective Heat and Mass Transfer

In most of the investigations discussed above, the evaporation and condensation processes involved a stagnant gas surrounding a levitated droplet. Because the microparticle is held quite tenaciously by the electrical fields of the quadrupole, it is possible to have a significant gas flow through the balance chamber, which permits the study of external convective heat and mass transfer. Kronig and Bruijsten (1951) appear to have been the first to develop a solution of the convective diffusion equation for heat and mass transfer from a sphere into a flowing stream at low Reynolds number. A regular perturbation technique failed, so they developed an *ad hoc* singular perturbation method to arrive at the following expression for the Sherwood number (Sh = $2K_G a/D_{ij}$, where K_G is the gas phase mass transfer coefficient) or Nusselt number (Nu = $2h_G a/k_j$, in which h_G is the gas phase heat transfer coefficient, and k_j is the thermal conductivity of the gas):

$$\text{Sh or Nu} = 2 + \frac{1}{2}\text{Pe} + \frac{581}{1920}\text{Pe}^2 + \cdots, \tag{133}$$

where the Peclet number for mass transfer is defined by Pe = $2aU_\infty/D_{ij}$, and for heat transfer Pe = $2aU_\infty/\alpha$. Here U_∞ is the uniform gas velocity upstream of the sphere, α is the thermal diffusivity of the surrounding medium, and creeping flow (Re $\ll$ 1) is assumed.

Using a somewhat more sophisticated singular perturbation method, Acrivos and Taylor (1962) obtained

$$\text{Sh or Nu} = 2 + \frac{1}{2}\text{Pe} + \frac{1}{4}\text{Pe}^2 \ln \text{Pe} + 0.03404\text{Pe}^2 + \frac{1}{16}\text{Pe}^3 \ln \text{Pe} + \cdots. \tag{134}$$

Relaxing the restriction of low Reynolds number, Rimmer (1968, 1969) used a matched asymptotic expansion technique to develop a solution in terms of Pe and the Schmidt number Sc (or Prandtl number Pr for heat transfer), where Sc = ν/D_{ij} and Pr = ν/α in which ν is the kinematic viscosity of the flowing fluid. His solution, valid for Pe < 1 and Sc = O(1), is

$$\text{Sh} = 2 + \frac{1}{2}\text{Pe} + \frac{1}{4}\text{Pe}^2 \ln \text{Pe} + f(\text{Sc})\text{Pe}^2 + \cdots, \tag{135}$$

and the function $f(\text{Sc})$ [or $f(\text{Pr})$] is

$$f(\text{Sc}) = -\frac{1}{4}\left[\frac{173}{160} + \ln 2 - \gamma - \frac{\text{Sc}^2}{2} + \frac{\text{Sc}}{4} - \left(1 + \frac{3}{2}\text{Sc} - \frac{\text{Sc}^3}{2}\right)\ln\left(1 + \frac{1}{\text{Sc}}\right)\right], \tag{136}$$

where γ is Euler's constant ($\gamma = 0.57722...$). In the limit $Sc \to \infty$, $f(Sc) = 0.03404$, which recovers the result of Acrivos and Taylor.

For large values of Pe and Re $\ll 1$, Levich (1962) solved the governing convective diffusion equation using a similarity solution method, which yielded

$$Sh = 1.008 Pe^{1/3}. \tag{137}$$

Acrivos and Goddard (1965) developed an asymptotic expansion for the large-Pe problem and obtained

$$Sh = 0.991 Pe^{1/3}[1 + 0.92 Pe^{-1/3} + O(Pe^{-2/3})]. \tag{138}$$

All of these solutions fail for $Pe = O(1)$, so Davis and his colleagues (Zhang and Davis, 1987; Taflin and Davis, 1987) performed evaporative mass transfer experiments over the Peclet number range $0.01 < Pe < 4$. using electrodynamically levitated droplets of hexadecane in flowing N_2 and He and dodecanol in N_2.

In this case the quasi–steady-state result, Eq. (104), is replaced by

$$a^2 = a_0^2 - \frac{Sh D_{ij} M_i p_i^0(T_\infty)}{\rho R T_\infty}(t - t_0), \tag{139}$$

which reduces to Eq. (104) for $Pe \ll 1$. Thus, the absolute value of the slope of a graph of a^2 versus time should increase as the gas velocity is increased.

Zhang and Davis used phase functions to determine the droplet radius as a function of time, and Taflin and Davis analyzed resonance spectra to obtain that information. In both cases the gas entered as a laminar jet through a hole in the bottom electrode, and the jet was calibrated to yield the gas velocity "seen" by the droplet as a function of volumetric flow rate through the balance. Typical results for a dodecanol droplet evaporating in nitrogen obtained by Taflin and Davis are presented in Fig. 40. For this particular run the nitrogen flow rate was set at 87 mL/min at the outset of the experiment. At about 26 minutes the flow rate was decreased to 39 mL/min, and then at about 54 minutes it was increased to 141 mL/min. The changes in slope are clearly seen in the figure, and the Sherwood number is simply determined as twice the ratio of the slope for evaporation with flow to the slope measured for stagnant gas. The Peclet number at each point was determined from the velocity U_∞ determined by the flowmeter calibration, the size determined by light-scattering, and the diffusivity determined from measurements in a stagnant gas by the methods discussed earlier.

Based on data of the type displayed in Fig. 40, Davis and his coworkers determined the Sherwood numbers for a number of droplet experiments, and their results are displayed in Fig. 41, a graph of the Sherwood number as a function of the Peclet number.

As expected, the large-Pe theory of Levich fails to predict the Sherwood

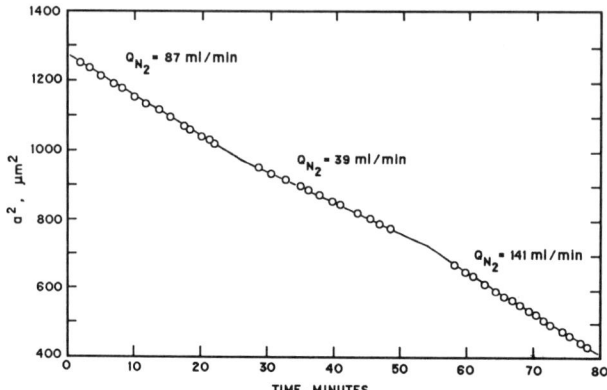

FIG. 40. The effect of gas flow rate on the evaporation rate of a dodecanol droplet in nitrogen at 295 K, from Taflin and Davis (1987).

numbers at these lower Peclet numbers, but the equations of Kronig and Bruijsten and Acrivos and Taylor are in good agreement with the data for Pe $\leqslant$ 1. For Pe $\geqslant$ 1, however, both equations diverge from the experimental data.

Zhang and Davis proposed an interpolation formula based on Churchill and Usagi's (1972) method of combining asymptotic solutions to obtain a correlation valid over the whole range of variables. Their correlation is

$$Sh = 2 + \left[\frac{1}{(0.5Pe + 0.3026Pe^2)^n} + \frac{1}{(1.008Pe^{1/3})^n}\right]^{-1/n}. \quad (140)$$

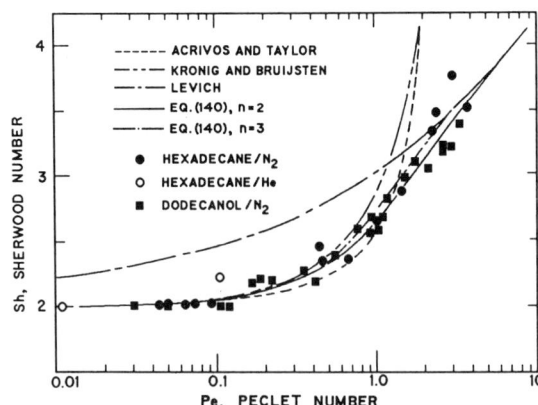

FIG. 41. A comparison among mass transfer equations and the droplet evaporation data of Zhang and Davis (1987) and Taflin and Davis (1987).

This correlation reduces to the solution of Kronig and Bruijsten for small Pe and recovers Levich's result for large Pe. As shown in Fig. 41, the best fit of the data corresponds to $n = 3$, but $n = 2$ fits the data nearly as well.

Finlayson and Olson (1987) used the Galerkin finite element numerical method to explore heat transfer to spheres at low to intermediate Reynolds numbers ($1 \leqslant \text{Re} \leqslant 100$) and for Prandtl numbers in the range 0.001–1,000. They found that the best correlation of their data was an interpolation formula of the form proposed by Zhang and Davis; their correlation is

$$\text{Nu} = 2 + \left[\frac{1}{\text{Pe}/2} + \frac{1}{(0.9\text{Pe}^{1/3}\text{Re}^{0.11})}\right]^{-1} \tag{141}$$

B. Particle Heating by Electromagnetic Energy Transfer

The study of microparticles at elevated temperatures requires particle heating, and this can be accomplished in at least three ways: (i) the chamber and gas surrounding the levitated particle can be heated to the desired temperature by conventional methods of a heating jacket around the balance and preheating of the gas entering the chamber, (ii) radiant heating of the chamber and particle, and (iii) electromagnetic heating of the particle by a laser beam. All of these methods are fraught with difficulty because of large convection effects within the chamber and the difficulty of controlling and measuring the particle temperature, but direct heating of the particle by electromagnetic radiation is a particularly attractive approach that was pioneered by Spjut et al. (1985) in their *electrodynamic thermogravimetric analyzer* (EDTGA). Bar-Ziv et al. (1989) published an extensive review of the application of the EDTGA to combustion studies.

1. *Electromagnetic Heating Theory*

Electromagnetic energy absorption by a small dielectric sphere can be analyzed by the formalism developed by Mie (1908) and Debye (1909). Kerker and Cooke (1973) developed expressions for electromagnetic energy absorption that were subsequently used by Lin (1975) and Akhtaruzzaman and Lin (1977) to develop closed-form formulas for the photophoretic velocity, radiometric force, and the mobility of an absorbing dielectric sphere in the free-molecule regime. Dusel et al. (1979) and Pluchino (1981) examined the effects of particle size and complex refractive index on the source function, and they graphically displayed the extremely asymmetrical behavior of the internal electromagnetic fields.

Allen et al. (1991) observed that when Raman spectra were being obtained for an octadecene droplet, the droplet evaporated faster when the laser intensity was increased. Recognizing that the increased evaporation rate was due to weak absorption of the electromagnetic energy, they analyzed the problem of electromagnetic heating of an absorbing droplet to show that the complex refractive index of the dielectric medium can be determined from evaporation measurements. An outline of the theory follows.

Transient heating effects occur for a small particle over a very short time after heating commences, and then quasi-steady state is reached. The quasi-steady temperature distribution within the sphere is governed by the thermal energy equation

$$\kappa_1 \left[\frac{1}{r^2} \frac{\partial}{\partial r} \left(r^2 \frac{\partial T}{\partial r} \right) + \frac{1}{r^2 \sin \theta} \frac{\partial}{\partial \theta} \left(\sin \theta \frac{\partial T}{\partial \theta} \right) + \frac{1}{r^2 \sin^2 \theta} \frac{\partial^2 T}{\partial \phi^2} \right] = -Q(r, \theta, \phi), \quad (142)$$

where κ_1 is the thermal conductivity of the particle, and the source function is given by Eq. (88). The heat source may also be written in terms of the absorption cross section as follows:

$$Q(r, \theta, \phi) = \frac{3\pi a^2 I_i Q_{abs}}{4\pi a^3}. \quad (143)$$

Note that the heat source is proportional to the irradiance of the incident beam.

Internal convection is precluded by the small size of the droplets. Observations of levitated droplets with diameters of order $50\,\mu m$ show no signs of internal convection when the droplets are heated with a laser. To observe internal convection it is necessary to have scattering sites such as crystallites within the droplet.

As indicated by Fig. 23 and Fig. 24, the source function can be highly asymmetrical. For the liquid droplet corresponding to Fig. 23, one would expect the internal temperature to be higher near the back and front of the sphere because of the spikes in the source function in those regions. As a result, the evaporation rate should be enhanced at the rear stagnation point and the front of the sphere. To calculate the evaporation rate when internal heating occurs, one must solve the full problem of conduction within the sphere coupled with convective heat and mass transport in the surrounding gas.

The interfacial boundary condition may be written in terms of the convective heat transfer of sensible heat, latent heat transfer due to evaporation, and, if the surface temperature is high enough, radiant heat transfer. Mathematically, the surface boundary condition is

$$-\kappa_1 \frac{\partial T}{\partial r}(a, \theta, \phi) = h[T(a, \theta, \phi) - T_\infty] - \lambda_{vap} J_i + e_1 \sigma_r [T(a, \theta, \phi)^4 - T_\infty^4], \quad (144)$$

where h is a heat transfer coefficient, J_i is the mass flux of vapor, λ_{vap} is the heat of vaporization, e_1 is the emissivity of the particle, σ_r is the Stefan–Boltzmann radiation constant, and T_∞ is the temperature in the surrounding medium far from the particle.

In the azimuthal and polar angle directions, the temperature distribution satisfies the periodicity conditions

$$T(r, 0, \phi) = T(r, 2\pi, \phi) \quad \text{and} \quad T(r, \theta, 0) = T(r, \theta, 2\pi).$$

For a quasi-steady process, the mass flux is obtained from Eq. (97), for $J_i = \rho_1 da/dt$, where ρ_1 is the droplet density. For $p_\infty = 0$ one obtains

$$J_i = \frac{D_{ij} M_i p_i^0(T_a)}{aRT_a}, \tag{145}$$

where T_a is the surface temperature.

The vapor pressure may be written in terms of the Clausius–Clapeyron equation,

$$\ln \frac{p_i^0(T_a)}{p_i^0(T_\infty)} = \frac{\lambda_{\text{vap}}}{R} \left(\frac{1}{T_a} - \frac{1}{T_\infty} \right). \tag{146}$$

Here the heat of vaporization is considered to be negative.

If the surface temperature does not differ greatly from the surrounding temperature, the highly nonlinear surface boundary condition may be simplified by linearizing the expression for the radiation flux and the Clausius–Clapeyron equation to yield the approximation

$$-\kappa_1 \frac{\partial T}{\partial r}(a) = h_e [T(a) - T_\infty] - \frac{\lambda_{\text{vap}} D_{ij} M_i p_i^0(T_\infty)}{aRT_\infty}, \tag{147}$$

where the effective heat transfer coefficient h_e is defined by

$$h_e = h + \frac{\lambda_{\text{vap}}^2 D_{ij} M_i p_i^0(T_\infty)}{aR^2 T_\infty^3} + 4 e_1 \sigma_r T_\infty^3. \tag{148}$$

Introducing nondimensional variables,

$$U = \frac{(T - T_\infty)}{T_\infty} \quad \text{and} \quad x = \frac{r}{a},$$

the conduction equation and interfacial boundary condition transform to

$$\frac{1}{x^2} \frac{\partial}{\partial x} \left(x^2 \frac{\partial U}{\partial x} \right) + \frac{1}{x^2 \sin \theta} \frac{\partial}{\partial \theta} \left(\sin \theta \frac{\partial U}{\partial \theta} \right) + \frac{1}{x^2 \sin^2 \theta} \frac{\partial^2 U}{\partial \phi^2} = -\frac{a^2}{\kappa_1 T_\infty} Q(x, \theta, \phi) \tag{149}$$

and

$$\frac{\partial U}{\partial x}(1, \theta, \phi) + \text{Bi}\, U(1, \theta, \phi) = A, \tag{150}$$

where A and Bi are parameters defined by

$$A = \frac{\lambda_{\text{vap}} D_{ij} M_i p_i^0(T_\infty)}{\kappa_1 R T_\infty^2} \quad \text{and} \quad \text{Bi} = \frac{h_e a}{\kappa_1}.$$

For a solid particle containing no volatile matter, $A = 0$, and h_e includes only h and the radiation term.

The solution of this system of equations may be written in terms of a Green's function

$$U(x, \theta, \phi) = \frac{A}{(1 + \text{Bi})} x$$
$$+ \int_{-\pi}^{\pi} \int_0^{\pi} \int_0^1 G(x, \theta, \phi; x', \theta', \phi')(F(x', \theta', \phi')x'^2 \sin \theta' \, dx' d\theta' d\phi', \tag{151}$$

where the Green's function is

$$G(x, \theta, \phi; x', \theta', \phi')$$
$$= \sum_{n=0}^{\infty} \frac{1}{4\pi} x'^n \left[x^{-n-1} + \frac{(n+1-\text{Bi})}{(n+\text{Bi})} x^n \right] P_n(\cos \theta^*) \quad (\text{for } x' < x). \tag{152}$$

Here

$$\cos \theta^* = \cos \theta \cos \theta' + \sin \theta \sin \theta' \cos(\phi - \phi'),$$

and the source function is

$$F(x', \theta', \phi') = \frac{2A}{(1 + \text{Bi})x'} + \frac{Q(x', \theta', \phi')}{\kappa_1 T_\infty x'^2}. \tag{153}$$

Once the surface temperature distribution has been computed, using Eq. (151), the droplet evaporation rate is given by integrating the local mass flux over the droplet surface,

$$\frac{da}{dt} = -\frac{D_{ij} M_i}{4\pi a \rho_1 R} \int_0^{2\pi} \int_0^{\pi} \frac{p_1^0(T_a)}{T_a} \sin \theta \, d\theta \, d\phi. \tag{154}$$

Allen et al. (1991) performed these computations for 1-octadecene droplets, and they measured the evaporation rate of the droplets as a function of laser power. To determine the absolute irradiance I_i of the laser beam, they also measured the force on the particle exerted by the laser beam using the techniques discussed above. The photon pressure force is given by Eq. (87), which involves the complex refractive index. The real component of the refractive index n was determined from optical resonance measurements, and the imaginary component was obtained iteratively. That is, they assumed a value of k, and computed the cross-sections C_{abs}, C_{sca}, C_{ext}, and C_{pr} using the equations presented earlier. Next, they computed the evaporation rate for this assumed value of k, and they compared the results with the experimental data. If the results did not agree, they adjusted k and repeated the exercise. Their results are presented in Table III.

Table III shows that the experimental and predicted evaporation rates are in good agreement at all beam intensities. There is some inconsistency at the highest power levels. It was difficult to maintain the droplet in the center of the laser beam at the highest power level, and the measured evaporation rate is somewhat low as a result of that problem. Additional computations demonstrate that the predicted evaporation rate is quite sensitive to the choice of the imaginary component of N, so the results suggest that this evaporation method is suitable for the determination of the complex refractive index of weakly absorbing liquids. For strong absorbers, the linearizations of the Clausius–Clapeyron equation and of the radiation energy loss term in the interfacial boundary condition may not be valid. In this event, a numerical solution of the governing equations is required. The structure of the source function, however, makes this a rather tedious task.

2. The EDTGA

Spjut and his associates heated levitated microparticles and measured their temperatures radiometrically. They levitated a microparticle in a bihyperboloidal quadrupole, and they irradiated the particle from opposite sides by splitting a cw CO_2 laser beam and directing the beams by means of mirrors to illuminate the particle symmetrically (Spjut et al., 1987). The particle temperature was measured by multiple-color pyrometry (Spjut, 1987; Spjut and Bolsaitis, 1987) using narrow-band infrared detectors. The temperature

TABLE III
RESULTS FOR AN OCTADECENE DROPLET EVAPORATING IN N_2 AT 297.2 K FOR COMPLEX
REFRACTIVE INDEX $N = 1.4521 + i1.3 \times 10^{-7}$[a]

Radius (μm)	Laser Power (W)	Beam Intensity (W/m^2)	Force (N)	Measured da^2/dt (μm^2/s)	Theoretical da^2/dt (μm^2/s)
21.08	1.2	8.85×10^7	1.37×10^{-10}	-0.0202	-0.0202
20.60	1.0	7.53×10^7	1.15×10^{-10}	-0.0206	-0.0202
19.64	0.75	5.10×10^7	7.06×10^{-11}	-0.0182	-0.0183
18.42	0.50	2.88×10^7	3.47×10^{-11}	-0.0174	-0.0175
23.55	low	low	—	-0.0167	-0.0167

[a]From Allen et al. (1991).

was obtained from the measured intensities I_λ by applying Planck's distribution law,

$$I_\lambda = \frac{2\varepsilon_\lambda C_1}{\lambda^5 [\exp(C_2/\lambda T) - 1]}, \qquad (155)$$

where C_1 and C_2 are constants, and ε_λ is the wavelength-dependent emissivity of the particle. The accuracy of the measurement depends on the accuracy of the intensity measurement and on knowledge of the spectral emissivity. The measurement of thermal transients requires a detector with a very fast response time to yield a reliable temperature.

Figure 42 shows the temperatures measured by two-color pyrometry for step changes in temperature compared with the true temperature and calculated temperatures based on the response characteristics of the detectors (time constant = 0.311 s). The response speed of the detectors in this case was too slow to follow the actual temperature decrease, but the temperature rise is reasonably well detected. Spjut and Bolsaitis reported that two-color temperatures are unreliable when the optical properties of the microparticle change during the experiment, but they showed that single-wavelength temperatures can yield consistent results and, with some caveats, are adequate for particle temperature measurement.

It should be pointed out that there is an upper limit of temperature for successful particle levitation, for charge loss occurs at elevated temperatures. When the particle temperature is sufficiently high, two mechanisms can cause the particle to lose charge. Thermionic emission and charge neutralization due to thermal ionization of the gas can both produce rapid charge loss. Bar-Ziv et al. (1989) discussed these phenomena, showing that the upper limit is approximately 1,500 K. Ward and Davis (1990) examined the analogous problem of charge loss due to gas-phase ionization by a radio-active source,

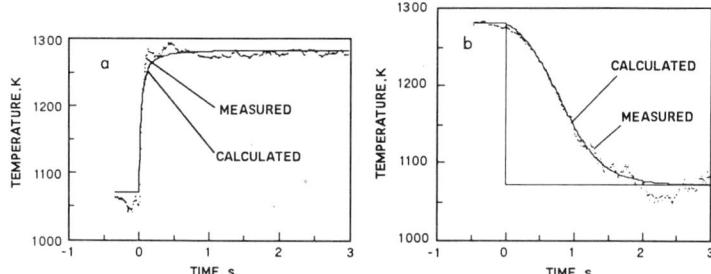

FIG. 42. The response to a step increase (a) and a step decrease (b) in temperature of a laser-heated particle, from Spjut and Bolsaitis (1987). The data were obtained using two-color pyrometry, and the calculated curves are based on the response characteristics of the detectors.

showing that the electrodes of the quadrupole compete with the particle for charge capture.

Although research on single particles at elevated temperatures is still in its infancy and difficulties abound, electromagnetic heating of levitated microparticles offers much opportunity for precise measurement of devolatilization rates and other phenomena, including gas/solid and gas/droplet chemical reactions.

V. Gas/Microparticle Chemical Reactions

There has been much less work done in the area of microparticle chemistry compared with studies of the physics of small particles, but this is an area of considerable interest to chemical engineers. The work of Matijevic and his colleagues on polymeric aerosols and of Rubel and Gentry on gas/droplet reactions was mentioned earlier, as was the application of fluorophores by Ward et al. (1987) to explore microparticle polymerization. The recent development of microparticle spectroscopic techniques makes it possible to follow chemical reactions between a reactive gas and a microparticle.

In addition to the gravimetric method used by Rubel and Gentry and the fluorescence measurements of Ward and his coworkers, two methods of following gas/particle chemical reactions have been developed recently.

A. Optical Resonance Spectroscopy

Taflin and Davis (1990) applied a new method, which might be termed *optical resonance spectroscopy*, to follow the reaction between bromine vapor and a levitated droplet of an olefin, 1-octadecene (OCT). The reaction is highly exothermic and, if carried out in bulk, easily overheats and produces undesired by-products in addition to the primary product, 1,2-dibromooctadecane (DBO). The reaction can be followed gravimetrically because it involves a large percentage mass change due to the molecular weight difference between OCT and DBO ($M_{OCT} = 252.49$ and $M_{DBO} = 412.30$). Furthermore, the reactant and product have different refractive indices ($n_{OCT} = 1.4448$ and $n_{DBO} = 1.4800$), and both are Raman-active. These attributes make the reaction a good model reaction to develop laboratory techniques.

Taflin (1988) recognized that the refractive index change associated with the OCT/Br_2 reaction would affect optical resonances of a reacting droplet, so he carried out bromination experiments with levitated droplets using the apparatus shown in Fig. 43.

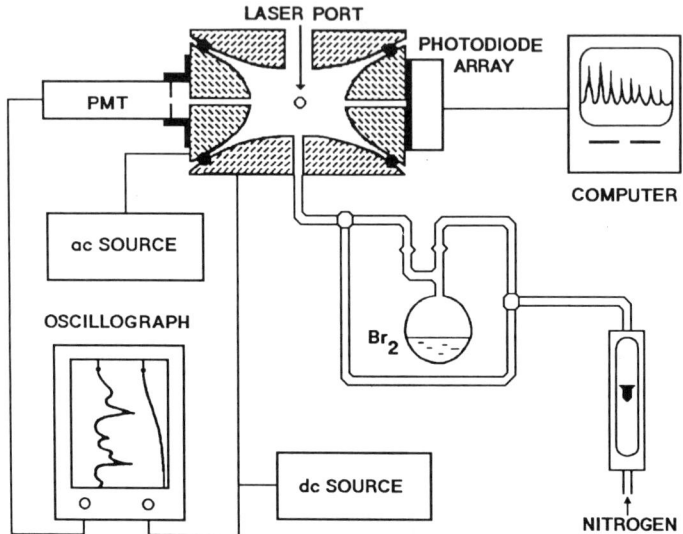

FIG. 43. The apparatus used by Taflin and Davis (1990) to study the bromination of an olefin droplet by optical resonance spectroscopy. Reprinted with permission from *J. Aerosol Sci.* 21, 73–86, Taflin, D. C., and Davis, E. J., Copyright © 1990, Pergamon Press plc.

Taflin trapped the droplet in a flow of nitrogen, then switched the nitrogen flow through the bromine generator shown in the figure. The dilute vapor/gas mixture flowed past the droplet, and the chemical reaction was followed by recording the levitation voltage, phase functions, and the optical resonance spectrum of the scattered light at right angles to the vertically polarized laser beam.

An optical resonance spectrum covering the duration of the reaction is presented in Fig. 44. Three stages of the experiment are shown: (i) prior to the introduction of bromine, slow evaporation of OCT occurs; (ii) after the introduction of bromine, it is absorbed and chemically reacts; and (iii) after completion of the reaction, the bromine flow is stopped, and desorption of dissolved bromine occurs together with very slow evaporation of the low-volatility product DBO. During the intermediate stage the resonance frequency changes significantly as the reaction proceeds. The reaction slows greatly as it nears completion, as evidenced by the decreasing resonance frequency late in the second stage.

Interpretation of the resonance spectrum is quite tedious because both the size and refractive index varied during the experiment. Phase function data obtained during the experiment provide an estimate of the size, and this estimate was used to initiate computations of the resonance spectrum from Mie theory to match the experimental spectrum. Once the refractive index

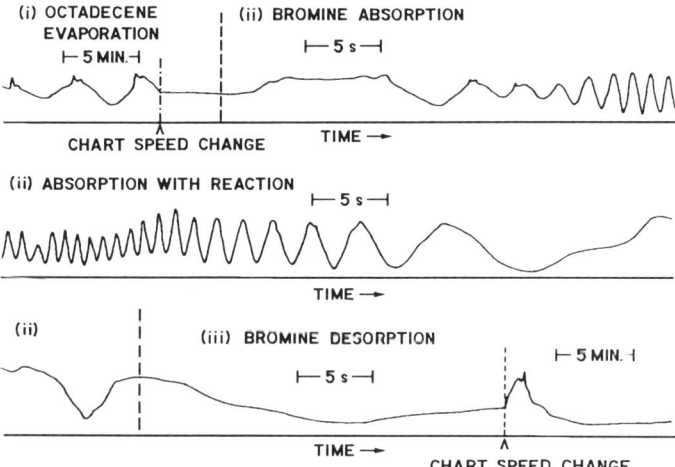

FIG. 44. An optical resonance spectrum from Taflin and Davis (1990) for the chemical reaction between an octadecene droplet and bromine vapor. Reprinted with permission from *J. Aerosol Sci.* **21**, 73–86, Taflin, D. C., and Davis, E. J., Copyright © 1990, Pergamon Press plc.

was determined by such matching procedures, the droplet composition was obtained from the relationship between refractive index and composition. The composition can also be calculated from the levitation voltage, and Fig. 45 compares the results obtained by the two methods.

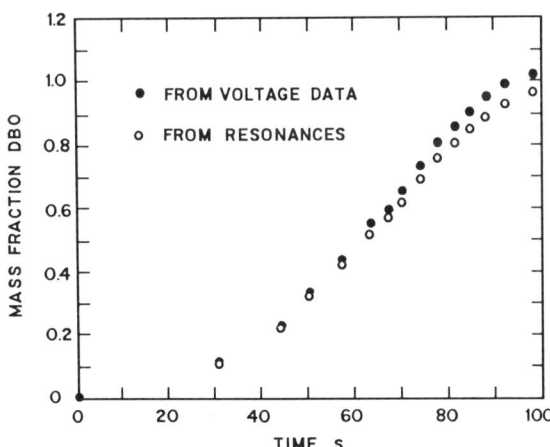

FIG. 45. Droplet composition data for a reacting droplet of octadecene based on the levitation voltage and analysis of the optical resonance spectrum. Reprinted with permission from *J. Aerosol Sci.* **21**, 73–86, Taflin, D. C., and Davis, E. J., Copyright © 1990, Pergamon Press plc.

84　　　　　　　　　　　E. JAMES DAVIS

The composition of the product determined from the levitation voltage data is slightly greater than that extracted by analysis of the optical resonance spectrum, but the results are generally in good agreement. The voltage data had to be corrected for the drag force exerted on the droplet by the gas flow, and this correction is subject to some error, so it is likely that the voltage data are somewhat less accurate than the results obtained from the light-scattering data.

A major problem with the experiments of Taflin is that the bromine composition varied during an experiment because of the relatively long space-time (ratio of the chamber volume to the gas flow rate) associated with the light-scattering chamber. Gas entering the chamber as a laminar jet mixed with the gas already in the chamber, which led to dilution of the inlet bromine concentration during much of the reaction time. The chamber had a volume of approximately 25 cm^3, and the gas flow rate ranged from 22 mL/min to 65 mL/min. Thus, a typical space-time was about one minute, which is of the order of the reaction time.

Although optical resonance spectroscopy is suitable for studying chemical reactions involving two reactants and a single product, it cannot be extended to more than two or three components because one cannot interpret the resonance spectrum uniquely to obtain the composition from the refractive index. Raman spectroscopy is an attractive alternative approach.

B. Raman Spectroscopy

Using Raman spectroscopy, Buehler (1991) examined the same reaction studied by Taflin, and he attempted to eliminate the problem of a large space-time by means of the apparatus shown in Fig. 13. The mesh electrodes permitted gas to flow through the dc electrodes, and a large flow rate could be achieved without having a high velocity, which tends to destabilize the particle. Furthermore, the gas passes through the chamber more as plug flow than as the mixed flow encountered by Taflin. As a result, the reactive-vapor and inert-gas mixture entering the chamber tended not to become diluted. The light-scattering chamber shown in Fig. 13 was coupled to an optical system and dual monochromator, shown in an overhead view in Fig. 46.

Buehler explored the possibility of following the olefin/Br$_2$ reaction via semi-continuous recording of Raman spectra. The C═C bond of the olefin and the C—Br bonds of the product both have well-defined Raman peaks, and Fig. 47 shows the relevant parts of the Raman spectrum for a 50 mass% mixture of 1-bromooctadecane and 1-octadecene. The large peak between the smaller peaks characteristic of the C═C and C—Br bonds is a weak line of the excitation source, which was an argon-ion laser operating at a nominal power of one watt at $\lambda = 488$ nm. The weak line corresponding to $\lambda = 507$ nm

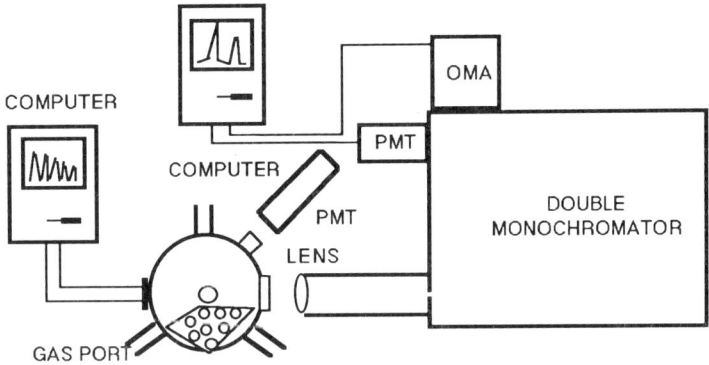

FIG. 46. An overhead view of the electrodynamic balance and Raman spectrometer developed by Buehler (1991) for gas/microparticle chemical reaction studies.

was used as a reference line. The C—Br bond has a peak at a wavenumber, v, of $650\,\text{cm}^{-1}$, and the C=C bond corresponds to $v = 1,650\,\text{cm}^{-1}$.

Thus, if the monochromator is set to a wavenumber of $650\,\text{cm}^{-1}$, the formation of the C—Br bond can be recorded as a function of time during a chemical reaction between bromine and an olefin. Buehler used this method to follow the bromine/octadecene reaction. It should be pointed out that the Raman scattering is a function of droplet size as well as composition, and unlike bulk Raman it is complicated by the morphological resonances

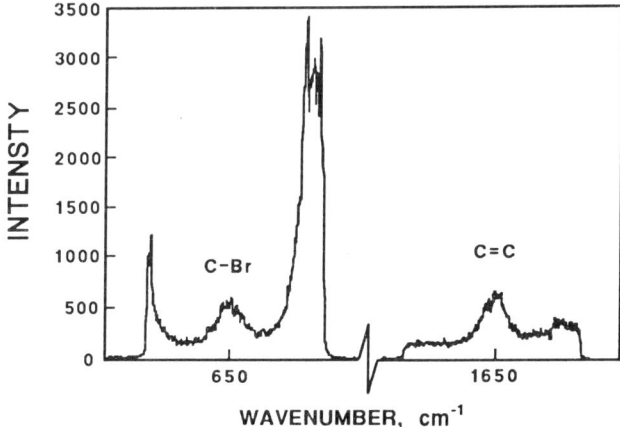

FIG. 47. A partial Raman spectrum for a binary droplet consisting of 50 mass % 1-octadecene and 1-bromooctadecane, from Davis and Buehler (1990).

demonstrated in Fig. 28. Buehler's objective was to determine if Raman spectroscopy can be used to monitor the droplet composition during the chemical reaction, taking into account the complications introduced by optical resonances.

Buehler used the single channel detector (PMT) shown in Fig. 46 with the monochromator set at $v = 650 \pm 2\,\text{cm}^{-1}$ to record the Raman signal for an integration time of one second. The PMT was cooled to 253 K by means of a solid-state cooler, and its output was processed by photon-counting electronics. The levitation voltage, the elastic morphological resonance spectrum, and the Raman signal were recorded simultaneously, and Fig. 48 displays the elastic and inelastic scattering signals for a bromine/octadecene reaction experiment.

Bromine entered the light-scattering chamber approximately at the start of the section of the oscillograph tracing shown. The absorption of bromine causes an increase in droplet size, which is associated with the first large peak in the upper tracing of Fig. 48. The two arrows located to the right of this first peak indicate when the dc levitation voltage was adjusted. This voltage adjustment, needed to balance the particle, moved the droplet in the laser beam and increased the scattering intensity. The sequence of elastic scattering peaks following bromine addition result from the changes in size and refractive index of the droplet, and the frequency of these peaks is seen to decrease as the reaction nears completion. The Raman signal in the lower tracing shows an increase in the intensity associated with the C—Br bond, and the fluctuations in the irradiance indicate the effects of the elastic morphological resonances on the Raman scattering. The levitation voltage

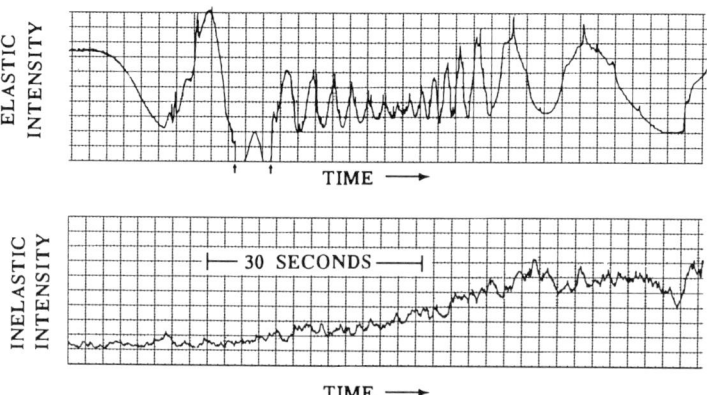

FIG. 48. The elastic resonance spectrum and the Raman signal during the bromination of 1-octadecene (from Buehler, 1991).

(not displayed in Fig. 48) showed a smooth increase in the droplet mass as reaction proceeded.

Since the Raman signal depends on the amount of the relevant chemical present, there should be a correspondence between the Raman irradiance associated with the product and the mass of the droplet as indicated by the dc voltage. Figure 49 compares the time-dependent raw data of V_{dc} and the output of the photon-counting processor for the Raman signal. Both sets of data are sigmoid curves increasing from the unreacted initial state of the droplet to complete conversion. The voltage data can be used to obtain the conversion quite directly, but, based on experiments with binary droplets, Buehler found that the Raman signal correlates better with the volume fraction of the chemical species than with its mass. Thus, one should not expect close agreement between the Raman signal and the levitation voltage, for the density of the droplet varied from $790 \, kg/m^3$ at the outset of the experiment to $1,048 \, kg/m^3$ when the reaction was complete. As a result of this large density variation, the volume fraction and mass fraction of either component are quite different during the time of the reaction. Furthermore, the Raman signal is affected by morphological resonances, which account for the peaks discernable in the Raman signal at about 19 s, 40 s, and 79 s. There are no corresponding voltage fluctuations.

The studies of Taflin and Buehler indicate that microparticle spectroscopic techniques can be used to follow gas/microparticle chemical reactions. The use of morphological resonances to determine the refractive index of a reacting droplet has limited applicability because there must be a unique relationship between composition and refractive index to allow the method to be used to follow chemical reactions. Raman spectroscopy has broader applications, but one must deal with morphological resonances if droplets are

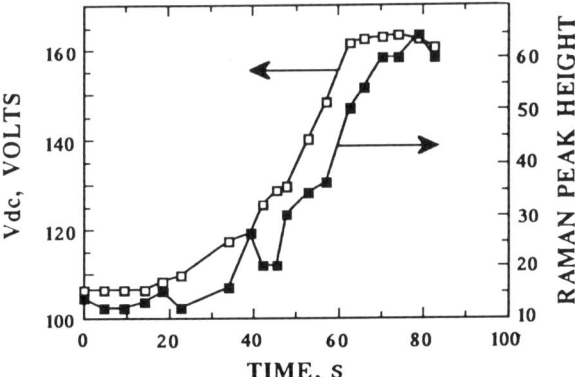

FIG. 49. The levitation voltage and the Raman peak height associated with the C—Br bond for the chemical reaction between 1-octadecene and bromine (from Buehler, 1991).

involved. Solid particles do not have well-defined resonance characteristics because of their nonuniformity. They rotate and "twinkle" in a laser beam. To minimize the effects of the nonuniformity on the inelastic scattering signal, one should average the signal over a time large compared with the rotational time constant.

Thus, provided that one takes into account the effects of morphological resonances on the Raman spectra of microspheres, it appears that qualitative and quantitative chemical analysis of microparticles can be achieved.

VI. Concluding Comments

This review of the chemistry and physics of microparticles and their characterization is by no means comprehensive, for the very large range of masses that can be studied with the electrodynamic balance makes it possible to explore the spectroscopy of atomic ions. This field is a large one, and Nobel laureates Hans Dehmelt and Wolfgang Paul have labored long in that fruitful scientific garden. The application of particle levitation to atmospheric aerosols, to studies of Knudsen aerosol phenomena, and to heat and mass transfer in the free-molecule regime would require as much space as this survey.

There are numerous applications to chemical engineering research currently under study in several laboratories in the United States and Europe, and the author hopes that this review will stimulate even more research. Microparticle chemical reaction studies are in their infancy, and there is much to be learned at the level of the single particle because internal diffusion can be eliminated as a rate-controlling process. Reactions at elevated temperatures are possible with the *caveat* that there is an upper limit above which charge-loss accelerates.

Although this overview of the electrodynamic balance and its applications has been directed to chemical engineers, it is hoped that physicists, chemists, atmospheric scientists, aerosol researchers, and environmental engineers will find something here to interest them.

Acknowledgment

The author is indebted to the National Science Foundation (CTS-8611779), the Department of Energy (Grants DE-FG21-87MC24214 and DE-FG22-89PC89790), the Petroleum Research Fund of the American Chemical Society (Grant No. 16422-AC7), and Philip Morris USA for grants in support of microparticle research. Special thanks go to Professors Milton Kerker and Egon Matijevic of Clarkson University for introducing the author to the *world of small dimensions* and the marvels of light scattering. Many of my students and former students made significant contributions to microparticle research and should be recognized. Among these are Dr. Randoph Chang, Professor Asit K. Ray, Dr. Ravindran Periasamy, Dr. Christopher Guzy, Dr. Daniel Taflin, Dr. Timothy Ward, Dr. Mark Buehler, Theresa Allen, John Fulton, Michael Bridges, Edward Chorbajian, and Deborah Hart.

Notation

a	Droplet radius, μm	J	Mass flux, kg/m²-s
a_n	Scattering coefficient	k	Wave number
A	Constant in Eq. (150)	k_B	Boltzmann constant, J/K
A^*	Constant in Eq. (121)	Kn	Knudsen number
b_n	Scattering coefficient	K	Constant in Eq. (131)
B^*	Constant in Eq. (121)	m	Mass, kg
$Bi = h_e a/\kappa_1$	Biot number	$m = N_i/N$	Refractive index ratio
c	Velocity of light, m/s	M	Molecular weight, kg/kmol
c_n	Internal field coefficient	**M**	Vector spherical harmonic function
C	Molar concentration, kmol/m³	**M**	Dipole moment
C_k	Cross-section, m²	n	Number density, m⁻³
C_0	Balance constant	N	Refractive index
d_n	Internal field coefficient	$Nu = 2ah_e/k$	Nusselt number
D_{ij}	Diffusion coefficient, m²/s	**N**	Vector spherical harmonic function
e	Charge on the electron, C		
e	Unit vector	p	Pressure, Pa
e_1	Emissivity	p^0	Vapor pressure, Pa
E	Electric vector, V/m	P	Total pressure, Pa
f	ac frequency	$P_n^m(x)$	Associated Legendre function
F	Force, N		
$F(x, \theta, \phi)$	Function defined by Eq. (153)	$Pe = 2aU_\infty/D_{ij}$	Peclet number for mass transfer
g	Gravitational constant, m/s²	$Pe = 2aU_\infty/\alpha$	Peclet number for heat transfer
G	Green's function	q	Coulombic charge, C
h	Planck constant, J/Hz	**q**	Heat flux, W/m²
h_k	Heat transfer coefficient, W/m²-K	$Q(x, \theta, \phi)$	Source function
		Q_k	Irradiation cross section, m²
H	Heaviside function		
H	Magnetic vector	r	Radial coordinate
i	Intensity, W/m²	R	Ideal gas constant
I	Intensity, W/m²	$Re = 2aU_\infty/v$	Reynolds number
j_n	Spherical Bessel function of the first kind	S_k	Scattering function
		S_{ij}	Slope of a² versus t plot

S'	Slope in Eq. (115)	ϕ_H	Correction factor in Eq. (102)
S	Poynting vector		
$Sc = \nu/D_{ij}$	Schmidt number	ϕ_M	Correction factor in Eq. (102)
$Sh = 2aK_G/D_{ij}$	Sherwood number		
t	Time, s	Φ_{ij}	Lennard–Jones potential
T	Temperature, K	γ_i	Activity coefficient
U	Dimensionless temperature	κ	Thermal conductivity, W/m-K
U_∞	Upstream velocity, m/s		
V	Voltage, V	λ	Wavelength of light, m
V_k	Molar volume, m³/kmol	λ_{vap}	Heat of vaporization, J/kmol
W	Electromagnetic energy flow, W	Λ	Mean free path, m
$x = r/a$	Dimensionless radial coordinate	μ	Viscosity, N-s/m²
		μ	Magnetic permeability, H/m
$x = 2\pi N a/\lambda$	Size parameter		
$\mathbf{x}$	Position vector	ν	Kinematic viscosity, m²/s
y	Transverse coordinate	θ	Polar angle
$y_n(x)$	Spherical Bessel function of the second kind	ρ	Density, kg/m³
		σ	Electrical conductivity
z	Mole fraction	σ_{lg}	Surface tension, N/m
z_0	Half distance between electrodes	σ_{ij}	Collision diameter
		σ_r	Stefan–Boltzmann radiation constant, W/m²-K⁴
$z_n(x)$	Spherical Bessel function		
$Z = z/z_0$	Dimensionless vertical position	τ	Dimensionless time
		ν_m	Frequency, s⁻¹
Greek		$\omega = 2\pi f$	Circular frequency, s⁻¹
α	Constant in Eq. (118)	Ω_{ij}	Collision integral
α	Thermal diffusivity, m²s	ξ	Extent of reaction in Eq. (23)
$\boldsymbol{\alpha}$	Polarizability tensor		
β	Field strength parameter	ψ	Generating function
β	Constant in Eq. (118)		
β'	Parameter in Eq. (115)	*Subscripts*	
χ	Electric susceptibility	a	Evaluated at the surface
δ	Drag parameter	abs	Absorbed energy
ε_{ij}	Lennard–Jones force constant	0	Reference state of initial state
ε_0	Permittivity of free space, F/m	1	Interior phase
		2	Exterior phase
ϕ	Potential	∞	Evaluated far from the surface
ϕ	Azimuthal angle		
ϕ_i	Parameter defined by Eq. (124)	i	Incident beam
		s	Scattered irradiation

References

Abbas, M. A., and Latham, J., *J. Fluid Mech.* **30**, 663 (1967).

Abramowitz, M., and Stegun, I. A., "Handbook of Mathematical Functions." U.S. Government Printing Office, Washington, D.C., 1964.

Acrivos, A., and Goodard, J. D., *J. Fluid Mech.* **23**, 273 (1965).
Acrivos, A., and Taylor, T. D., *Phys. Fluids* **5**, 387 (1962).
Aden, A. L., and Kerker, M., *J. Appl. Phys.* **22**, 1242 (1951).
Akhtaruzzaman, A. F. M., and Lin, S. P., *J. Colloid Interface Sci.* **61**, 170 (1977).
Allen, T. M., Taflin, D. C., and Davis, E. J., *Ind. Eng. Chem. Res.* **29**, 682 (1990).
Allen, T. M., Buehler, M. F., and Davis, E. J., *J. Colloid Interface Sci.* **142**, 343 (1991).
Anderson, A., "The Raman Effect, Vol. 1: Principles." Marcel Dekker, New York, 1971.
Arnold, S., *J. Aerosol Sci.* **10**, 49 (1979).
Arnold, S., and Amani, Y., *Opt. Lett.* **5**, 242 (1980).
Arnold, S., Amani, Y., and Orenstein, A., *Rev. Sci. Instrum.* **51**, 1202 (1980).
Arnold, S., Neuman, M., and Pluchino, A. B., *Opt. Lett.* **9**, 4 (1984).
Ashkin, A., and Dziedzic, J. M., *Phys. Rev. Lett.* **38**, 1351 (1977).
Ashkin, A., and Dziedzic, J. M., *Appl. Opt.* **20**, 1803 (1981).
Ashkin, A., and Dziedzic, J. M., *Science* **235**, 1517 (1987).
Ashkin, A., Dziedzic, J. M., Bjorkholm, J. E., and Chu, S., *Opt. Lett.* **11**, 288 (1986).
Ataman, S., and Hanson, D. N., *Ind. Eng. Chem. Fundam.* **8**, 833 (1969).
Bar-Ziv, E., Jones, D. B., Spjut, R. E., Dudek, D. R., Sarofim, A. F., and Longwell, J. P., *Combust. Flame* **75**, 81 (1989).
Benner, R. E., Barber, P. W., Owen, J. F., and Chang, R. K., *Phys. Rev. Lett.* **44**, 475 (1980).
Berg, T. G. O., Trainor, R. J., Jr., and Vaughan, U., *J. Atmos. Sci.* **27**, 1173 (1970).
Bird, R. B., Stewart, W. E., and Lightfoot, E. N., "Transport Phenomena." Wiley, New York, 1960.
Bohren, C. F., and Huffman, D. R., "Absorption and Scattering of Light by Small Particles." John Wiley & Sons, New York, 1983.
Bohren, C. F., and Hunt, A. J., *Can. J. Phys.* **55**, 1930 (1977).
Born, M., and Wolf, E., "Principles of Optics, Electromagnetic Theory of Propagation Interference and Diffraction of Light," 6th ed. Pergamon Press, Oxford, 1980.
Bridges, M. A., M. S. thesis, University of Washington, Seattle (1990).
Brock, J. R., *J. Colloid Sci.* **17**, 768 (1962).
Buehler, M. F., Ph.D. dissertation, University of Washington, Seattle (1991).
Buehler, M. F., Allen, T. M., and Davis, E. J., *J. Colloid Interface Sci.* **146**, 343 (1991).
Chang, R., and Davis, E. J., *J. Colloid Interface Sci.* **54**, 352 (1976).
Chapman, S., and Cowling, T. G., "The Mathematical Theory of Nonuniform Gases," 3rd ed. Cambridge University Press, Cambridge, United Kingdom, 1970.
Chew, H., McNulty, P. J., and Kerker, M., *Phys. Rev. A* **13**, 396 (1976).
Chuchottaworn, P., Fujinami, A., and Asano, K., *J. Chem. Eng. Japan* **16**, 18 (1983).
Churchill, S. W., and Usagi, R., *AIChE J.* **18**, 1121 (1972).
Chylek, P., Ramaswamy, V., Ashkin, A., and Dziedzic, J. M., *Appl. Opt.* **22**, 2302 (1983).
Cohen, M. D., Flagan, R. C., and Seinfeld, J. H., *J. Phys. Chem.* **91**, 4563 (1987a).
Cohen, M. D., Flagan, R. C., and Seinfeld, J. H., *J. Phys. Chem.* **91**, 4575 (1987b).
Cohen, M. D., Flagan, R. C., and Seinfeld, J. H., *J. Phys. Chem.* **91**, 4583 (1987c).
Cooper, D. W., *Aerosol Sci. Tech.* **5**, 287 (1986).
Davis, E. J., *Langmuir* **1**, 379 (1985).
Davis, E. J., *ISA Trans.* **26**, 1 (1987).
Davis, E. J., in "Optical Methods for Ultransensitive Detection and Analysis: Techniques and Applications," Vol. 1435 (B. L. Fearey, ed.). *Proc. SPIE*, p. 216.
Davis, E. J., and Buehler, M. F., *Mat. Res. Soc. Bull.* **15**, 26 (1990).
Davis, E. J., and Chorbajian, E., *Ind. Eng. Chem. Fundam.* **13**, 272 (1974).
Davis, E. J., and Periasamy, R., *Langmuir* **1**, 373 (1985).
Davis, E. J., and Ray, A. K., *J. Chem. Phys.* **67**, 414 (1977).
Davis, E. J., and Ray, A. K., *J. Aerosol Sci.* **9**, 411 (1978).
Davis, E. J., and Ray, A. K., *J. Colloid Interface Sci.* **75**, 566 (1980).

Davis, E. J., Ravindran, P., and Ray, A. K., *Chem. Eng. Commun.* **5**, 251 (1980).
Davis, E. J., Zhang, S. H., Fulton, J. H., and Periasamy, R., *Aerosol Sci. Tech.* **6**, 273 (1987).
Davis, E. J., Ward, T. L., Rodenhizer, D. G., Jenkins, R. W., Jr., and McRae, D. D., *Part. Sci. Tech.* **6**, 169 (1988).
Davis, E. J., Buehler, M. F., and Ward, T. L., *Rev. Sci. Instrum.* **61**, 1281 (1990).
Debye, P., *Ann. Phys.* **30**, 57 (1909).
Dehmelt, H. G., and Major, F. G., *Phys. Rev. Lett.* **8**, 213 (1962).
Dehmelt, H. G., and Walls, F. L., *Phys. Rev. Lett.* **21**, 127 (1968).
Doyle, A., Moffett, D. R., and Vonnegut, B., *J. Colloid Sci.* **19**, 136 (1964).
Dusel, P. W., Kerker, M., and Cooke, D. D., *J. Opt. Soc. Am.* **69**, 55 (1979).
Eisenklam, P., Arunachalam, S. A., and Weston, J. A., "11th Symp. (Intl.) on Combustion," p. 715. The Combustion Institute, New York, 1966.
Finlayson, B. A., and Olson, J. W., *Chem. Eng. Commun.* **58**, 431 (1987).
Fletcher, H., *Phys. Rev.* **33**, 81 (1911).
Frickel, R. H., Shaffer, R. E., and Stamatoff, J. B., Report No. ARCSL-TR-77041, Chemical Systems Laboratory, Aberdeen Proving Ground, Maryland, 1978.
Fuller, E. N., Schettler, P. D., and Giddings, J. C., *Ind. Eng. Chem.* **58**, 19 (1966).
Fulton, J. H., M. S. thesis in Chemical Engineering, University of Washington, Seattle (1985).
Fung, K. H., and Tang, I. N., *Appl. Opt.* **27**, 206 (1988a).
Fung, K. H., and Tang, I. N., *Chem. Phys. Lett.* **147**, 509 (1988b).
Fung, K. H., and Tang, I. N., *Chem. Phys. Lett.* **163**, 560 (1989a).
Fung, K. H., and Tang, I. N., *J. Colloid Interface Sci.* **130**, 219 (1989b).
Fung, K. H., and Tang, I. N., *Appl. Spectrosc.* **45**, 734 (1991).
Gmitro, J. I., and Vermeulen, T., *AIChE J.* **10**, 740 (1964).
Hadamard, M. J., *Compt. Rendus* **152**, 1735 (1911).
Hightower, R. L., and Richardson, C. B. *Atmos. Environ.* **22**, 2587 (1988).
Hirschfelder, J. O., Bird, R. B., and Spotz, E. L. *Chem. Revs.* **44**, 205 (1949).
Hirschfelder, J. O., Curtiss, C. F., and Bird, R. B., "Molecular Theory of Gases and Liquids." Wiley, New York, 1967.
Ingebrethsen, B. J., and Matijevic, E., *J. Colloid Interface Sci.* **100**, 1 (1984).
Ingebrethsen, B. J., Matijevic, E., and Partch, R. E., *J. Colloid Interface Sci.* **95**, 228 (1983).
Iwamoto, T., Itoh, M., and Takahashi, K., *Aerosol Sci. Tech.* **15**, 127 (1991).
Jackson, J. D., "Classical Electrodynamics." Wiley, New York, 1975.
Jacobsen, S., and Brock, J. R., *J. Colloid Sci.* **20**, 544 (1965).
Jeans, J. H., "The Dynamical Theory of Gases." Dover, New York, 1954.
Kennard, E. H., "Kinetic Theory of Gases." McGraw-Hill, New York, 1938.
Kerker, M., "The Scattering of Light and Other Electromagnetic Radiation." Academic Press, New York, 1969.
Kerker, M., *Aerosol Sci. Tech.* **1**, 275 (1982).
Kerker, M., and Cooke, D. D., *Appl. Opt.* **12**, 1378 (1973).
Kerker, M., and Druger, S. D., *Appl. Opt.* **18**, 1172 (1979).
Koningstein, J. A., "Introduction to the Theory of the Raman Effect." D. Reidel Publishing Co., Dordrecht, 1972.
Kronig, R., and Bruijsten, J., *Appl. Sci. Res.* **A2**, 439 (1951).
Kurtz, C. A., and Richardson, C. B., *Chem. Phys. Lett.* **109**, 190 (1984).
Lettieri, T. R., and Preston, R. E., *Opt. Commun.* **54**, 348 (1985).
Levich, V. G., "Physicochemical Hydrodynamics." Prentice-Hall, Englewood Cliffs, 1962.
Lin, H.-B., and Campillo, A. J., *Appl. Opt.* **24**, 422 (1985).
Lin, S. P., *J. Colloid Interface Sci.* **51**, 66 (1975).
Liu, B. Y. H., and Ahn, K., *Aerosol Sci. Tech.* **6**, 215 (1987).
Lorenz, L., *Videnskab. Selskab. Skrifter* **6**, 1 (1890).

Maloney, D. J., and Spann, J. F., "Proc. 22nd Intl. Symp. Combustion," p. 1999. Combustion Institute, (1988).
Maxwell, J. C., "Collected Scientific Papers," Vol. 11, p. 625. Cambridge University Press, Cambridge, United Kingdom, 1890.
Mie, G., *Ann. Phys.* **25**, 377 (1908).
Millikan, R. A., *Phys. Rev. Ser. I* **32**, 349 (1911).
Millikan, R. A., "Electrons (+ and −), Protons, Photons, Neutrons, and Cosmic Rays." The University of Chicago Press, Chicago, 1935.
Müller, A., *Ann. Phys.* **6**, 206 (1960).
Nakamura, K., Partch, R. E., and Matijevic, E., *J. Colloid Interface Sci.* **99**, 118 (1984).
National Research Council, "Frontiers in Chemical Engineering." National Academy Press, Washington, D.C., 1988.
Okuyama, K., Kousaka, Y., Tohge, N., Yamamoto, S., Wu, J. J., Flagan, R. C., and Seinfeld, J. H., *AIChE J.* **32**, 2010 (1986).
Oster, G., and Nishijima, Y., *J. Amer. Chem. Soc.* **78**, 1581 (1956).
Oster, G., and Yang, N.-L., *Chem. Rev.* **68**, 125 (1968).
Partch, R. E., Matijevic, E., Hodgson, A. W., and Aiken, B. E., *J. Polym. Sci., Polym. Chem. Ed.* **21**, 961 (1983).
Partch, R. E., Nakamura, K., Wolfe, K. J., and Matijevic, E., *J. Colloid Interface Sci.* **105**, 560 (1985).
Paul, W., and Raether, M., *Z. Phys.* **140**, 262 (1955).
Philip, M. A., Gelbard, F., and Arnold, S., *J. Colloid Interface Sci.* **91**, 507 (1983).
Pluchino, A. B., *Appl. Opt.* **20**, 2986 (1981).
Pope, M., *J. Chem. Phys.* **36**, 2810 (1962a).
Pope, M., *J. Chem. Phys.* **37**, 1001 (1962b).
Pope, M., Arnold, S., and Rozenshtein, L., *Chem. Phys. Lett.* **62**, 589 (1979).
Preston, R. E., Lettieri, T. R., and Semerjian, H. G., *Langmuir* **1**, 365 (1985).
Qian, S.-X., and Chang, R. K., *Phys. Rev. Lett.* **56**, 926 (1986a).
Qian, S.-X., and Chang, R. K., *Opt. Lett.* **11**, 371 (1986b).
Qian, S.-X., Snow, J. B., and Chang, R. K., *Opt. Lett.* **10**, 499 (1985).
Qian, S.-X., Snow, J. B., Tzeng, H.-M., and Chang, R. K., *Science* **231**, 486 (1986).
Ravindran, P., Davis, E. J., and Ray, A. K., *AIChE J.* **25**, 966 (1979).
Ray, A. K., Davis, E. J., and Ravindran, P., *J. Chem. Phys.* **71**, 582 (1979).
Ray, A. K., Johnson, R. D., and Souyri, A., *Langmuir* **5**, 133 (1989).
Ray, A. K., Souyri, A., Davis, E. J., and Allen, T. M., *Appl. Opt.*, **30**, 3974 (1991a).
Ray, A. K., Devakottai, B., Souyri, A., and Huckaby, J. L., *Langmuir* **7**, 525 (1991b).
Rayleigh, Lord, *Philos. Mag.* **4**, 107 (1871).
Rayleigh, Lord *Philos. Mag.* **14**, 184 (1882).
Reed, L. D., *J. Aerosol Sci.* **8**, 123 (1977).
Rhim, W. K., Chung, S. K., Trihn, E. H., and Elleman, D. D., *Mater. Res. Soc. Symp. Proc.* **87**, 329 (1987a).
Rhim, W. K., Chung, S. K., Hyson, M. T., and Elleman, D. D., *Mater. Res. Soc. Symp. Proc.* **87**, 103 (1987b).
Richardson, C. B., and Hightower, R. L., *Atmos. Environ.* **21**, 971 (1987).
Richardson, C. B., and Spann, J. F., *J. Aerosol Sci.* **15**, 563 (1984).
Richardson, C. B., Jefferts, K. B., and Dehmelt, H. G., *Phys. Rev.* **165**, 80 (1968).
Richardson, C. B., Hightower, R. L., and Pigg, A. L., *Appl. Opt.* **25**, 1226 (1986).
Richardson, C. B., Pigg, A. L., and Hightower, R. L., *Proc. Roy. Soc. Lond. A* **422**, 319 (1989).
Rimmer, P. L., *J. Fluid Mech.* **32**, 1 (1968).
Rimmer, P. L., *J. Fluid Mech.* **35**, 827 (1969).
Rosenblatt, P., and LaMer, V. K., *Phys. Rev.* **70**, 385 (1946).

Rubel, G. O., and Gentry, J. W., *J. Aerosol Sci.* **15**, 661 (1984a).
Rubel, G. O., and Gentry, J. W., *J. Phys. Chem.* **88**, 3142 (1984b).
Rybczynski, D. P., *Bulletin Intern. Acad. Sci. Cracovie* **A403**, 40 (1911).
Sageev, G., Seinfeld, J. H., and Flagan, R. C., *Rev. Sci. Instrum.* **57**, 933 (1986).
Schmitt, K. H., Z. *Naturforsch.* **14a**, 870 (1959).
Schweiger, G., *Part. Charact.* **4**, 67 (1987).
Schweiger, G., *Aerosol Sci. Tech.* **12**, 1016 (1990a).
Schweiger, G., *Opt. Lett.* **15**, 156 (1990b).
Schweiger, G., *J. Raman Spectrosc.* **21**, 165 (1990c).
Schweiger, G., *J. Aerosol Sci.* **21**, 483 (1990d).
Schweizer, J. W., and Hanson, D. N., *J. Colloid Interface Sci.* **35**, 417 (1971).
Sitarski, M., *Particulate Sci. Tech.* **5**, 193 (1987).
Sloane, C. S., and Elmoursi, A. A., "Conf. Record 1987 IEEE Industry Applications Mtg., Part II, p. 1568, IEEE Publ. Services, New York, 1987.
Snow, J. B., Qian, S.-X., and Chang, R. K., *Opt. Lett.* **10**, 37 (1985).
Spann, J. F., and Richardson, C. B., *Atmos. Environ.* **19**, 819 (1985).
Spjut, R. E., *Opt. Eng.* **26**, 467 (1987).
Spjut, R. E., and Bolsaitis, P. P., *Mat. Res. Soc. Symp. Proc.* **87**, 295 (1987).
Spjut, R. E., Sarofim, A. F., and Longwell, J. P., *Langmuir* **1**, 355 (1985).
Spjut, R. E., Elliott, J. F., and Bolsaitis, P. P., *Mat. Res. Soc. Symp. Proc.* **87**, 95 (1987).
Straubel, H., *Z. Elektrochem.* **60**, 1033 (1956).
Taflin, D. C., Ph.D. dissertation, University of Washington, Seattle (1988).
Taflin, D. C., and Davis, E. J., *Chem. Eng. Commun.* **55**, 199 (1987).
Taflin, D. C., and Davis, E. J., *J. Aerosol Sci.* **21**, 73 (1990).
Taflin, D. C., Zhang, S. H., Allen, T., and Davis, E. J., *AIChE J.* **34**, 1310 (1988).
Taflin, D. C., Ward, T. L., and Davis, E. J., *Langmuir* **5**, 376 (1989).
Tang, I. N., and Munkelwitz, H. R., *J. Colloid Interface Sci.* **98**, 430 (1984).
Tang, I. N., and Munkelwitz, H. R., *J. Colloid Interface Sci.* **141**, 109 (1991).
Thurn, R., and Kiefer, W., *Appl. Spectros.* **38**, 78 (1984a).
Thurn, R., and Kiefer, W., *J. Raman Spectros.* **15**, 411 (1984b).
Thurn, R., and Kiefer, W., *Appl. Opt.* **24**, 1515 (1985).
van de Hulst, H. C., "Light Scattering by Small Particles." Dover, New York, 1981.
Wagner, P. E., in "Topics in Current Physics Vol. 29 Aerosol Microphysics II Chemical Physics of Microparticles." W. H. Marlow, ed., Springer-Verlag, Berlin, 1982.
Ward, T. L., and Davis, E. J., *J. Aerosol Sci.* **21**, 875 (1990).
Ward, T. L., Zhang, S. H., Allen, T., and Davis, E. J., *J. Colloid Interface Sci.* **118**, 343 (1987).
Ward, T. L., Davis, E. J., Jenkins, R. W., Jr., and McRae, D. D., *Rev. Sci. Instrum.* **60**, 414 (1989).
Weast, R. C., and Selby, S. M., "Handbook of Chemistry and Physics." Chem. Rubber Co., Boca Raton, 1988.
Weiss-Wrana, K., *Astron. Astrophys.* **126**, 240 (1983).
Wiscombe, W. J., *Appl. Opt.* **19**, 1505 (1980).
Wuerker, R. F., Shelton, H., and Langmuir, R. V., *J. Appl. Phys.* **30**, 342 (1959).
Wyatt, P. J., and Phillips, D. T., *J. Colloid Interface Sci.* **39**, 125 (1972).
Yang, W., and Carleson, T. E., AIChE Annual Meeting, Chicago (1990).
Zhang, S. H., and Davis, E. J., *Chem. Eng. Commun.* **50**, 51 (1987).

DETAILED CHEMICAL KINETIC MODELING: CHEMICAL REACTION ENGINEERING OF THE FUTURE

Selim M. Senkan

Department of Chemical Engineering
University of California at Los Angeles
Los Angeles, California

I.	Introduction and Overview	95
II.	Computational Quantum Chemistry	101
III.	Thermochemistry	111
IV.	Estimation of Thermochemistry by Conventional Methods	113
V.	Estimation of Thermochemistry by Quantum Mechanics	126
VI.	Rate Processes	131
VII.	Chemical Processes	132
VIII.	Estimation of Chemical Rate Parameters by Conventional Methods	134
	A. Unimolecular Reactions	134
	B. Simple Fission Reactions	136
	C. Complex Unimolecular Reactions	139
	D. Bimolecular Reactions	144
	E. Direct Metathesis Reactions	145
	F. Association Reactions	149
IX.	Estimation of Rate Parameters by Quantum Mechanics	152
X.	Energy-Transfer-Limited Processes	160
XI.	Heterogeneous Reactions	172
XII.	An Example: Detailed Chemical Kinetic Modeling of the Oxidation and Pyrolysis of CH_3Cl/CH_4	175
	Acknowledgments	190
	Note	190
	References	190

I. Introduction and Overview

Chemical reaction engineering education and research over the past decades emphasized transport phenomena, as evident from the contents of textbooks and research papers published in this field (see also Carberry, 1988, and Dudukovic, 1987, and references therein, for the assessment of this situation). These developments, however, were achieved at the expense of paying similar attention to the underlying chemistry and chemical kinetics that are also important, especially for the development of new chemical manufacturing processes. At present it is common to see in chemical reaction engineering a reaction represented simply by A → B, devoid of any specifics of the underlying chemistry. Furthermore, this *oversimplification* in chemistry is almost always coupled with an *oversophistication* in mathematical analysis of of the associated transport phenomena.

However, this trend is changing because of the growing sophistication of the chemical industry and because of the need to invent and implement new manufacturing processes for the industry to remain productive and competitive. For example, because of increasing concerns over production flexibility, selectivity, quality control, safety, and environmental protection, it is no longer acceptable to rely solely on simple (empirical) chemical models. Empirical models, although continuing to be useful, have long been recognized to be of limited utility because of their narrow range of applicability and lack of both interpretive and predictive capabilities.

In contrast, in detailed chemical kinetic modeling (DCKM) the underlying chemistry of the manufacturing process is integrated into the process model. This is accomplished first by developing a comprehensive description of the associated chemical transformations in terms of *irreducible* chemical events or *elementary reactions*. The DCKMs developed are then combined with appropriate transport models for the simulation of the overall process. It must be noted that DCKM is a fundamentally different approach from complex reaction modeling, which is being used in chemical reaction engineering for data correlation purposes. In complex reaction modeling, reaction intermediates often correspond to stable species derived primarily from experimental observations; thus, their reactions do not correspond to elementary processes. Consequently, complex reaction modeling is of little interpretive or predictive utility.

In principle, detailed mechanisms comprise large, putatively complete sets of elementary reactions for which independent rate coefficient parameters, frequently expressed in the form

$$k = AT^n \exp(-E/RT), \qquad (1)$$

are available from direct measurements or estimable from theoretical considerations. Estimations are based on the judicious application of theories of rate processes in a conventional manner, as well as in conjunction with modern quantum chemical calculations, and by using analogies between similar types of reactions. At present we rely on all of these complementary methods.

Our ability to develop detailed chemical kinetic models has improved considerably over the past decade because of simultaneous developments in a number of related areas. First, with the availability of fast computers, together with efficient numerical algorithms to handle stiff differential equations, an increasing complexity of problems can now be coded and solved rapidly without making any fundamental assumptions. For example, it is no longer necessary to invoke the pseudo-steady-state assumption to treat reactive radicals or intermediates in modeling the kinetics of complex chemical processes in an effort to obtain analytical expressions (see, for example, Butt, 1980; Carberry, 1976). In fact, one can argue that no conclusion based on the assumption of pseudo-steady-state conditions or through the neglect of reactions presumed to be unimportant in a process would be acceptable in the absence of a detailed modeling work that proves the validity of these assumptions.

Recently, a number of very efficient and transportable software packages have become available for the solution of stiff differential equations involved in detailed chemical kinetic modeling (see, for example, Hindmarsh, 1980; Petzold, 1982; Caracatsios and Steward, 1985). Consequently, the actual solution of equations no longer limits the modeling process. Instead, the limiting factor today is the availability of reliable and fundamentally based chemical reaction mechanisms.

Second, the quality and quantity of data on the thermochemistry of species and on the kinetics and mechanisms of individual elementary reactions, especially those in the gas phase, have improved significantly over the past decade because of advances made in experimental methods. In particular, the thermochemistry of a large number of species is now available in convenient tabulations (see, for example, Pedley *et al.*, 1986; Chase *et al.*, 1985; Benson, 1976; Bahn, 1973). Similarly, the rates and mechanisms of a large number of elementary reactions have also been evaluated and documented (see, for example, Berces and Marta, 1988; Tsang and Hampson, 1986; Kerr and Drew, 1987; Basevich, 1987; DeMore *et al.*, 1985; Westbrook and Dryer, 1984; Hanson and Salimian, 1984; Warnatz, 1984; Kondratiev, 1972; Kerr and Moss, 1981; Baulch *et al.*, 1981; Westley, 1976, 1980; Benson and O'Neal, 1970). In addition, mature empirical methods are now in place to estimate thermochemistry and reaction rate parameters from theoretical considerations (see, for example, Gaffney and Bull, 1988; Benson, 1976). Although considerable progress also has been made in heterogeneous reactions, the available data are more limited, and the theoretical framework to estimate thermochemical and rate parameters is less well developed.

Third, with recent advances made in theoretical and computational quantum mechanics, it is possible to estimate thermochemical information via electronic structure calculations (Dewar, 1975; Dunning *et al.*, 1988). Such a capability, together with the transition state theory (TST) (Eyring, 1935), also allows the determination of the rate parameters of elementary reactions from first principles. Our ability to estimate activation energy barriers is particularly significant because until the advent of computational quantum mechanics, no fundamentally based method was available for the determination of this important variable.

Finally, although DCKMs comprise a large number of elementary reactions, which is necessary to keep the range of applicability of these models as broad as possible and to preserve the elementary nature of the mechanisms, experience with these models has indicated that only a small fraction of the reactions may be important under a given set of conditions (Westbrook and Dryer, 1984). Consequently, sections of detailed mechanisms can be developed, tested, and modified for the continued improvement of DCKMs.

From a practical standpoint, the extent of "detail" in detailed chemical kinetic mechanisms is a strong function of the objectives of the effort. As a general guideline, however, the number of species and the associated chemical reactions increase with increasing complexity—e.g., the number of atoms—of the species involved. For example, to model the gas-phase kinetics of oxidative pyrolysis of CH_3Cl and to describe the formation of C_2 hydrocarbon products, more than 150 reactions and 35 species were considered (Karra and Senkan, 1988b). In contrast, to model soot formation in C_2H_2 combustion, a mechanism consisting of about 500 reactions and more than 140 species was necessary (Frenklach et al., 1986). Subsequently, the size of larger mechanisms may be reduced to capture the major features of these processes (see, for example, Jones and Lindstedt, 1988; Paczko et al., 1986). However, this is neither necessary nor desirable, especially when fast computers and efficient codes are available to analyze large amounts of numerical data. In fact, the major problem in detailed chemical kinetic modeling today is just the opposite, i.e., the mechanisms developed may not be detailed enough, and potentially important elementary reactions may have been overlooked.

Mechanism reduction, nevertheless, may be necessary in some applications—for example, to model multidimensional reactive flows. Even the fastest computers today cannot handle such problems using detailed mechanisms in a reasonable time frame. It must be recognized, however, that models that utilize reduced mechanisms would have a far narrower range of applicability than the ones that use comprehensive reaction mechanisms. Furthermore, models that are based on reduced mechanisms cannot be expected to be valid outside the limits set in the mechanism reduction step.

Since detailed chemical kinetic mechanisms involve the participation of a large number of species in a large number of elementary reactions, *sensitivity* and *reaction path* analyses are also essential elements of DCKM. Sensitivity analysis provides a means to assess the limits of confidence we must put on our model predictions in view of uncertainties that exist in reaction rate parameters and thermochemical and thermophysical data utilized, as well as the initial and boundary conditions used in the modeling work. Through reaction path analysis, major reaction pathways responsible for the production and consumption of each species can be identified. Powerful formalisms have been developed to undertake sensitivity analysis in detailed chemical kinetic modeling (see, for example, Tilden et al., 1981; Siegneur et al., 1982; Rabitz et al., 1983; Hwang, 1983). In addition, efficient and transportable computer codes to implement these techniques have also been developed (Lutz et al., 1988; Caracatsios and Steward, 1985; Kramer et al., 1982a, b).

Sensitivity and reaction path analyses also represent techniques to identify rank-order of importance of reactions in the mechanism. Important reactions can then be isolated and studied with greatest scrutiny, without wasting scarce resources on unimportant ones. As new and improved data on important elementary reactions become available, they are reincorporated into the evolving mechanisms. Revised mechanisms are then resubjected to sensitivity and reaction path analyses for the redetermination of the rank-order of importance of reactions.

In Fig. 1, various elements involved with the development of detailed chemical kinetic mechanisms are illustrated. Generally, the objective of this effort is to predict macroscopic phenomena, e.g., species concentration profiles and heat release in a chemical reactor, from the knowledge of fundamental chemical and physical parameters, together with a mathematical model of the process. Some of the fundamental chemical parameters of interest are the thermochemistry of species, i.e., standard state heats of formation ($\Delta H_f(T_0)$), and absolute entropies ($S(T_0)$), and temperature-dependent specific heats ($c_p(T)$), and the rate parameter constants A, n, and E, for the associated elementary reactions (see Eq. (1)). As noted above, evaluated compilations exist for the determination of these parameters. Fundamental physical parameters of interest may be the Lennard–Jones parameters ($\varepsilon/\kappa, \sigma$), dipole moments ($\mu$), polarizabilities ($\alpha$), and rotational relaxation numbers (z_{rot}) that are necessary for the calculation of transport parameters such as the viscosity (μ) and the thermal conductivity (k) of the mixture and species diffusion coefficients (D_{ij}). These data, together with their associated uncertainties, are then used in modeling the macroscopic behavior of the chemically reacting system. The model is then subjected to sensitivity analysis to identify its elements that are most important in influencing predictions.

When the objective of the modeling effort is to develop and validate a reaction mechanism, the major uncertainty in the model must reside in the detailed chemical kinetic mechanism. Under these conditions, the process must be studied either under transport-free conditions, e.g., in plug-flow or stirred-tank reactors, or under conditions in which the transport phenomena can be modeled very precisely, e.g., under laminar flow conditions. This way,

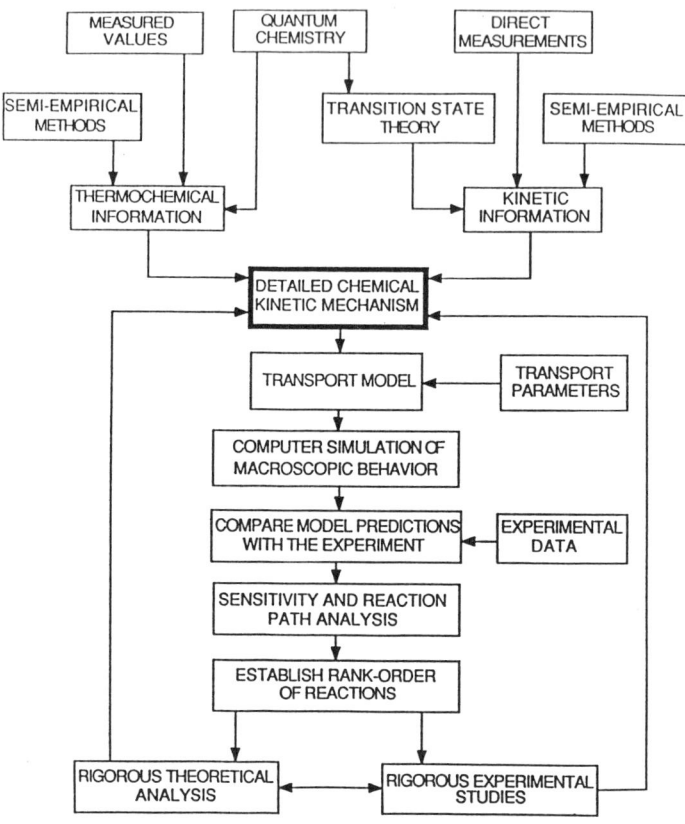

computer predictions become primarily influenced by the uncertainties in the reaction mechanism.

In this communication we focus on issues related to the development of chemical kinetic mechanisms, since modeling of the associated transport phenomena is more mature, and comprehensive discussions are available in a number of advanced textbooks (see, for example, Seinfield, 1986; Rosner, 1986; Dixon-Lewis, 1984; Toong, 1983; Bird et al., 1960). Although we principally discuss the development of detailed mechanisms for homogeneous gas-phase processes, the approach taken can be extended to describe processes in condensed phases and in heterogeneous media such as those observed in living systems (Neta, 1988) and to heterogeneous catalysis (Dumesic et al., 1987; Zhdanov et al., 1988; Boudart and Djega-Mariadassou, 1984; Allara and Edelson, 1977; Baetzold and Somorjai, 1976).

We must also note that detailed chemical kinetic modeling successfully is being used to describe macroscopic phenomena in the stratosphere and troposphere (see, for example, Filby, 1988, and Seinfeld, 1986, and references therein), and in combustion (see Miller and Fisk, 1987; Westbrook and Dryer, 1984, and references therein). The importance of such a modeling approach has also been recognized in the manufacture of useful chemicals by partial oxidation of hydrocarbons (Pitchai and Klier, 1986; Karra and Senkan, 1987), in the thermal cracking of hydrocarbons (Dente and Ranzi, 1983; Willems and Froment, 1988a, b) and in the manufacture of microelectronic devices (Coltrin *et al.*, 1986).

The material presented here is organized in the following manner: First, some of the fundamentals associated with computational quantum mechanics are discussed, since this approach plays a central role in the estimation of both thermochemical and reaction rate parameters. Second, methods to estimate thermochemical properties using both conventional and quantum chemical methods are presented. Third, a working treatment of rates of elementary reactions is presented based on the thermodynamic formulation of the transition state theory (TST), followed by the discussion of methods for the estimation of rate parameters, again using both conventional and quantum mechanical methods. Fourth, a review of reactions that are influenced by energy transfer processes is presented because of their importance in many gas-phase reaction systems. Fifth, a brief TST treatment of heterogeneous reactions is presented to illustrate the similarities involved with gas-phase reactions. Finally, as an example for detailed chemical kinetic modeling, the development and the application of the mechanism of oxidation and pyrolysis of CH_3Cl/CH_4 in flames and flow reactors are discussed.

II. Computational Quantum Chemistry

To predict thermochemistry and reaction rate parameters, the total energy (ε_T) of a collection of atoms must be determined as a function of their geometrical configuration in space. This requires the determination of interactions among all the atoms involved, and these interactions collectively define a potential energy hypersurface (PES) in $3N - 6$ dimensions, where N is the number of atoms. On this hypersurface the minima correspond to stable species or molecules, and the saddle points represent transition states. A chemical reaction can then be described by a movement from one minimum to another, and the difficulty of the reaction, i.e., the activation energy, corresponds to the height of the barrier separating these minima. In Figure 2, an isometric projection diagram for a hypothetical PES is presented showing the three configurations of a triatomic molecule ABC in which the

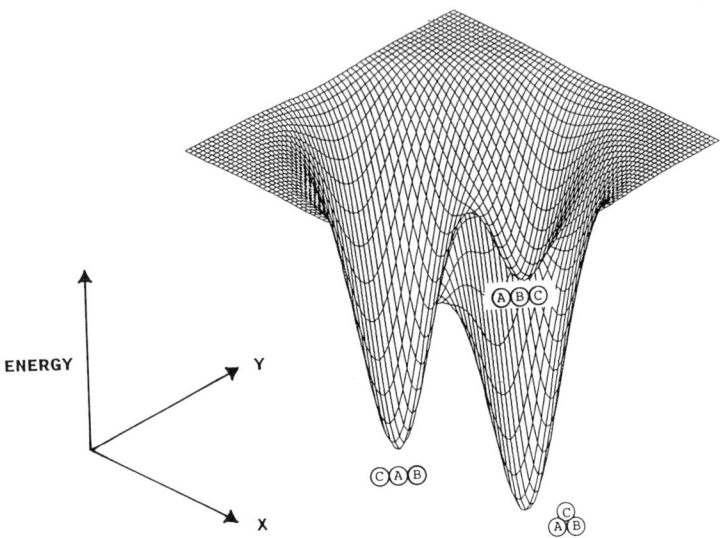

FIG. 2. Potential energy surface of a hypothetical collection of atoms.

BC bond is fixed. It should be noted that PESs represent the energy of the collection of atoms at *absolute zero*, i.e., they provide no information on the translational, vibrational, or rotational energies that are normally present. The latter forms of energy clearly must be considered to determine thermochemistry under conditions of practical interest.

The determination of ε_T requires that Schrödinger's equation for the collection of atoms must be solved with all the associated nuclei and electrons in three dimensions. For a molecule with N nuclei and n electrons, the time-independent Schrödinger's equation is given by

$$\left[-\sum_{j=1}^{N} (\hbar^2/2M_j)\nabla_j^2 - (\hbar^2/2m)\sum_{j=1}^{n} \nabla_j^2 - \sum_{i=1}^{n}\sum_{j=1}^{N} Z_j e^2/(4\pi\varepsilon_0|R_j - r_i|) \right.$$

$$+ (1/2)\sum_{i=1}^{n}\sum_{j=i+1}^{n} e^2/(4\pi\varepsilon_0|r_i - r_j|)$$

$$\left. + (1/2)\sum_{i=1}^{N}\sum_{j=i+1}^{N} Z_i Z_j e^2/(4\pi\varepsilon_0|R_i - R_j|) \right] \Psi = \varepsilon_T \psi, \tag{2}$$

where $\hbar^2 = h^2/4\pi^2$, in which h is Planck's constant, e is the electronic charge, ∇_j^2 is the Laplacian operator centered around j, R_i and r_j are the position vectors for the nuclei i and the electron j, respectively, ε_0 is the permittivity of vacuum, m is the mass of the electron, M_j is the mass of the jth nucleus, and Ψ

is the total wave function. In shorthand notation Schrödinger's equation can be written as

$$H\Psi = (T_N + T_e + V_{Ne} + V_{ee} + V_{NN})\Psi = \varepsilon_T \Psi, \tag{3}$$

where H is the Hamiltonian operator, T and V are the kinetic and potential energy operators, and subscripts N and e refer to nuclear and electronic components of T and V, respectively.

For the hydrogen atom, which has one nucleus (charge $+e$) and one electron ($-e$), Eq. (2) can be reduced into the following familiar form:

$$-[(\hbar^2/2\mu)\nabla^2 - e^2/(4\pi\varepsilon_0 r)]\Psi(r, \theta, \phi) = \varepsilon_T \Psi(r, \theta, \phi), \tag{4}$$

where ∇^2 is the Laplacian operator in spherical coordinates (r, θ, ϕ), r is the distance between the nucleus and the electron, and μ is the reduced mass, $\mu = mM/(m + M)$. Exact solutions of Eq. (4) are available and are presented in Table I. It should be noted that these exact solutions provide valuable insights into the electronic structure of larger atoms, since they can be viewed as being made from the hydrogen atom by suitably adding electrons and protons. In particular, hydrogen-atom–like wavefunctions clearly illustrate the angular dependency of electronic orbitals, as shown in Fig. 3a.

These insights have also been extremely useful in the molecular orbital (MO) theory of chemical bonding. For example, in Fig. 3b the molecular

TABLE I
THE HDROGENLIKE-ATOM WAVE FUNCTIONS $\psi = R(r)Y(\theta, \varphi)$ FOR THE $n = 1$ AND $n = 2$ SHELLS[a]

n	l	m	$R(r)$	$Y(\theta, \varphi)$	Wave-Function Symbol
1	0	0	$2\left(\dfrac{Z}{a_0}\right)^{3/2} e^{-Zr/a_0}$	$\left(\dfrac{1}{4\pi}\right)^{1/2}$	$1s$
2	0	0	$\left(\dfrac{Z}{2a_0}\right)^{3/2}\left(2 - \dfrac{Zr}{a_0}\right) e^{-Zr/2a_0}$	$\left(\dfrac{1}{4\pi}\right)^{1/2}$	$2s$
2	1	0	$\dfrac{1}{\sqrt{3}}\left(\dfrac{Z}{2a_0}\right)^{3/2}\left(\dfrac{Zr}{a_0}\right) e^{-Zr/2a_0}$	$\left(\dfrac{3}{4\pi}\right)^{1/2} \cos\theta$	$2p_z$
2	1	$+1$	$\dfrac{1}{\sqrt{3}}\left(\dfrac{Z}{2a_0}\right)^{3/2}\left(\dfrac{Zr}{a_0}\right) e^{-Zr/2a_0}$	$\left(\dfrac{3}{8\pi}\right)^{1/2} \sin\theta\, e^{-i\varphi}$	$2p_x, 2p_y$
2	1	-1	$\dfrac{1}{\sqrt{3}}\left(\dfrac{Z}{2a_0}\right)^{3/2}\left(\dfrac{Zr}{a_0}\right) e^{-Zr/2a_0}$	$\left(\dfrac{3}{8\pi}\right)^{1/2} \sin\theta\, e^{+i\varphi}$	

[a] $a_0 = h^2/4\pi^2 m e^2 = 0.529$ Å. Z is the effective nuclear charge, which, for the hydrogen atom, has the value 1.

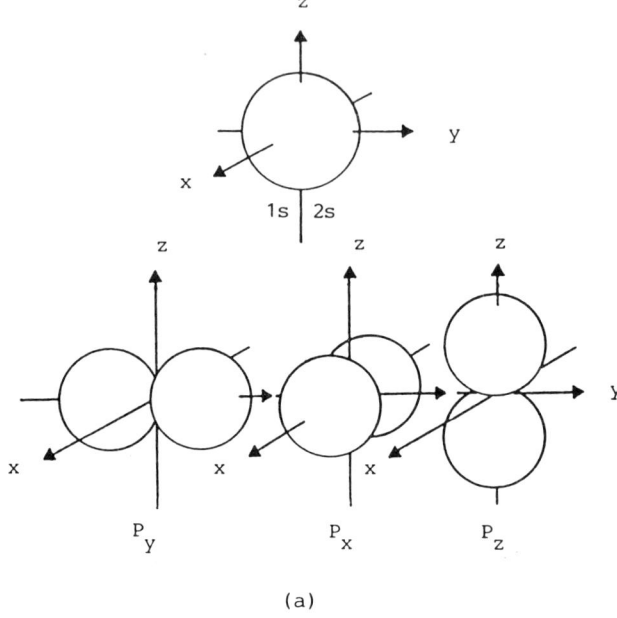

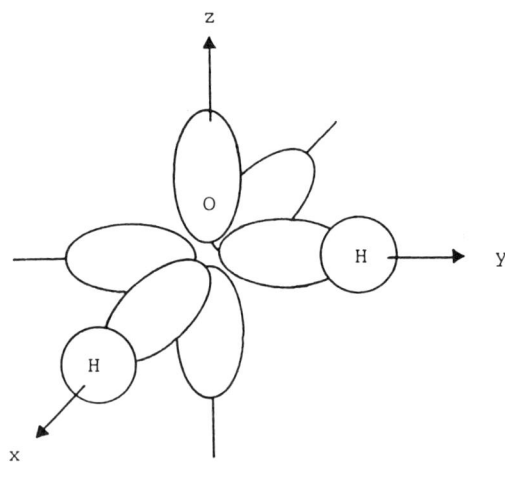

Fig. 3. (a) H-atom–like orbitals $1s$, $2s$, $2p_x$, $2p_y$, $2p_z$ showing angular dependency; (b) the structure of the water molecule.

structure of H_2O is illustrated, showing the utilization of the two p orbitals of oxygen (e.g., p_x and p_y) by the two hydrogen atoms. Although the H–O–H bond angle is expected to be 90 degrees based on the hydrogen-atom-like orbitals, the actual bond angle has been determined to be about 104 degrees because of H–H repulsion (Pauling, 1960). Similarly, the molecular structures of C_2H_2, C_2H_4, and CH_4, shown in Figs. 4a–c, can also be explained in terms of different extents of hybridization of the s and p orbitals.

Since the exact solution of Schrödinger's equation for multi-electron, multi-nucleus systems turned out to be impossible, efforts have been directed towards the determination of approximate solutions. Most modern approaches rely on the implementation of the Born–Oppenheimer (BO) approximation, which is based on the large difference in the masses of the electrons and the nuclei. Under the BO approximation, the total wavefunction can be expressed as the product of the electronic (ψ) and nuclear (η) wavefunctions, leading to the following electronic and nuclear Schrödinger's equations:

$$(T_e + V_{Ne} + V_{ee} + V_{NN})\psi = \varepsilon_e \psi, \quad (5)$$

$$(T_N + \varepsilon_e)\eta = \varepsilon_T \eta \quad (6)$$

where ε_e is the electronic energy. Once the electronic Schrödinger equation is

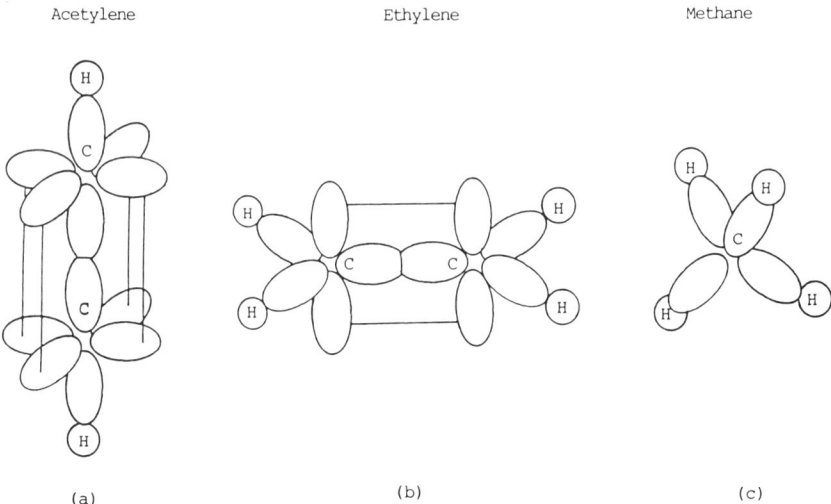

FIG. 4. Molecular orbitals showing the hybridization of s and p orbitals. (a) Acetylene (C_2H_2, sp hybridization); (b) ethylene (C_2H_4, sp^2 hybridization), and (c) methane (CH_4, sp^3 hybridization).

solved numerically or analytically, the total energy of the molecule can then be obtained from the solution of the nuclear equation (6).

For a fixed nuclear configuration, V_{NN} would be constant; thus, the electronic Schrödinger equation can be rewritten as the following:

$$(T_e + V_{Ne} + V_{ee})\psi = E_e\psi, \qquad (7)$$

where

$$E_e = \varepsilon_e - V_{NN}. \qquad (8)$$

The nuclear configuration can then be changed to establish new potential fields V_{Ne} and V_{NN}, which in return leads to a new set of wavefunctions, and thus to a new electronic energy and a new total energy. By repeating this procedure for a large number of nuclear configurations, the potential energy surface, such as that shown in Fig. 2, can be generated.

Under the Born–Oppenheimer approximation, two major methods exist to determine the electronic structure of molecules: The valence bond (VB) and the molecular orbital (MO) methods (Atkins, 1986). In the valence bond method, the chemical bond is assumed to be an electron pair at the onset. Thus, bonds are viewed to be distinct atom–atom interactions, and upon dissociation molecules always lead to neutral species. In contrast, in the MO method the individual electrons are assumed to occupy an orbital that spreads the entire nuclear framework, and upon dissociation, neutral and ionic species form with equal probabilities. Consequently, the charge correlation, or the avoidance of one electron by others based on electrostatic repulsion, is overestimated by the VB method and is underestimated by the MO method (Atkins, 1986). The MO method turned out to be easier to apply to complex systems, and with the advent of computers it became a powerful computational tool in chemistry. Consequently, we shall concentrate on the MO method for the remainder of this section.

In the MO approach molecular orbitals are expressed as a linear combination of atomic orbitals (LCAO); atomic orbitals (AO), in return, are determined from the approximate numerical solution of the electronic Schrödinger equation for each of the parent atoms in the molecule. This is the reason why hydrogen-atom–like wavefunctions continue to be so important in quantum mechanics. Mathematically, MO–LCAO means that the wavefunctions of the molecule containing N atoms can be expressed as

$$\psi_{MO} = \sum_{k=1}^{N} C_k \phi_k, \qquad (9)$$

where ϕ_k are the approximate atomic wavefunctions (i.e., wavefunctions centered around the kth atom), and C_k are the LCAO expansion coefficients. The direct implementation of the MO–LCAO to systems consisting of many nuclei and many electrons turned out to be difficult because of the three-

DETAILED CHEMICAL KINETIC MODELING 107

dimensional nature of orbitals, especially in evaluating the contributions of electronic repulsions to the ε_T. These difficulties are partially removed by the use of the self-consistent field (SCF) method (Daudel et al., 1983).

In the SCF approach, each electron is assumed to move in an average field due to nuclei and the remaining electrons. Consequently, electronic repulsions are formally considered in the procedure. For example, consider a hypothetical atom (nuclear charge Z) containing two electrons as shown in Fig. 5. The electrons occupy orbitals ϕ_1 and ϕ_2, with corresponding electron densities of $-e\phi_1^2$ and $-e\phi_2^2$, respectively. According to the SCF approach, the potential seen by electron 2 would be

$$V_2 = -Ze^2/(4\pi\varepsilon_0|R - r_2|) + e^2/(4\pi\varepsilon_0)\int \phi_1^2(r_1)/(r_1 - r_2)\,d\tau, \tag{10}$$

where the second term represents the electronic repulsion integral in which $d\tau_1$ is the volume element in the orbital ϕ_1. To determine ϕ_2 the following Schrödinger equation must be solved:

$$[-(\hbar^2/2m)\nabla_2^2 + V_2]\phi_2 = \varepsilon_2\phi_2, \tag{11}$$

or, in shorthand notation, the following Hartree–Fock (H–F) equation:

$$H_i^F \phi_2 = \varepsilon_2 \phi_2. \tag{12}$$

However, as evident from these equations, ϕ_1 must be known to determine ϕ_2, and vice versa; thus, an iterative calculation would be needed. For the general case of N nuclei and n electrons, the resulting H–F expression for the wave function ϕ_i will be

$$H_i^F \phi_i = \varepsilon_i \phi_i. \tag{13}$$

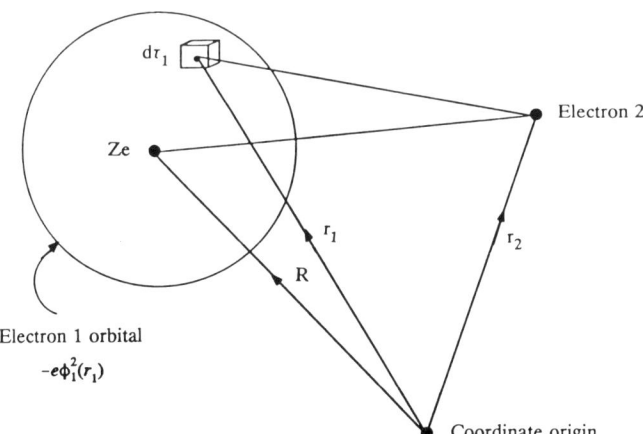

FIG. 5. Schematic diagram for one nucleus (Ze) and two electrons for the discussion of SCF.

Since the exact solution of the Hartree–Fock equation for molecules also proved to be impossible, numerical methods approximating the solution of the Schrödinger's equation at the HF limit have been developed. For example, in the Roothan–Hall SCF method, each SCF orbital is expressed in terms of a linear combination of fixed orbitals or *basis sets* (φ_i). These orbitals are fixed in the sense that they are not allowed to vary as the SCF calculation proceeds. From n basis functions, new SCF orbitals are generated by

$$\psi_k = \sum_{i=1}^{n} C_{k,i} \varphi_i, \tag{14}$$

and for each k, a vector of C_k values are obtained. The C_ks are then determined from the solution of the matrix Hartree–Fock equations (Daudel et al., 1983):

$$H^F C_k = \varepsilon_k S C_k, \tag{15}$$

where S represents the matrix of electron overlap integrals in which

$$S_{ij} = \int \varphi_i \varphi_j \, d\tau. \tag{16}$$

Because the calculation of H^F also involves the evaluation of a large number of one- and two-electron integrals, the need for computational resources increase with increasing number of electrons. As an approximate guideline, computational requirements increase with the fourth power of the number of electrons present in the system.

In minimal basis sets, each atom is represented by a single orbital of each type. For example, oxygen is represented by $1s$, $2s$, $2p_x$, $2p_y$, and $2p_z$ orbitals only. In double zeta basis sets, twice the functions in the minimum basis sets are used. Extended basis sets generally refer to sets that make use of functions that are more than the minimum basis set.

Among the earlier basis sets have been the Slater-type orbitals (STO). The STO sets make use of the $\exp(-ar)$ type of functions, where r represents distance from the center of the nucleus, because of their initial use in the SCF calculations for atoms (see Table I). Most of the current SCF calculations, however, make use of gaussian-type orbitals (GTO), in which basis sets are constructed using functions of the form $\exp(-br^2)$. Although gaussian functions have improper features, e.g., they have the wrong behavior at $r = 0.0$ and they decay too fast at large r, their form helps enormously in the numerical integration of electron repulsion integrals (Lowe, 1978).

At present two parallel approaches are being undertaken to determine the PES or the ε_T: The *ab initio* and *semiempirical* quantum mechanical methods. *Ab initio* methods attempt to undertake the just-mentioned calculations rigorously without relying on the use of any information other than the basis sets (φ_i); to increase the resulting accuracy of the calculations large basis sets are used (Dunning et al., 1988). Although *ab initio* methods hold a great promise for the future, they presently suffer from the fact that they are computationally very demanding. Consequently, *ab initio* methods have been applied to systems containing a limited number of atoms of the first row in the periodic table (Dunning et al., 1988). When larger systems are investigated, empirical corrections often are applied to the results of *ab initio* calculations to account for errors that seem to occur as a result of basis-set truncation. We shall discuss these corrections briefly below.

Semiempirical methods, on the other hand, utilize minimum basis sets to speed up computations, and the loss in rigor is compensated by the use of experimental data to reproduce important chemical properties, such as the heats of formation, molecular geometries, dipole moments, and ionization potentials (Dewar, 1976; Stewart, 1989a). As a result of their computational simplicity and their chemically useful accuracy, semiempirical methods are widely used, especially when large molecules are involved (see, for example, Stewart, 1989b; Dewar et al., 1985; Dewar, 1975).

Most present-day semiempirical methods are based on the idea of the neglect of differential overlap (NDO) of inner electrons developed by Pople and co-workers (see, for example, Pople and Beveridge, 1970; Dewar, 1975). NDO-type approximations generally result in a decrease in computational resource requirements that are $1/100$ to $1/1000$ of the corresponding *ab initio* methods.

Once the electronic Schrödinger equations are solved, the total energy of the collection of atoms (ε_T) is obtained by summing all the electronic energies, ε_k, and the nuclear repulsions in the following manner:

$$\varepsilon_T = \sum_{k=1}^{n} e_k + \sum_{i=1}^{N} \sum_{j=i+1}^{N} Z_i Z_j e^2 / (4\pi\varepsilon_0 |R_i - R_j|). \qquad (17)$$

The heat of formation can then be obtained from this total energy via

$$\Delta H_f = \varepsilon_T - \sum_{k=1}^{n} \varepsilon_k^A + \sum_{i=1}^{N} \Delta H_{fi}^A, \qquad (18)$$

where ε_k^A and ΔH_{fi}^A are the electron energies and the heats of formation of individual atoms, respectively. Clearly, this approach requires the accurate knowledge of ΔH_{fi}^A, which may or may not be available. However, as such data become available—for example, as a result of new improved experiments or via high-level *ab initio* calculations—ΔH_f can be determined with greater accuracy.

Alternate methods, each involving some degree of empiricism, also exist for the determination of ΔH_f. One popular approach is to use isodesmic reactions which are reactions in which the products and reactants contain the same number of bonds of each type. The following is an example of an isodesmic reaction that involves the breakage and formation of sp^3 C–Cl bond in alkanes:

$$CH_3 + C_2H_5Cl \rightleftharpoons CH_3Cl + C_2H_5. \quad (19)$$

By choosing a suitable isodesmic reaction, the heat of formation of the new species can be determined from the calculated value of the heat of reaction and the thermochemistry of the remaining species, which must also be known. This approach provides an empirical correction in the form of cancellation of correlation energy that accompany the formation and breakage of specific types of bonds.

Other, more elaborate approaches involve corrections that are applied to individual bonds (Melius, 1990; Melius and Binkley, 1986) and groups (Wiberg, 1984). These methods are based on the analogies of empirical schemes for estimating heats of formation as sums of contributions by individual bonds and groups (Benson, 1976). However, it must be recognized that while these methods are of considerable value, when applicable, they are restricted to molecules built entirely from standard bonds or groups. We shall discuss bond and group additivity methods later.

A method involving the use of atom equivalents for relating *ab initio* energies to heats of formation has also been suggested (Ibrahim *et al.*, 1985 and 1987). In this approach, atom equivalents were determined based on the environment in which they reside; consequently, this method corresponds to an intermediate method between group and bond additivity methods.

Following the determination of the minimum energy, i.e., the optimal geometrical configuration of the atoms, a force calculation can also be undertaken to determine the vibrational frequencies of the molecule. The entropy and specific heat of the molecule can then be calculated via the use of statistical mechanics in a straightforward manner by considering contributions from translational, rotational, and vibrational degrees of freedom. We shall discuss the use of statistical mechanics to determine entropy and heat capacity later. Once the translational, rotational, and vibrational energies of the molecule are determined, they can be used to correct ΔH_f to reflect its true value under conditions of practical interest, i.e., at a temperature other than absolute zero.

A number of highly sophisticated public domain software packages are available to undertake both *ab initio* and semiempirical quantum mechanical calculations. Among the most popular *ab initio* packages are the GAUSSIAN88 (Firsch, 1988) and GAMESS (Schmidt *et al.*, 1987). Semiempirical programs include the MNDO (Dewar and Thiel, 1977), AM1 (Dewar *et al.*, 1985), and MNDO-PM3 (Stewart, 1989a), all of which are available in MOPAC (Stewart, 1987). These modern packages contain procedures to calculate optimized molecular geometries and thermochemical data, as well as spectroscopic information of molecules using built-in basis sets.

In summary, computational quantum mechanics has reached such a state that its use in chemical kinetics is possible. However, since these methods still are at various stages of development, their routine and direct use without carefully evaluating the reasonableness of predictions must be avoided. Since *ab initio* methods presently are far too expensive from the computational point of view, and still require the application of empirical corrections, semiempirical quantum chemical methods represent the most accessible option in chemical reaction engineering today. One productive approach is to use semiempirical methods to build systematically the necessary thermochemical and kinetic-parameter data bases for mechanism development. Following this, the mechanism would be subjected to sensitivity and reaction path analyses for the determination of the rank-order of importance of reactions. Important reactions and species can then be studied with greatest scrutiny using rigorous *ab initio* calculations, as well as by experiments.

III. Thermochemistry

An accurate knowledge of the thermochemical properties of species, i.e., $\Delta H_f(T_0)$, $S(T_0)$, and $c_p(T)$, is essential for the development of detailed chemical kinetic models. For example, the determination of heat release and removal rates by chemical reaction and the resulting changes in temperature in the mixture requires an accurate knowledge of ΔH_f and c_p for each species. In addition, reverse rates of elementary reactions are frequently determined by the application of the principle of microscopic reversibility, i.e., through the use of equilibrium constants, K_{eq}. Clearly, to determine K_{eq}, the knowledge of ΔH_f and S for all the species appearing in the reaction mechanism would be necessary.

The knowledge of thermochemistry is also important in early parts of mechanism development. For example, by examining the heats of reaction of competing elementary processes, unlikely reaction paths can be identified *a priori* and eliminated from further consideration. To illustrate this, consider the following simple unimolecular decomposition processes for CH_3Cl:

$$CH_3Cl \rightleftharpoons CH_3 + Cl \quad (\Delta H_r = 84\,\text{kcal/mol}), \quad (20)$$

$$CH_3Cl \rightleftharpoons CH_2Cl + H \quad (\Delta H_r = 100\,\text{kcal/mol}). \quad (21)$$

Since both of these reactions proceed through similar transition states, their rate constants have similar pre-exponential factors, and the ΔH_r corresponds to activation energies. Consequently, at process temperatures, i.e., $T \approx 1000°C$ or less, the ratio of the rate coefficient of reaction (20) to that of (21) will be the order 10^4 or more. Therefore, C–Cl bond scission, i.e., reaction (20), will always dominate the unimolecular decomposition of CH_3Cl.

From thermodynamics, the following relationships are also available for the thermochemical quantities:

$$K_{eq} = k_f/k_b = \exp(-\Delta G_r(T)/RT), \quad (22)$$

$$\Delta G_r(T) = \Delta H_r(T) - T\Delta S_r(T), \quad (23)$$

$$\Delta H_r(T) = \Delta H_r(T_0) + \int_{T_0}^{T} \Delta c_p(T)\,dT, \quad (24)$$

$$\Delta S_r(T) = \Delta S_r(T_0) + \int_{T_0}^{T} (\Delta c_p(T)/T)\,dT. \quad (25)$$

That is, the equilibrium constant (K_{eq}) for an elementary reaction at any temperature is related to the ratio of the rates for the forward (k_f) and backward (k_b) rate coefficients. The K_{eq} can be determined from the knowledge of the standard state ΔH_f and S of all the species participating in the elementary reaction, and the temperature-dependent specific heats.

Accumulated data to date indicates that K_{eq} is relatively insensitive to the details of the structure of species. Consequently, additivity rules and model compounds, in conjunction with statistical mechanics, can readily be used to estimate entropies and specific heats of molecules and radicals. These estimates generally are accurate within ± 1.0 cal/mol-K, leading to uncertainties in rate and equilibrium constants well within $\pm 20\%$.

The most important uncertainty associated with the determination of K_{eq} is related to our ability to predict heats of formation of species. At present, ΔH_f values for stable species can be predicted within ± 5 kcal/mol using various forms of additivity principles, provided these rules are applicable. Estimations based on semiempirical quantum mechanics are more general and can be as accurate. Although *ab initio* calculations can be more accurate, they are computationally prohibitive. For radical species, the associated uncertainties in ΔH_f generally are larger.

When the uncertainty associated with ΔH_f is ± 5 kcal/mol, rate and equilibrium constants can be estimated within a factor of 10 at process temperatures, i.e., 500–1,500 K. This level of accuracy may be acceptable for preliminary mechanism development work and for the identification of important reactions in a DCKM. However, it would clearly be desirable to know ΔH_f within ± 1 kcal/mol, which would lead to the determination of rate and equilibrium constants that are accurate within a factor of two. Since this level of accuracy is very close to the limits of accuracy of most experimental measurements, improvements in ΔH_f are often difficult. Consequently, computational quantum chemistry holds a great promise for the accurate determination of ΔH_f.

At present, a considerable amount of thermochemical data exists on stable, as well as some free radical, species (see, for example, Chase et al., 1985;

Pedley et al. 1986; Benson, 1976). These compilations represent a valuable starting point to estimate thermochemistry of analogous new compounds and to check the validity of the theoretical approaches used to estimate such properties.

IV. Estimation of Thermochemistry by Conventional Methods

These techniques include methods that are highly empirical in nature, such as those based on various forms of additivity principles, as well as those that are based on the use of statistical mechanical calculations. Although the latter methods represent precise tools to determine S and c_p from molecular properties, they are of little utility for the prediction of ΔH_f.

The bond additivity rule is the first-order formalism to estimate thermochemistry of stable molecules. Bond contributions to thermochemical properties for some select bonds are given in Table II. The application of this

TABLE II

BOND CONTRIBUTIONS TO ΔH_f, S, AND $c_p{}^a$ (FROM BENSON, COPYRIGHT © 1976 JOHN WILEY & SONS, INC.)

Bond	c_p	S_{int}	ΔH_f	Bond	c_p	S_{int}	ΔH_f
C–H	1.7	12.9	−3.8	S–S	5.4	11.6	−6.0
C–D	2.1	13.6	−4.7	C_d–C^b	2.6	−14.3	6.7
C–C	1.9	−16.4	2.7	C_d–H	2.6	13.8	3.2
C–F	3.3	16.9	−52.5	C_d–F	4.6	18.6	−39.0
C–Cl	4.6	19.7	−7.4	C_d–Cl	5.7	21.2	−5.0
C–Br	5.1	22.6	2.2	C_d–Br	6.3	24.1	9.7
C–I	5.5	24.6	14.1	C_d–I	6.7	26.1	21.7
C–O	2.7	−4.0	−12.0	>CO–H^c	4.2	26.8	−13.9
O–H	2.7	24.0	−27.0	>CO–C	3.7	−0.6	−14.4
O–D	3.1	24.8	−27.9	>CO–O	2.2	9.8	−50.5
O–O	4.9	9.1	21.5	>CO–F	5.7	31.6	−77.0
O–Cl	5.5	32.5	9.1	>CO–Cl	7.2	35.2	−27.0
C–N	2.1	−12.8	9.3	ϕ–H^d	3.0	11.7	3.25
N–H	2.3	17.7	−2.6	ϕ–C^d	4.5	−17.4	7.25
C–S	3.4	−1.5	6.7	(NO$_2$)–O^d	—	43.1	−3.0
S–H	3.2	27.0	−0.8	(NO)–O^d	—	35.5	9.0
C_ϕ–C_ϕ	(biphenyl)	—	10.0	C_d–C_d	—	—	7.5

^{a}Units: c_p and S in cal/mol-K, ΔH_f in kcal/mol.
^{b}C_d is vinylic carbon.
^{c}>CO– represents carbonyl carbon.
^{d}NO and NO$_2$ considered as univalent terminal groups, ϕ hexavalent unit.

method is quite straightforward; for example, the heat of formation of CH_3Cl at 298 K can be calculated as follows:

$$\Delta H_f(298) = 3\Delta H_f(C\text{---}H) + \Delta H_f(C\text{---}Cl)$$
$$= 3(-3.8) + (-7.4) = -18.8 \text{ kcal/mol}. \quad (26)$$

This is within 5% of the experimentally measured value of -19.6 kcal/mol, and no corrections are needed for the ΔH_f in this example. Similarly, the intrinsic entropy for CH_3Cl can be determined as follows:

$$S_{int}(298) = 3S(C-H) + S(C-Cl) = 3(12.90) + (19.70) = 58.4 \text{ cal/mol-K}. \quad (27)$$

The absolute entropy of any species is given by

$$S_{abs} = S_{int} + R \ln(\eta) - R \ln(\sigma) + R \ln(g_e), \quad (28)$$

where η, σ, and g_e ($g_e = 2s + 1$, where s is the total spin) are the number of optical isomers, the total symmetry numbers, and electronic degeneracies, respectively. Applying this correction to the S_{int} for CH_3Cl ($\eta = 1$, $\sigma = 3$, and $g_e = 1$), $S_{abs}(298)$ can be can be determined to be 56.2 cal/mol-K, which is very close to the experimental value of 56.0 cal/mol-K.

Symmetry number (σ) is the total number of independent permutations of identical atoms or groups in a molecule that can be arrived by simple rotations. Both the internal (σ_{int}) and external (σ_{ext}) symmetries must be considered in establishing the symmetry number, where

$$\sigma = (\sigma_{int}) \times (\sigma_{ext}). \quad (29)$$

Symmetry numbers for a number of prototype molecules are presented in Table III.

The next, and most useful, level of sophistication to estimate thermochem-

TABLE III
TOTAL SYMMETRY NUMBERS FOR MOLECULES

σ	Species
1	HCl, HCN, N_2O, CH_3OH, CH_3CHO
2	H_2, N_2, CO_2, H_2O, C_2H_2, CH_2O, C_2H_4O
3	NH_3, CH_3Cl, PCl_3, AsH_3, SiH_3Br
4	C_2H_4, B_2H_6, C_4H_8 (cyclobutane)
6	SO_3, C_3H_6 (cyclopropane)
12	CH_4, C_6H_6 (benzene)
18	All n-alkanes

ical properties involves group additivity rules, in which a group is defined as a polyvalent atom with all of its ligands. Some examples are given below:

$$CH_3OH \Rightarrow C(O)(H)_3 + O(C)(H), \qquad (30)$$

$$n\text{-}C_3H_8 \Rightarrow 2C(C)(H)_3 + C(C)_2(H)_2, \qquad (31)$$

$$1\text{-}C_4H_8 \Rightarrow C(C)(H)_3 + C(C)(C_d)(H)_2 + C_d(H)(C) + C_d(H)_2 \qquad (32)$$

As evident from these examples, thermochemistry of all linear alkanes can be determined by the knowledge of the group properties of $C(C)(H)_3$ and $C(C)_2(H)_2$. To describe all alkanes, two additional groups, $C(C)_3(H)$ and $C(C)_4$, and their properties are required. Although the data base needed to utilize group additivity rules fully is considerable, a large body of information has accumulated over the years rendering group additivity a method of choice in estimating thermochemistry today. In Table IV, some group contributions to ΔH_f, S, and c_p are illustrated. Space limitations prohibit the listing of all groups. Therefore, the reader is referred to standard tabulations in literature (see, for example, Benson, 1976; Stein et al., 1977; Stein and Fahr, 1985).

The application of group additivity is also straightforward. For example, Table V illustrates the application of this method for the estimation of the thermochemical properties of $1\text{-}C_4H_8$. Computerized versions of this approach are also available (Martinez, 1973; Seaton et al., 1974).

It should be recognized, however, that although bond and group additivity rules represent useful tools when applicable, they are restricted to molecules built from standard bonds and groups. Consequently, when new molecules are encountered, for which available bond and group values do not apply, their thermochemistry must be determined either experimentally or via computational quantum mechanics.

In the model compound approach, the thermochemistry of new species in a given series or class of compounds is determined relative to a similar model compound, with corrections made to account for differences. In estimating entropies and specific heats, differences in symmetry, masses, vibrations, moments of inertia, rotational barriers, and electronic degeneracies are accounted for in accordance with the formulations derived using statistical mechanics. In estimating the heats of formation, contributions due to the presence of isomers (e.g. *cis*, *trans*, *gauche*), and nongroup interactions (e.g., strain due to ring formation, resonance stabilization) must be considered. Corrections to ΔH_f due to isomeric forms usually are minor.

From statistical mechanics, the following relationships are available for the entropy and specific heat (McQuarrie, 1976; Laidler, 1987):

$$S = R \ln Q + R[\partial \ln Q / \partial \ln T]_V, \qquad (33)$$

$$c_v = R/T^2 [\partial^2 (\ln Q) / \partial (1/T)^2]_V, \qquad (34)$$

TABLE IV
Group Contributions to ΔH_f, S, and c_p*

Group	$\Delta H_f(298)$	$S_{int}(298)$	$c_p(300)$	$c_p(500)$	$c_p(800)$
$C-(H)_3(C)$	−10.20	30.41	6.19	9.40	13.02
$C-(H)_2(C)_2$	−4.93	9.42	5.50	8.25	11.07
$C-(H)(C)_2$	−1.90	−12.07	4.54	7.17	9.31
$C-(C)_4$	0.50	−35.10	4.37	7.36	8.77
$C_d-(H)_2$	6.26	27.61	5.10	7.51	10.07
$C_d-(H)(C)$	8.59	7.97	4.16	5.81	7.65
$C_d-(C)_2$	10.34	−12.70	4.10	4.99	5.80
$C_d-(C_d)(H)$	6.78	6.38	4.46	6.75	8.35
$C_d-(C_d)(C)$	8.88	−14.6	4.40	5.93	6.50
$C_d-(C_B)(H)$	6.78	6.38	4.46	6.75	8.35
$C_d-(C_B)(C)$	8.64	−14.6	4.40	5.93	6.50
$C_d-(C_t)(H)$	6.78	6.38	4.46	6.75	8.35
$C-(C_d)(C)(H)_2$	−4.76	9.80	5.12	8.32	11.22
$C-(C_d)_2(H)_2$	−4.29	10.2	4.7	8.4	11.3
$C-(C_d)(C_B)(H)_2$	−4.29	10.2	4.7	8.4	11.3
$C-(C_t)(C)(H)_2$	−4.73	10.30	4.95	7.93	10.86
$C-(C_B)(C)(H)_2$	−4.86	9.34	5.84	8.98	11.49
$C-(C_d)(C)_2(H)$	−1.48	−11.69	4.16	7.34	9.46
$C-(C_t)(C)_2(H)$	−1.72	−11.19	3.99	6.85	9.10
$C-(C_B)(C)_2(H)$	−0.98	−12.15	4.88	7.90	9.73
$C-(C_d)(C)_3$	1.68	−34.72	3.99	7.43	8.92
$C-(C_B)(C)_3$	2.81	−35.18	4.37	8.09	9.19
$C_t-(H)$	26.93	24.7	5.27	6.49	7.47
$C_t-(C)$	27.55	6.35	3.13	3.81	4.60
$C_t-(C_d)$	29.20	6.43	2.57	3.50	5.34
$C_t-(C_B)$	29.20	6.43	2.57	3.50	5.34
$C_B-(H)$	3.30	11.53	3.24	5.46	7.54
$C_B-(C)$	5.51	−7.69	2.67	3.68	4.96
$C_B-(C_d)$	5.68	−7.80	3.59	4.38	5.28
$C_B-(C_t)$	5.68	−7.80	3.59	4.38	5.28
$C_B-(C_B)$	4.96	−8.64	3.33	4.89	5.76
C_a	34.20	6.0	3.9	4.7	5.3
$C_{BF}-(C_B)_2(C_{BF})$	4.8	−5.0	3.0	4.2	5.2
$C_{BF}-(C_B)(C_{BF})_2$	3.7	−5.0	3.0	4.2	5.2
$C_{BF}-(C_{BF})_3$	1.5	1.4	2.0	3.5	4.7

C_d represents a double-bonded C atom, C_t a triple bonded C-atom, C_B a C atom in a benzene ring, and C_a an allenic C atom. By convention, group values for $C-(X)(H)_3$ will always be taken as those for $C-(C)(H)_3$ when X is any other polyvalent atom such as C_d, C_t, C_B, O, and S. C_{BF} represents a carbon atom in a fused ring system such as naphthalene, anthracene, etc. $C_{BF}-(C_{BF})_3$ represents the group in graphite.

*See Benson 1976 for a comprehensive list of groups and group contributions. Table reprinted with permission, Copyright © 1976 John Wiley & Sons, Inc.

TABLE V
ESTIMATION OF THERMOCHEMISTRY BY THE GROUP ADDITIVITY RULES:
$1-C_4H_8[CH_3CH_2CHCH_2]$

Group	ΔH_f	S	$c_p(300)$	$c_p(800)$
C(C)(H)$_3$	−10.2	30.4	6.2	13.0
C(C)(Cd)(H)$_3$	−4.8	9.8	5.1	11.2
Cd(H)(C)	8.6	8.0	4.2	7.7
Cd(H)$_2$	6.3	27.6	5.1	10.1
Sum	−0.1	75.8	20.6	42.0
Measured	0.0	73.0	19.0	41.8

where $c_p = c_v + R$ for an ideal gas, and Q is the total molar partition function, which can be expressed as the product of partition functions for each of the degrees of freedom of the system shown below:

$$Q = Q_{\text{trans}} \times Q_{\text{rot}} \times Q_{\text{vib}} \times Q_{\text{elec}}. \tag{35}$$

Explicit expressions for these partition functions are listed in Table VI, together with the corresponding expressions for S and c_v.

For a molecule with N atoms, its $3N$ degrees of freedom would be split into three translational degrees of freedom (corresponding to x-, y-, and z-directions), and three rotational degrees of freedom for nonlinear molecules and two for linear ones. Therefore, $3N - 6$ and $3N - 5$ vibrational degrees of freedom exist for nonlinear and linear molecules, respectively. Vibrational frequencies can be obtained from convenient tabulations (see, for example, Shimanouchi, 1972; Chase et al., 1985).

Since both S and c_v depend on the logarithm of Q, they can be evaluated as a sum of contributions for each degree of freedom. For example, for a nonlinear molecule, the following expressions represent its S and c_p:

$$S = R\{\ln[M^{3/2}T^{5/2}/P] - 1.165\} + R\{\ln(T^{3/2}(I_xI_yI_z)^{1/2}/\sigma) + 134.68\}$$

$$+ R\left\{\sum_{i=1}^{3N-6} [-\ln(1 - \exp(-1.44v_i'/T)) + 1.44v_i'/(T(\exp(-1.44v_i'/T) - 1))]\right\}$$

$$+ R\{\ln g_e\},$$

$$c_p = R + R\{3/2\} + R\{3/2\}$$

$$+ R\left\{2.07 \sum_{i=1}^{3N-6} [(v_i'/T)^2 \exp(1.44v_i'/T)/(\exp(1.44v_i'/T) - 1)^2]\right\}$$

$$+ R\{g_e(E_e/RT)^2 \exp(-E_e/RT)/[1 + g_e \exp(-E_e/RT)]^2\}, \tag{37}$$

where the units of S and c_p are those of the R used in the calculations. The

TABLE VI
CALCULATION OF ENTROPY AND HEAT CAPACITY FROM STATISTICAL MECHANICS

Motion	Partition Function	Absolute Entropy	Heat Capacity
Translation	$\left(\dfrac{2\pi mkT}{h^2}\right)^{3/2} V$	$R\left\{\ln\left(\dfrac{M^{3/2}T^{5/2}}{P}\right) - 1.165\right\}$	$\dfrac{3}{2}R$
Rotation, linear	$\dfrac{8\pi^2 IkT}{\sigma h^2}$	$R\left\{\ln\left(\dfrac{IT}{\sigma}\right) - 91.71\right\}$	R
Rotation, nonlinear	$\dfrac{\pi^{7/2}\left(\dfrac{8kT}{h^2}\right)^{3/2}(I_X I_Y I_Z)^{1/2}}{\sigma}$	$R\left\{\ln\left(\dfrac{T^{3/2}(I_X I_Y I_Z)^{1/2}}{\sigma}\right) + 134.68\right\}$	$\dfrac{3}{2}R$
Vibration	$\dfrac{1}{1 - \exp\left(-\dfrac{h\nu_i'}{kT}\right)}$	$R\left\{-\ln\left(1 - \exp\left(-\dfrac{1.44\nu_i'}{T}\right)\right) + \dfrac{1.44\nu_i'}{T(\exp\left(-\dfrac{1.44\nu_i'}{T}\right) - 1)}\right\}$	$2.07\left(\dfrac{\nu_i'}{kT}\right)^2 \dfrac{\exp\left(\dfrac{1.44\nu_i'}{T}\right)}{\left(\exp\left(\dfrac{1.44\nu_i'}{T}\right) - 1\right)^2}$
Electronic	$g_e \exp\left(-\dfrac{E_e}{kT}\right)$	$R\ln(g_e)$	$R\left[\dfrac{g_e\left(\dfrac{E_e}{kT}\right)^2 \exp\left(-\dfrac{E_e}{kT}\right)}{1 + g_e\exp\left(-\dfrac{E_e}{kT}\right)}\right]^2$

symbols in Eqs. (36) and (37), as well as those shown in Table VI, are defined as the following: c is the speed of light in vacuum, 2.99×10^{12} cm/s; g_e is the electronic degeneracy; h is Planck's constant, 6.626×10^{-27} erg-s/molecule; $I(\text{mr}^2)$ is the moment of inertia of a linear molecule in g-cm^2; $I_x = \Sigma m_i(y_i^2 + z_i^2)$, $I_y = \Sigma m_i(x_i^2 + z_i^2)$, and $I_z = \Sigma m_i(x_i^2 + y_i^2)$ are the moments of inertia about the center of gravity in g-cm^2; k is Boltzmann's constant, 1.38×10^{-16} erg/molecule-K; m_i is the mass of the atom, in g; M is the molecular weight, in amu; P is the pressure, in atm; T is the temperature, in K; R is the gas constant, 1.987 cal/mol-K; r is the radius of rotation, in cm; σ is the rotational symmetry number, $v' = v/c$, in cm^{-1}; and v is the vibrational frequency, in s^{-1}. It should be noted that electronic contributions to c_p generally are insignificant, except for molecules with low-lying electronic states, such as ClO$_2$ and NO$_2$. Internal rotations, which occur around single bonds, can also contribute S and c_p. Empirical corrections for this are discussed in Benson (1976).

To illustrate the application of statistical mechanical formulas, consider the determination of S and c_p for a nonlinear molecule such as H$_2$O at 298 K and 1 atm. The geometry of the H$_2$O molecule and its atomic coordinates, relative to the center of gravity, are presented in Fig. 6, together with its three moments of inertia I_x, I_y, and I_z. The symmetry number for H$_2$O is 2, and its vibrational frequencies are 3,657, 1,595, and 3,756 cm^{-1} (Shimanouchi, 1972). The use of Eq. (36) leads to the following estimate for S:

$$S = 34.6(S_{\text{trans}}) + 10.42(S_{\text{rot}}) + 0.0(S_{\text{vib}}) + 0.0(S_{\text{ele}})$$
$$= 45.02 \text{ cal/mol-K (experimental value 45.1 cal/mol-K.} \quad (38)$$

Similarly, for the heat capacity, Eq. (37) can be used to show that

$$c_p = 1.987 + 2.98(c_{v,\text{trans}}) + 2.98(c_{v,\text{rot}}) + 0.0(c_{v,\text{vib}}) + 0.0(c_{v,\text{elec}})$$
$$= 7.95 \text{ cal/mol-K (experimental value 8.0 cal/mol-K)} \quad (39)$$

Both these results favorably compare with experimental measurements of these thermochemical properties. It is important to note that vibrational contributions to both S and c_p are negligible in this example. However, for molecules containing heavy atoms, i.e., with low vibrational frequencies, and those with a large number of atoms, i.e., with many degrees of vibrational freedom, as well as at high temperatures, vibrations would be expected to contribute significantly to both S and c_p.

As is evident from the preceding example, statistical mechanics is particularly useful for determining S and c_p from molecular properties, when such information is available. The translational contribution to entropy depends on the mass or the molecular weight, which is known precisely. However, the determination of rotational entropy requires the knowledge of the geometry of the molecule, including bond distances, bond angles, and dihedral angles,

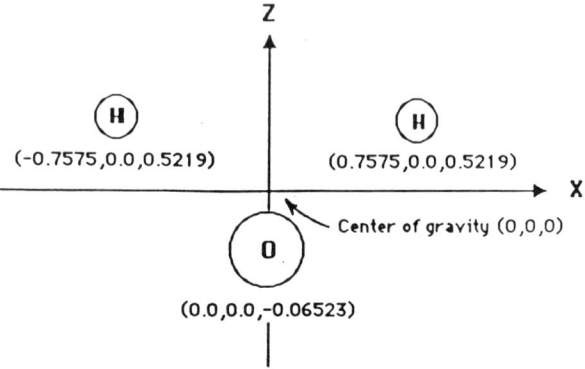

$$I_X - \sum_{i-1}^{n} m_i(Y_i^2+Z_i^2) - m_o(Y_o^2+Z_o^2) + 2m_H(Y_H^2+Z_H^2)$$

$$- \frac{1}{N_\cdot}[16(-0.06523\times10^{-8})^2 + 2(0.5291\times10^{-8})^2\,]$$

$$- 1.0175\times10^{-40} \quad (g\text{-}cm^2)$$

$$I_Y - \sum_{i-1}^{n} m_i(Z_i^2+X_i^2) - m_o(Z_o^2+X_o^2) + 2m_H(Z_H^2+X_H^2)$$

$$- \frac{1}{N_\cdot}[16(-0.06523\times10^{-8})^2 + 2((0.5291\times10^{-8})^2 + (0.7575\times10^{-8})^2)\,]$$

$$- 2.9229\times10^{-40} \quad (g\text{-}cm^2)$$

$$I_Z - \sum_{i-1}^{n} m_i(X_i^2+Y_i^2) - m_o(X_o^2+Y_o^2) + 2m_H(X_H^2+Y_H^2)$$

$$- \frac{1}{N_\cdot}[2(0.7575\times10^{-8})^2\,]$$

$$- 1.9054\times10^{-40} \quad (g\text{-}cm^2)$$

$$I_X I_Y I_Z - 1.0175\times2.9229\times1.9054\times(10^{-40})^3$$

$$- 5.6668\times10^{-120} \quad (g^3\text{-}cm^6)$$

$(N_\cdot - 6.023\times10^{23})$

FIG. 6. Geometry of the H_2O molecule and its moments of inertia.

which may not be known accurately. Similarly, contributions to entropy and heat capacity from vibrational degrees of freedom require the knowledge of the $3N-6$ characteristic frequencies of the molecule, which are also geometry-dependent, and thus may not be known with certainty. Fortunately, however, both S and c_p are not very sensitive to molecular geometry and vibrational frequencies. Consequently, we can estimate S and c_p with reasonable accuracy without precise knowledge of the molecular structure using analogous compounds (Benson 1976). We should note that molecular geometries can be determined from minimum energy considerations using quantum mechanics, and we shall discuss these issues further later.

Although we may not be able to use statistical mechanical results directly to estimate S and c_p, because of our lack of knowledge of precise molecular geometry and vibrational frequencies, we still can make use of these formulations to make systematic corrections to the thermochemistry of model compounds or to estimate these properties at different conditions. For example, based on the statistical mechanical formulas presented in Table VI, a working equation for estimating entropies from model compounds can be derived as

$$S_n = S_m + 3/2R\ln(M_n/M_m) + R\ln(\sigma_m/\sigma_n)$$
$$- R\ln(\eta_m/\eta_n) - R\ln(g_{e-m}/g_{e-n}), \qquad (40)$$

where the subscripts n and m refer to the properties of the new and model compound, respectively, and with the assumption that moments of inertia remain unchanged. Thus, the entropy of $CH_3CH_2CH_2OH$ ($\sigma = 3$) can be derived from $CH_3CH_2CH_2CH_3$ ($\sigma = 18$), since the masses of CH_3 and OH are close to one another, that is:

$$S_{C_3H_7OH} = S_{C_3H_8} + 3/2R\ln(16/15) + R\ln(18/3)$$
$$= 64.6 + 0.19 + 3.6 = 68.4\,\text{cal/mol-K}. \qquad (41)$$

This result compares favorably with the value 67.3 cal/mol-K for the experimentally determined entropy for propanol.

It is also possible to estimate the entropies of cyclic compounds from corresponding chains. The net entropy loss for most cases is of the order 3–5 cal/mol-K per rotor, or for each of the $n - 1$ atoms of the ring. For example, the S_{int} for c-C_3H_6 can be derived from open chain n-C_3H_8 ($S_{int} = 70.8$ cal/mol-K) by reducing the S_{int} of the latter by about $2(5) = 10$ cal/mol-K, resulting in a value of about 60.8 cal/mol-K; the experimentally determined S_{int} for c-C_3H_6 is virtually the same.

Differences in specific heats can be obtained in a similar fashion. Since translational and rotational contributions to c_p at elevated temperatures are minor, the differences to be accounted for are entirely due to vibrational effects. The most effective way to accomplish this is to identify the incremental contribution of each atom or group to c_p, and add or subtract this value from the model compound. For example, the c_p for C_2H_5Cl can be calculated by comparing it to C_2H_6 ($c_p(298) = 12.7\,\text{cal/mol-K}$). Since the differences in c_p are entirely due to the replacement of an H atom by Cl in a normal alkane, we can estimate this effect by considering CH_4 and CH_3Cl. In the case of the CH_4–CH_3Cl system, $\Delta c_p = 1.2\,\text{cal/mol-K}$. Applying this correction to the c_p of C_2H_6 results in the estimated heat capacity of $C_2H_5Cl = 12.7 + 1.2 = 13.9\,\text{cal/mol-K}$, which is within 8% of the experimentally determined value of $15.0\,\text{cal/mol-K}$ for the c_p of C_2H_5Cl. Clearly, some discrepancy is expected because of differences in the masses of CH_3 and C_2H_5 that lead to different vibrational frequencies.

As stated above, the thermochemistry of free radicals can also be estimated by the group additivity method, if group values are available. With the exception of a few cases reported in Benson (1976), however, such information presently does not exist. Therefore, we rely on the model compound approach (for S and c_p) and bond dissociation energy (BDE) considerations and computational quantum mechanics for the determination of the heats of formation of radicals.

Since free radicals can be viewed as being derived from a (closest) stable molecule by the removal of an atom, again only differences must be sought in establishing their thermochemistry. In addition to the symmetry considerations noted earlier, spin corrections clearly are required to establish the thermochemistry of free radicals. For example, $C_2H_5 (\sigma = 6, s = \tfrac{1}{2})$ is derived from $C_2H_6 (\sigma = 18)$; therefore, the entropy of C_2H_5 can be calculated to be

$$S_{C_2H_5} = S_{C_2H_6} + 3/2 R \ln(29/30) + R \ln(18/6) + R \ln(1/(1/2))$$
$$= 58.4 \text{ cal/mol-K}. \tag{42}$$

This is in good agreement with the experimentally determined value of 58.0 cal/mol-K.

The heats of formation of radicals can be estimated most efficiently by using information on bond dissociation energies, which is a property that approximately can be transferred between similar species. For example, available data indicate that in alkanes (except CH_4) all the primary C–H bond strengths are about 98 kcal/mol, all the secondary C–H bond strengths are about 95 kcal/mol, and all tertiary C–H bond strengths are 92 kcal/mol. Similarly, all C–Cl bond strengths in alkanes are about 80 kcal/mol. The ΔH_f of radicals can then be determined from a corresponding dissociation reaction, in which $\Delta H_r \approx \text{BDE}$, and from the known heats of formation of the remaining species in the reaction. In Table VII, dissociation energies for a large number of bonds are presented together with the heats of formation for the corresponding radicals. For example, consider the estimation of the heat of formation of CH_3, which can be determined from the reaction $CH_3Cl \rightleftharpoons CH_3 + Cl$, where ΔH_r is about 84 kcal/mol (see Eq. (20)). Therefore, the heat of formation of CH_3 can be determined as the following:

$$\Delta H_f(CH_3) = BDE + \Delta H_f(CH_3Cl) - \Delta H_f(Cl)$$
$$= 84 - 18.8 - 29.0 = 36 \text{ kcal/mol}. \qquad (43)$$

This value compares quite well with the experimental value of 35 kcal/mol.

Certain special bond types require particular attention for the determination of ΔH_f. The bond dissociation energies of first pi-bonds in C–C, such as those observed in C_2H_4, are about 60 kcal/mol; the second C–C pi-bonds, such as those observed in C_2H_4, have a strength of about 70 kcal/mol. The pi-bond in the C O system, such as those observed in aldehydes, is in excess of 150 kcal/mol. Consequently, once formed, —C=O almost exclusively leads to CO production in oxidation processes. The proper accounting of the strengths of these bonds is essential for correctly estimating heats of formation. For example, consider the decomposition of C_2H_5:

$$C_2H_5 \rightleftharpoons C_2H_4 + H \qquad (\Delta H_r = 38 \text{ kcal/mol}). \qquad (44)$$

The reason for this low heat of reaction is that while a primary C–H bond (98 kcal/mol) is being broken, a C–C pi-bond (60 kcal/mol) is simultaneously being formed.

Strain energy (E_s) is a major correction that must be applied to all cyclic compounds. Since additivity rules have been the major tool to estimate heats of formation, strain energy often represents the discrepancy between these estimates and the experimentally determined ΔH_f. Available data suggest that strain energy is relatively constant in groups of rings, and decreases with increasing number of members in rings: E_s values are about 25–30 kcal/mol for three- and four-membered rings, and about 6 kcal/mol for five-membered rings (Benson, 1976).

Resonance-stabilized compounds also represent special cases. For example, in C_3H_5 the discrepancy between experimental (41 kcal/mol) and group-additivity estimated (51 kcal/mol) ΔH_f is about -10 kcal/mol, and this is because of the stabilization induced by the delocalization of the pi electrons. In the case of benzene, the discrepancy between experimental ΔH_f (19.8 kcal/mol) and group additivity estimates (52.8 kcal/mol) is -33.0 kcal/mol (similar analysis using bond additivity rules leads to a resonance energy of -37 kcal/mol). These results suggest that resonance stabilization is primarily responsible for the lower heat of formation of benzene, since strain energy in six-membered rings, including those with double bonds, is likely to be small (i.e., 6 kcal/mol or less). In Table VIII, resonance stabilization energies observed in a variety of molecules are presented based on bond additivity considerations (Pauling, 1960). These values must be added to the heats of formation determined using the bond

TABLE VII
BOND DISSOCIATION ENERGIES FOR SELECT SPECIES (ALL UNITS IN KCAL/MOL) (FROM WENTRUP, COPYRIGHT © 1984 JOHN WILEY & SONS, INC.)

R—X		(52.1)* H	(34.0) CH$_3$	(25.7) C$_2$H$_5$	(20.8) n-C$_3$H$_7$	(17.8) i-C$_3$H$_7$	(7.5) t-C$_4$H$_9$	(18.9) F	(29.0) Cl	(26.7) Br	(25.5) I	(9.0) OH	(3.4) OCH$_3$	(47.2) NH$_2$
H—	(52.1)	104.2	104	98	98	95	92	135.8	103.1	87.4	71.3	119	105	110
CH$_3$—	(34)	104	88.2	84.8	85.2	84.2	81.8	108.8	83.6	69.8	56.1	91.1	81.4	86.7
C$_2$H$_5$—	(25.7)	98	84.8	81.8	81.6	80.4	77.9	107.1	80.8	67.6	53.2	90.9	80.8	84.3
n-C$_3$H$_7$—	(20.8)	98	85.2	81.6	81.5	80.4	77.5	107.3	80.1	68.0	53.4	91.0	81.0	84.8
i-C$_3$H$_7$—	(17.8)	95	84.2	80.4	80.4	78.2	74.2	106.1	80.4	68	53.1	91.9	81.4	85
t-C$_4$H$_9$—	(7.5)	92	81.8	77.7	77.5	74.2	68.8	108.4	80.2	66.1	50.4	91.2	80.6	83.6
△	(61.3)	100.7	89.6	86.2	86.3	85.4	82.7	110.9	85.4	73.7	58.6	97.5	87.8	91.2
◇	(51.2)	96.5	85.8	82.5	82.5	81.6	79	107.2	81.7	69.9	54.9	93.7	84	87.4
⬠	(24.3)	94.8	83.6	80.3	80.5	79.6	77	105.2	79.7	67.9	52.9	91.3	82	85.4
⬡	(13.9)	95.5	84.9	80.6	80.9	80.4	77.8	106	82	68.7	51.6	91.3	82.8	86.2
⬢	(12.2)	92.5	81.5	78.2	78.2	77.3	74.7	102.9	77.4	65.6	50.6	89.4	79.7	83.1
H$_2$C=CH—	(68.4)	108	97.5	94.3	94.5	92.8	90.4	118.7	88.8	76.4	63	100.4	88.2	
⬡—CH$_3$	(77.7)	110	99.7	96.3	96.6	94.5	90.6	123.9	94.5	79.2	64.4	109.7	98.4	104.1

Structure	ΔH_f													
H₂C=CH—CH₂— / H₂C=CH—CH— / CH₃	(41.4)	88.6	75.6	72.4	72.2	71.5	68.3	97.7	71.2	57.2	44.1	80	70.3	75.2
(methylcyclopentene)	(30.4)	82.5	71.0	67.9	67.2	65.8	64.4	93.5	67.8	55.4	40.5	79.3	68.8	72.4
HC≡C—CH₂— / H₂C=CH	(38.4)	82.3	71.6	67.8	67.2	65.6	61.6	93.5	67.8	55.4	40.5	79.3	68.8	72.4
—CH— / H₂C=CH	(86.2)	93.9	81.1	77.5	77.4	76.3	73.4	103.2	76.7	63.9	49.3	86.9	76.9	80.7
H₂C=CH	(48.3)	75.2	64.2	60.4	60.4	58.2	54.2	86.1	60.4	48	33.1	71.9	61.4	65
CH₂— \\ C=C / H (trans) / H₂C=CH	(48.3)	82.3	69.5	65.9	65.8	64.7	61.8	91.6	65.1	52.3	37.7	75.3	65.3	69.1
(cyclopentadiene)	(61)	81.2												
(methylcyclohexene)	(44)	69.8	59	55.2	55.2	53	49	80.9	55.2	42.8	27.9	66.7	56.2	59.8
(methylcyclohexadiene)	(44)	70.1	59.3	55.5	55.5	53.3	49.3	81.2	55.5	43.1	28.2	67	56.5	60.1
(methylcycloheptatriene)	(65)	73.2	61.2	57.9	57.9	57	54.4	81.9	56.4	44.6	29.6	68.4	58.7	62.1
(phenyl)—CH₂—	(44.9)	85	71.8	68.7	69	67.8	65	96.3	69.4	54.7	40	77.9	66.9	71.9

*Values in parentheses are heats of formation.

TABLE VIII
RESONANCE STABILIZATION ENERGIES OF SELECT MOLECULES (PAULING, 1960)

Molecule	Resonance Energy^a, kcal/mol
Benzene, C_6H_6	37
Naphthalene, $C_{10}H_8$	75
Anthracene, $C_{14}H_{10}$	105
Phenanthrene, $C_{14}H_{10}$	110
Biphenyl, $C_{12}H_{10}$	79
Cyclopentadiene, C_5H_6	4
Styrene, $C_6H_5CHCH_2$	42
Phenylacetylene, C_6H_5CCH	47
Phenol, C_6H_5OH	44
Aniline, $C_6H_5NH_2$	42
Pyridine, C_5NH_5	43
Acids, RCOOH	28
Esters, RCOOR'	24

^{a}Based on bond additivity considerations.

additivity principle for the estimation of ΔH_f of species that are expected to exhibit resonance stabilization.

In summary, bond and group additivity rules, as well as the model compound approach, in conjunction with statistical mechanics, represent useful tools for the estimation of thermochemical properties. However, their utility for the determination of thermochemistry of new classes of compounds is limited, especially with regard to the determination of ΔH_f. For new classes of compounds, we must resort to experiments, as well as to computational quantum mechanical methods.

V. Estimation of Thermochemistry by Quantum Mechanics

As we noted earlier, theoretical determination of heats of formation requires the knowledge of the total energy (ε_T) of interaction of electrons and nuclei among various atoms in a molecule or radical. These interactions collectively define a potential energy hypersurface (PES), and the minima on the PES corresponds to the ΔH_f of stable geometrical arrangements representing molecules or radicals (see Fig. 2 again). In mathematical terms, this can be stated as

$$\partial(\varepsilon_T)/\partial(g_i) = 0.0, \tag{45}$$

$$\partial^2(\varepsilon_T)/\partial(g_i)^2 > 0.0, \quad \text{for a molecule or radical,} \tag{46}$$

where g_i is any geometrical variable. The above relationships can then be used to locate such minima on the PES. Although the procedure is difficult to implement in $3N - 6$ dimensions, efficient computational algorithms are increasingly becoming available to locate stationary points in multidimensional spaces (Daudel et al., 1983). Additional problems are faced in locating transition states, which will be discussed later (Dewar et al., 1984).

A very effective method to expedite the determination of stationary points, and thus to estimate ΔH_f, is to use chemical intuition, i.e., our approximate knowledge of molecular structure. For this, we first must describe the geometry of the molecule by approximately specifying bond distances, bond angles, and dihedral angles, and then undertake local energy minimization calculations to determine the optimum geometry and ε_T. The heat of formation can then be determined from ε_T by using the methods discussed earlier in this article. When more than one geometrical configuration is possible, such as in the case of isomers, we must repeat this procedure by starting with the approximate geometry of each likely configuration. In cases when we do not know the structure of the molecule, we may start with a number of likely structures based on reasonable initial guesses, and then determine the optimal geometry and energy for each starting configuration.

Since energy minimization calculations are expedited by the specification of an initial structure that is as close to the optimum geometry as possible, approximate radii for a variety of atoms, and bond angles for some common orbitals for simple molecules are summarized in Table IX.

In order to illustrate these ideas, let us consider the determination of the thermochemistry of C_4H_8, which has five stable isomers. These isomers and their experimental heats of formation are trans-$CH_3CHCHCH_3$ ($\Delta H_f(\exp) = -3.0$ kcal/mol), cis-$CH_3CHCHCH_3$ ($\Delta H_f(\exp) = -1.9$ kcal/mol), $CH_2CHCH_2CH_3$ ($\Delta H_f(\exp) = -0.2$ kcal/mol), $CH_2C(CH_3)_2$ ($\Delta H_f(\exp) = -4.3$ kcal/mol), and cyclic c-C_4H_8 ($\Delta H_f(\exp) = 6.8$ kcal/mol). The geometries of these isomers are illustrated in Fig. 7, where large circles represent the carbon atoms. In Table X, the optimum geometry of trans-$CH_3CHCHCH_3$ is described in standard z-matrix form (i.e., term of bond distances, bond angles, and dihedral angles) where the carbon atoms are numbered from 1 to 4 and the hydrogen atoms from 5 to 12. Calculated heats of formation of the isomers of C_4H_8, determined by using the MNDO formalism (Dewar and Thiel, 1976), are also presented in Fig. 7, together with the values for S and c_p. These calculations typically took about 20 min cpu time on a DEC MicroVax-II (1 mips machine) computer for each of the isomers.

As evident from these results, the predictions of the thermochemistry of various C_4H_8 isomers, with the exception of cyclic C_4H_8, are in reasonable accord with the available experimental data, i.e., the agreements in ΔH_f are all within ± 3 kcal/mol and the results exhibit *correct trends*. In the case of

TABLE IX
Approximate Atomic Radii for a Variety of Atoms and Bond Angles (Pauling, 1960)

Atom	Radius, Å
H	0.300
Li (single bond)	1.225
B (single bond)	0.810
(double bond)	0.710
(triple bond)	0.640
C (single bond)	0.772
(double bond)	0.667
(triple bond)	0.603
N (single bond)	0.740
(double bond)	0.620
(triple bond)	0.550
O (single bond)	0.740
(double bond)	0.620
(triple bond)	0.550
F (single bond)	0.720
(double bond)	0.600
Na (single bond)	1.572
Mg (single bond)	1.364
Al (single bond)	1.248
Si (single bond)	1.170
(double bond)	1.070
(triple bond)	1.000
P (single bond)	1.100
(double bond)	1.000
(triple bond)	0.930
S (single bond)	1.040
(double bond)	0.940
(triple bond)	0.870
Cl (single bond)	0.990
(double bond)	0.890
Br (single bond)	1.140
(double bond)	1.040
I (single bond)	1.330
(double bond)	1.230

Bond angles for some common bonds (Pauling, 1960)

Bond	Geometry	Approximate Bond Angle, Degree
sp^3 examples: CH_4, SiH_4	tetrahedral	109
sp^2 examples: C_2H_4, C_6H_6	planar	120
sp examples: C_2H_2, HCN	colinear	180

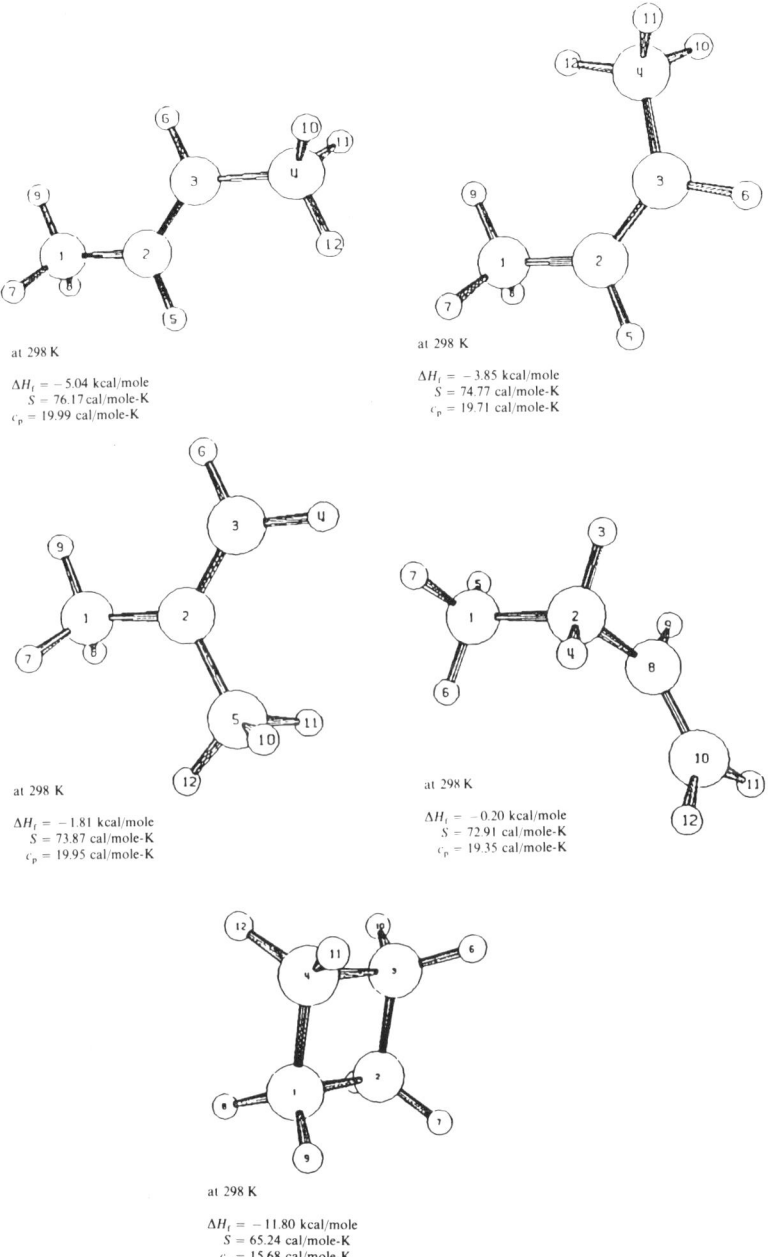

FIG. 7. Geometries and thermochemistry of C_4H_8 isomers calculated by the MNDO formalism (experimental ΔH_f are presented in the text).

TABLE X
OPTIMUM Z-MATRIX REPRESENTATION OF THE GEOMETRY OF trans-C_4H_8 OBTAINED BY MNDO[a]

Atom	Bond Length		Bond Angle		Dihedral Angle		Atom Number		
C	0.000000	0	0.000000	0	0.000000	0	0	0	0
C	1.496489	1	0.000000	0	0.000000	0	1	0	0
C	1.346135	1	126.297735	1	0.000000	0	2	1	0
C	1.496498	1	126.287045	1	−179.932162	1	3	2	1
H	1.095886	1	120.405953	1	0.020908	1	2	3	4
H	1.095862	1	120.394279	1	179.953803	1	3	2	4
H	1.110660	1	110.323728	1	58.939711	1	1	2	5
H	1.110757	1	110.283059	1	−59.949980	1	1	2	5
H	1.108807	1	113.259967	1	179.488580	1	1	2	5
H	1.110782	1	110.311462	1	58.929937	1	4	3	6
H	1.110717	1	110.303494	1	−59.981965	1	4	3	6
H	1.108831	1	113.239837	1	179.500392	1	4	3	6

[a]See Figure 7 for the ball and stick representation of this molecule.

cyclobutane, the strain associated with the formation of the ring structure clearly was underestimated by the calculations, although the entropy and specific heat are consistent with the structure. The determinations of accurate strain, as well as resonance stabilization energies, remain as important challenges in both semiempirical and *ab initio* quantum mechanical calculations (Stewart, 1989a; Ibrahim et al., 1985), necessitating the continued use of empirical corrections.

It should be noted again that, although we can determine the thermochemistry of C_4H_8 using the theoretically more pure *ab initio* methods, they are far too slow; thus, their application to systems as large as C_4H_8 is a costly proposition (Dunning et al., 1988; Dewar and Storch, 1985). In addition, as we noted previously, empirical corrections still are needed to extract accurate heats of formation from *ab initio* calculations (Melius, 1990; Wiberg, 1984; Lowe 1978).

The accuracy of MNDO-PM3, MNDO, and AM1 semiempirical quantum mechanical methods for the estimation of heats of formation, dipole moments, ionization potentials, and molecular geometries have recently been evaluated (Stewart, 1989b). For the MNDO-PM3 method, the average difference between predicted and experimental heats of formation for 657 compounds was 7.8 kcal/mol over an energy range of 350 kcal/mol, i.e., within about 2%. Other methods performed less well for the same group of survey compounds. The average difference reported was typical for free radical species, and the errors generally were much less for the case of closed-shell (stable) compounds.

As a final note, we should state that our experience with computational quantum mechanics suggests that although calculated energies may not be individually accurate, the resulting errors often are systematic in nature and lead to the establishment of useful quantitative trends. Once an accurate heat of formation for a species in a class of compounds becomes available, the established trends can then be calibrated for the determination of more accurate heats of formation.

In summary, computational quantum mechanics methods represent powerful new tools for the estimation of thermochemistry. However, their routine use clearly must be avoided, as there still are unresolved limitations of these methods. Consequently, we must continue to rely on conventional methods, experiments, and chemical intuition for the estimation of thermochemical properties.

VI. Rate Processes

It is convenient initially to classify elementary reactions either as energy-transfer–limited or chemical reaction-rate–limited processes. In the former class, the observed rate corresponds to the rate of energy transfer to or from a species either by intermolecular collisions or by radiation, or intramolecularly due to energy transfer between different degrees of freedom of a species. All thermally activated unimolecular reactions become energy-transfer–limited at high temperatures and low pressures, because the reactant can receive the necessary "activation" energy only by intermolecular collisions.

Chemical processes, in contrast, are processes that are not limited by rates of energy transfer. In thermal processes, chemical reactions occur under conditions in which the statistical distribution of molecular energies obey the Maxwell–Boltzmann form, i.e., the fraction of species that have an energy E or larger is proportional to $\exp(-E/RT)$. In other words, the rates of intermolecular collisions are rapid enough that all the species become "thermalized" with respect to the bulk gas mixture (Golden and Larson, 1984; Benson, 1976).

The transition state theory (TST) developed by Eyring and co-workers has been shown to be extremely useful to describe both the qualitative and quantitative features of chemical processes in the gas and condensed phases (Eyring, 1935; Glasstone et al., 1941). As we shall discuss below, TST also plays a central role in the determination of rate parameters by quantum mechanics.

At present two major approaches exist for the implementation of the TST theory. In the first approach, partition functions are used to describe the chemical equilibrium between the reactant(s) and the transition state (Laidler, 1987; Moore and Pearson, 1981). The partition function for the transition state is modified to account for the reaction coordinate. This approach readily allows the rationalization of the various temperature coefficients observed the preexponential factors in experimental data (Moore and Pearson, 1981); however, this method provide no insights for the expected activation energies for the reactions.

In the second approach, the chemical equilibrium between the reactant(s) and the transition state is expressed in terms of conventional thermodynamic functions, i.e., enthalpy and entropy changes. This method is easier to implement and provides useful insights for estimating both the preexponential factors and the activation energies. Consequently, we shall utilize the thermodynamic formulation of the TST in this paper.

Since the development of fundamental working equations for energy-transfer–limited reactions requires the knowledge of the rates of chemical processes, we will discuss the latter first.

VII. Chemical Processes

A chemical reaction can be viewed as occurring via the formation of an excited state that can be any one of the degrees of freedom of the collection of N atoms. That is, translational, rotational, vibrational, and electronic excitation can lead to a chemical reaction. We often do not need to consider explicitly the quantized nature of rotational and vibrational energies in practical applications because of time scale considerations. For example, when a chemical reaction proceeds via a vibrationally excited state, in which the average lifetime typically is about $3 \times 10^{-7}(T)^{0.5}/P$s, where T is in Kelvins and P is in atmospheres (Nicholas, 1976), the explicit treatment of quantum energy distribution in vibrational states would not be necessary when the experimental time scale is greater than 1 ms, at 1,000 K and 1 atm. This is because each species in the gas phase undergoes collisions every 3×10^{-11} $\sqrt{T/P}$ s; thus, the experimental time scale would be about 100 times longer than the lifetime of the vibrationally excited state. In other words, a thermal equilibrium would exist between the reactants and the surrounding bath gases, and the thermal energy distribution function can be used to estimate the concentration of activated states.

Most elementary chemical reactions can be categorized as unimolecular or bimolecular events. However, further phenomenological classification is useful for the development of detailed chemical kinetic models. This way, rate parameters for new reactions can be estimated rapidly and reliably by analogy to similar reactions in the same phenomenological class. In addition, the number of different elementary reactions that must separately be treated is reduced. It must be recognized, however, that exceptional cases

DETAILED CHEMICAL KINETIC MODELING 133

will always arise, thus the classification presented here must be used with caution.

In Fig. 8, the energy diagrams are presented for a number of prototypical elementary reactions. These diagrams essentially correspond to two-dimensional cross-sections of the potential energy surfaces described previously and shown in Fig. 2. In Figs. 8a and 8b, the respective energy diagrams for simple and complex unimolecular reactions are presented. Similarly, Figs.

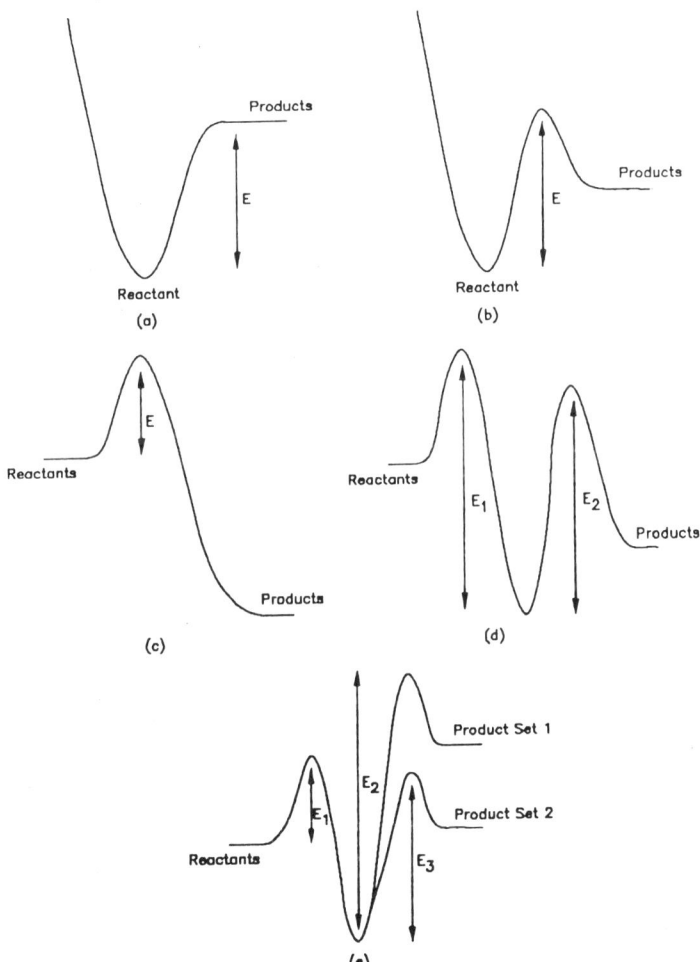

FIG. 8. (a) Energy diagrams for simple unimolecular decompositions: (b) complex unimolecular decompositions; (c) simple bimolecular (metathesis) reactions; (d) complex bimolecular reactions when one product channel is open; and (e) when two product channels are open.

8c and 8d illustrate the energetics of simple bimolecular (metathesis) and complex bimolecular reactions, in which the latter is indicated to proceed with the formation of a bound intermediate and when only one product channel is accessible. In Fig. 8e, the energy diagram for a complex bimolecular reaction is shown when two product channels are accessible, and the energy diagrams for the two channels have been superimposed. It is important to note that with the exception of bimolecular metathesis, all the other chemical reaction processes have the potential for being energy-transfer–limited; thus, their rates may depend not only on temperature, but also on pressure and the nature of the collision partner. The specific features for each of these diagrams will be discussed later.

VIII. Estimation of Chemical Rate Parameters by Conventional Methods

A. Unimolecular Reactions

These are the simplest chemical events in which an energized molecule undergoes a chemical reaction in complete isolation from other molecules. The necessary energy for reaction may accumulate in the molecule as a consequence of intermolecular collisions (thermal activation) or light absorption (photochemical activation), or as a result of energy release associated with the formation of a chemical bond in a *prior* reaction process (chemical activation). Once energy is imparted to a molecule, it is rapidly distributed among its vibrational and rotational energy levels. Consequently, energized molecules take many configurations before reaction or de-energization. If any one of these configurations corresponds to the localization of sufficient energy for reaction along the reaction coordinate, then the reaction occurs.

Since the rates of energization and de-energization depend upon the frequency of collisions, unimolecular reactions are, in principle, density- or "pressure"-dependent. That is, the rates of these reactions may be limited by the rates of energy transfer to or from the molecule by molecular collisions. Consequently, we need to know not only the high-pressure rate coefficients for unimolecular reactions, but also the extent of "fall-off" from the high-pressure limit as a function of the density of the system. Theoretical methods are available to make "fall-off" corrections, and they will be discussed later. In this section we limit our discussion of the rates of unimolecular reactions in the "high-pressure" regime where energy transfer processes are not rate-limiting.

DETAILED CHEMICAL KINETIC MODELING   135

The thermodynamic formulation of the transition state theory (TST), as applied to a unimolecular reaction described symbolically by

$$A \rightleftharpoons A^\ddagger \to P, \qquad (47)$$

leads to

$$-d[A]/dt = (kT/h)\exp(\Delta S^\ddagger/R)\exp(-\Delta H^\ddagger/RT)[A]\,\text{mol/cm}^3\text{-s} \qquad (48)$$

where [A] is the concentration of A in mol/cm^3, t is the time in s, k is the Boltzmann constant (1.38×10^{-16} erg/molecule-K), h is Planck's constant (6.63×10^{-27} erg-s/molecule), T is the absolute temperature in K, $\Delta S^\ddagger$ and $\Delta H^\ddagger$ are the entropy and enthalpy changes associated with the formation of the transition state, and R is the gas constant in units consistent with the units of S and E. The experimental activation energy E is then related to enthalpy change by $E = \Delta H^\ddagger + RT$. Since both S and H are temperature-dependent, $\Delta H^\ddagger$ and $\Delta S^\ddagger$ can further be expressed as

$$\Delta H^\ddagger(T) = \Delta H^\ddagger(T_0) + \langle \Delta c_p^\ddagger \rangle (T - T_0), \qquad (49)$$

$$\Delta S^\ddagger(T) = \Delta S^\ddagger(T_0) + \langle \Delta c_p^\ddagger \rangle \ln(T/T_0), \qquad (50)$$

where $\langle \Delta c_p^\ddagger \rangle$ represents the average specific heat change associated with the formation of the transition state at temperature T. Substituting the preceding relations into the previous rate expression results in

$$-d[A]/dt = (kT/h)(T/T_0)^{\langle \Delta c_p^\ddagger/R \rangle} \exp[1 - \langle \Delta c_p^\ddagger \rangle (T - T_0)/RT]$$
$$\times \exp(\Delta S^\ddagger(T_0)/R)\exp(-E/RT)[A]. \qquad (51)$$

From this expression, it is clear that a complex temperature dependency is an expected consequence of the TST analysis. When $\langle \Delta c_p^\ddagger \rangle$ is small relative to the gas constant R, the following familiar rate expression can be obtained:

$$-d[A]/dt = (kTe/h)\exp(\Delta S^\ddagger/R)\exp(-E/RT)[A]\,\text{mol/cm}^3\text{-s}, \qquad (52)$$

where e is the base of the natural logarithm (2.718). Because the activation energies of unimolecular reactions are generally large, the $\exp(-E/RT)$ term dominates the temperature dependency of their rates. Consequently, Arrhenius plots of unimolecular reactions generally do not exhibit much curvature under process conditions; thus, two-parameter rate expressions adequately represent such data.

It can be shown that the numerical value of the group (kTe/h) in the preceding expression is in the range $10^{13.5} - 10^{14.0}\,\text{s}^{-1}$ at process temperatures, i.e., temperatures in the range 500–1,500 K. Therefore, unless a unimolecular reaction proceeds through a "tight" transition state, i.e., $\Delta S^\ddagger = 0.0$, the pre-exponential factors of these reactions are expected to deviate from this range.

It is also useful further to classify unimolecular reactions as simple or complex, based on whether a single or multiple bonds are being broken or formed.

B. SIMPLE FISSION REACTIONS

These reactions are characterized by the formation of two radicals as a consequence of the cleavage of a single bond in an energized molecule, concomitant with an increase in bond length and entropy, i.e., $\Delta S^\ddagger > 0.0$. In addition, activation energies of these reactions generally correspond to the heat of reaction, and this is shown in Fig. 8a. In accordance with the principle of detailed balancing, the activation energy of the reverse process, i.e., radical–radical recombination, will be zero.

Simple fissions are responsible for the initiation of radical chain reactions, and some examples are

$$CH_4 \rightleftharpoons CH_3 + H, \quad (53)$$

$$CH_3Cl \rightleftharpoons CH_3 + Cl, \quad (54)$$

$$SiH_4 \rightleftharpoons SiH_3 + H, \quad (55)$$

$$Si_2Cl_6 \rightleftharpoons SiCl_3 + SiCl_3, \quad (56)$$

$$C_2H_5Cl \rightleftharpoons CH_3 + CH_2Cl. \quad (57)$$

In Table XI, experimental rate coefficients for a number of prototype

TABLE XI
HIGH-PRESSURE RATE COEFFICIENTS FOR ELEMENTARY REACTIONS[a]

Reaction Type	$\log_{10}(A)$	E kcal/mol	$\Delta H_r(298\ K)$ kcal/mol
Unimolecular Reactions ($k = A \exp(-E/RT)$, s^{-1})			
Simple Fission Reactions			
$CH_4 \rightleftharpoons CH_3 + H$	16.0	105	105
$CH_3Cl \rightleftharpoons CH_3 + Cl$	15.4	83.4	83.5
$CH_3OH \rightleftharpoons CH_3 + OH$	15.9	89.9	92.0
$C_2H_5Cl \rightleftharpoons C_2H_5 + Cl$	16.2	67.4	68.6
$C_2H_6 \rightleftharpoons C_2H_5 + H$	16.1	98.0	98.0
$C_2H_4 \rightleftharpoons C_2H_3 + H$	16.3	110	108
$CH_2CHCH_2CH_3 \rightleftharpoons CH_2CHCHCH_3 + H$	15.1	82.4	83.5
$C_6H_6 \rightleftharpoons C_6H_5 + H$	16.2	108	110
$SiH_4 \rightleftharpoons SiH_3 + H$	15.5	93	92
$H_2SiSiH_2 \rightleftharpoons SiH_2 + SiH_2$	16.0	59.0	68.0
$C_2H_6 \rightleftharpoons CH_3 + CH_3$	16.9	89.4	89.8
$C_2Cl_6 \rightleftharpoons CCl_3 + CCl_3$	17.8	68.4	69.7
$CH_2CO \rightleftharpoons CH_2 + CO$	14.5	71.1	78.0

TABLE XI Continued

Reaction Type	$\log_{10}(A)$	E kcal/mol	$\Delta H_r(298\text{ K})$ kcal/mol
$CH_3CHO \rightleftharpoons CH_3 + CHO$	15.3	79.2	84.9
$C_3H_8 \rightleftharpoons C_2H_5 + CH_3$	15.6	83.7	85.5
$n\text{-}C_4H_{10} \rightleftharpoons C_2H_5 + C_2H_5$	16.3	83.0	81.9
$CHCCH_2CH_2CCH \rightleftharpoons CHCCH_2 + CHCCH_2$	14.8	61.9	63.3
Recommendation	15.5	ΔH_r	
Complex Fission Reactions			
Radical fissions			
$C_2H_5 \rightleftharpoons C_2H_4 + H$	13.6	40.9	38.7
$C_2H_3 \rightleftharpoons C_2H_2 + H$	12.5	38.3	35.9
$C_2HCl_4 \rightleftharpoons C_2HCl_3 + Cl$	13.4	20.5	21.9
$C_2Cl_5 \rightleftharpoons C_2Cl_4 + Cl$	13.4	15.5	17.2
$n\text{-}C_3H_7 \rightleftharpoons C_3H_6 + H$	14.0	37.3	37.0
Recommendation	13.2	ΔH_r	
$CH_3CO \rightleftharpoons CH_3 + CO$	12.4	16.7	14.2
$n\text{-}C_3H_7 \rightleftharpoons CH_3 + C_2H_4$	13.0	32.9	26.4
$CH_3CHCH \rightleftharpoons CH_3 + C_2H_2$	13.1	33.4	26.2
$CH_3CH_2CH_2CH_2 \rightleftharpoons C_2H_5 + C_2H_4$	12.7	27.8	22.4
$CH_2CHCHCH \rightleftharpoons C_2H_3 + C_2H_2$	13.4	43.9	40.4
Recommendation	13.0	$\Delta H_r + 5$	
Molecular Fissions			
Three- and four-center eliminations			
$C_2H_5Br \rightleftharpoons C_2H_4 + HBr$	13.5	53.9	18.6
$C_2H_5Cl \rightleftharpoons C_2H_4 + HCl$	13.5	56.6	20.2
$C_2H_3Cl \rightleftharpoons C_2H_2 + HCl$	13.5	69.0	24.5
$1,2\text{-}C_2H_4Cl_2 \rightleftharpoons C_2H_3Cl + HCl$	13.6	58.0	15.0
$C_2Cl_6 \rightleftharpoons C_2Cl_4 + Cl_2$	13.7	54.0	30.0
$C_2HCl_5 \rightleftharpoons C_2Cl_4 + HCl$	14.0	59.0	9.2
$C_2H_6 \rightleftharpoons C_2H_4 + H_2$	13.5	65.0	33.1
Recommendation	13.5	$\frac{1}{3}$(sum of bond or $\Delta H_r + 25$ energies broken)	
Isomerization reactions			
Cis–trans isomerizations			
$CHDCHD\ c \rightleftharpoons t$	13.0	65.0	0.0
$CH_3CHCHCH_3\ c \rightleftharpoons t$	13.8	63.0	0.0
$CHClCHCl\ c \rightleftharpoons t$	13.4	56.9	0.0
$AHCCHA\ c \rightleftharpoons t$	12.8	42.8	0.0
$CH_3CHCHCOOCH_3\ c \rightleftharpoons t$	13.2	57.8	0.0
Recommendation	13.0	π-bond energy $+4.0$	
Atom migrations			
$1\text{-}C_3H_7 \rightleftharpoons 2\text{-}C_3H_7$	12.4	34.0	0.0
$2\text{-}C_3H_7 \rightleftharpoons 1\text{-}C_3H_7$	13.0	38.0	0.0
Recommendation	13.0	35.0 for 1,2 shift 25.0 for 1,3 shift	

TABLE XI Continued

Reaction Type	$\log_{10}(A)$	E kcal/mol	$\Delta H_r(298\text{ K})$ kcal/mol
Cyclization and decyclizations			
$c\text{-}C_3H_6 \rightleftharpoons C_3H_6$	15.2	65.0	-27.6
1,2-$(CH_3)_2C_3H_4 \rightleftharpoons$ noncyclic products (all)	14.0	62.0	—
$CH_3\text{-}c\text{-}C_3H_5 \rightleftharpoons n\text{-}C_4H_8$	15.4	65.0	-26.0
$C_2H_4O \rightleftharpoons CH_3CHO$	14.2	57.0	-26.9
Recommendation	15.0	60.0	

Bimolecular reactions ($k = AT^2 \exp(-E/RT)$, cm^3/mol-s)

Reaction Type	$\log_{10}(A)$	E kcal/mol	$\Delta H_r(298\text{ K})$ kcal/mol
Atom metathesis reactions			
$Cl + CH_4 \rightleftharpoons CH_3 + HCl$	6.71	1.6	
$H + CH_4 \rightleftharpoons H_2 + CH_3$	8.85	11.6	0.0
$H + C_2H_6 \rightleftharpoons H_2 + C_2H_5$	9.15	9.7	-4.4
$Cl + C_2H_6 \rightleftharpoons HCl + C_2H_5$	9.05	1.0	-3.0
$H + H_2O \rightleftharpoons H_2 + OH$	8.95	20.5	15.0
$O + OH \rightleftharpoons O_2 + H$	8.35	0.0	-16.8
$H + CH_3CHO \rightleftharpoons H_2 + CH_3CO$	8.46	4.2	—
Recommendation	8.50	$F_A \times F_C$, see below	
$NO_2 + CO \rightleftharpoons NO + CO_2$	8.54	31.9	-55.0
$CH_3 + C_2H_6 \rightleftharpoons CH_4 + C_2H_5$	6.55	10.8	-4.7
$CH_3 + CH_3CHO \rightleftharpoons CH_3CO + CH_4$	5.95	6.1	5.9
$CH_3 + C_6H_6 \rightleftharpoons CH_4 + C_6H_5$	5.55	9.6	-5.9
$CH_4 + C_6H_5 \rightleftharpoons CH_3 + C_6H_6$	6.65	11.1	-12.9
$CH_3 + CCl_4 \rightleftharpoons CH_3Cl + CCl_3$	6.65	9.1	
$OH + CH_4 \rightleftharpoons CH_3 + H_2O$	6.19	2.5	
Recommendation	7.00	$F_A \times F_C$, see below	

Form of the Reaction $A + BC \rightleftharpoons AB + C$

Group or Atom	F (Benson, 1976)
H	3.00
Cl	0.57
O	2.15
OH	1.30
NH_2	1.30
HO_2	1.70
CHO	1.55
CH_3	3.50
C_2H_5	2.85

Reaction Type	$\log_{10}(A)$	E kcal/mol	$\Delta H_r(298\text{ K})$ kcal/mol
Radical–radical metathesis			
$C_2H_5 + C_2H_5 \rightleftharpoons C_2H_4 + C_2H_6$	7.46	0.0	-63.0
$C_2H_3 + H \rightleftharpoons C_2H_2 + H_2$	8.35	0.0	-66.3
Recommendation	8.00	0.0	

TABLE XI Continued

Reaction Type	$\log_{10}(A)$	E kcal/mol	$\Delta H_r(298\ K)$ kcal/mol
Molecule–molecule metathesis			
$C_2H_4 + C_2H_4 \rightleftharpoons C_2H_3 + C_2H_5$	9.15	72.0	70.
$C_2H_6 + C_2H_4 \rightleftharpoons C_2H_5 + C_2H_5$	9.53	63.0	63.
Recommendation	Use microscopic reversibility		
Complex bimolecular reactions ($k = A\ \exp(-E/RT)$, cm^3/mol-s)			
$CH_3 + O_2 \rightleftharpoons CH_2O + OH$	13.7	34.6	−53.0
$CH_3 + O_2 \rightleftharpoons CH_3O + O$	13.2	28.7	29.0
$C_2H_3 + O_2 \rightleftharpoons CH_2O + CHO$	12.0	−0.25	−87.0
$C_2Cl_3 + O_2 \rightleftharpoons COCl_2 + COCl$	12.5	−0.83	−99.5
$C_3H_3 + O_2 \rightleftharpoons C_2H_2O + CHO$	10.5	2.87	−62.1
$HO_2 + NO \rightleftharpoons OH + NO_2$	12.3	−0.48	−6.85
Recommendation	None		

[a]Use the recommendations to estimate rate coefficients for analogous reactions.

reactions are presented, together with their enthalpy changes at 298 K and recommendations for the estimation of such parameters for analogous reactions. Experimental values were obtained from the conventional sources (e.g., Kerr and Drew, 1987; Warnatz, 1984; Kondratiev, 1972). It should be noted that the recommendations stated in Table XI are somewhat cruder, albeit quicker and easier than the methods of Benson (1976). Consequently, they are of considerable utility for the development of preliminary DCKMs. An inspection of Table XI reveals that high-pressure Arrhenius preexponential factors (As) for simple fission reactions are in a remarkably narrow range of $10^{15}-10^{16}$/s, in spite of the major differences in the nature of reactants involved. Since kTe/h is of the order $10^{13.5}-10^{14.0}$ at process temperatures, transition states in simple fission reactions clearly must be "loose", and $\Delta S^{\ddagger}$ must be positive and is of the order 9–15 e.u. The parameters presented in Table XI are particularly useful because they constitute a reasonable starting point to estimate the rate parameters of a large variety of analogous reactions and to check the reasonableness of numerical predictions using theoretical calculations.

C. Complex Unimolecular Reactions

These reactions are characterized by the concerted breakage and formation of multiple bonds with the formation of stable as well as radical products. Complex unimolecular reactions can be categorized into radical, molecular, and isomerization reactions, and rate parameters of some prototype reactions are also presented in Table XI. Because of the need to break and form multiple bonds, bond lengths in the transition state are not expected to stretch as much as those observed in simple fission reactions. Thus, complex unimolecular reactions are expected to proceed via a "tight" transition state, i.e., $\Delta S^{\ddagger}$ should be close to zero.

Examples of radical fissions include

$$C_2H_5 \rightleftharpoons C_2H_4 + H, \tag{58}$$

$$C_2H_3 \rightleftharpoons C_2H_2 + H, \tag{59}$$

$$C_2HCl_4 \rightleftharpoons C_2HCl_3 + Cl, \tag{60}$$

$$C_4H_9 \rightleftharpoons C_2H_4 + C_2H_5. \tag{61}$$

As evident from these examples, radical fissions typically involve the scission of a C–H, C–X, or C–C bond, coupled with the formation of double or triple bonds. As seen in Table XI, the preexponential A factors are generally clustered around the order of 10^{13}/s, consistent with the tight model for the transition state. It is interesting to note that activation energies of radical fissions follow, but are higher than, the enthalpies of reaction. In atom scissions, the experimental activation energy appears to be 1–2 kcal/mol larger than the heat of reaction. When the scission of polyatomic species is concerned, the observed activation energies are larger than the heat of reaction by about 4–5 kcal/mol. These differences can be explained by considering the additional strain energy required for the formation of the transition state in the latter case.

It is important to recognize that although the A factors of complex unimolecular reactions are about 10^2–10^3 smaller than those for simple bond fission reactions, their activation energies also are considerably lower. Consequently, complex radical decompositions can be dominant processes in the propagation steps of free radical chain reactions. The energy diagram for a complex unimolecular reaction is shown in Fig. 8b. An important feature of these reactions is that the reverse processes also exhibit a nonzero activation energy. In contrast, the reverses of simple unimolecular reactions almost always have zero activation energies.

The mechanisms of large radical decompositions can be estimated by applying the "β-scission" rule (McNesby et al., 1956; Westbrook, 1988). According to this rule, alkyl-type radicals decompose primarily by the breakage of a bond that is one bond away from the site of the radical. This is because isomerizations based on intramolecular transfer of atoms are energetically difficult, while the formation of a double bond makes the reaction process energetically more favorable. For example, the decomposition of 1-C_3H_7 would occur essentially by the following mechanism:

$$CH_3CH_2CH_2 \rightleftharpoons CH_3 + C_2H_4 \quad (\Delta H = 24 \text{ kcal/mol}), \tag{62}$$

and to a lesser extent via

$$CH_3CH_2CH_2 \rightleftharpoons CH_3CHCH_2 + H \quad (\Delta H = 34 \text{ kcal/mol}), \quad (63)$$

and not by

$$CH_3CH_2CH_2 \rightleftharpoons CH_3CH_2 + CH_2 \quad (\Delta H = 98 \text{ kcal/mol}). \quad (64)$$

That is, the CH_2 biradical formation process is far too endothermic to be of any significance compared to reactions (62) and (63). On the other hand, 2-C_3H_7 decomposes primarily by the following CH_3CHCH_2–H bond scission:

$$CH_3CHCH_3 \rightleftharpoons CH_3CHCH_2 + H \quad (\Delta H = 40 \text{ kcal/mol}), \quad (65)$$

because the following biradical formation reaction is too endothermic:

$$CH_3CHCH_3 \rightleftharpoons CH_3CH + CH_3 \quad (\Delta H = 90 \text{ kcal/mol}). \quad (66)$$

For radicals with more than one β site available, the breakage of the bond that results in the formation of largest radical product is usually preferred.

Molecular reactions are generally more difficult to treat because of the complexity of the possible transition states. The most widely studied complex molecular reaction class is HX elimination from halogenated hydrocarbons. These reactions proceed primarily via the formation of polar, four-centered "tight" transition states, and examples include

$$C_2H_5Cl \rightleftharpoons C_2H_4 + HCl, \quad (67)$$
$$C_2H_3Cl \rightleftharpoons C_2H_2 + HCl. \quad (68)$$

Three-centered reactions such as

$$CHCl_3 \rightleftharpoons CCl_2 + HCl \quad (69)$$

also appear possible, with the formation of biradical CCl_2. In addition, five- and six-centered molecular elimination reactions are also known to exist (Benson, 1976).

Experimental rate parameters for a variety of four-centered molecular reactions are also presented in Table XI. Although the activation energies and heats of reaction do not appear to exhibit a simple relationship similar to the other types of unimolecular reactions, there are some subtle relationships. First, for reactions involving cyclic transition states, activation energies correspond not only to the endothermicity of the overall reaction, but also to the strain energy associated with the formation of the transition state. As we discussed previously, strain energies generally are in the range 25–30 kcal/mol for three- and four-membered rings, and decrease considerably to values below 10 kcal/mol for larger rings. Consequently, for three- and four-center elimination reactions, activation energies are expected to be related to the heat of reaction by

$$E = \Delta H_r + (25-30) \text{ kcal/mol.} \tag{70}$$

Experimental activation energies presented in Table XI indeed are consistent with this observation. Furthermore, we can postulate that molecular eliminations that proceed with the formation of five-membered or larger cyclic transition states should exhibit activation energies that are much closer to the heats of reaction (O'Neil and Benson, 1967).

Second, activation energies for four-center molecular reactions are also structure-sensitive, and thus should have a less direct relationship with the enthalpy of the overall reaction (Benson and Hougen, 1965). Most notably, the activation energies for these reactions strongly depend on the polarizability of the various groups or atoms that participate in the formation of the transition state. As discussed recently, the more polarizable the group or atom, the lower the enthalpy of formation of the transition state, and thus the activation energy for the reaction (Russel et al., 1988a).

Since polarizabilities of groups or atoms are also manifested in bond strengths, some form of Hirschfelder's rule can also be developed to treat these types of reactions (Laidler, 1987). For example, the activation energies for many molecular elimination reactions seem to be correlated with one-third of the sum of the dissociation energies of the bonds that are being broken. This empirical relationship seem to hold well for a large variety of HX elimination reactions (see Table XI).

Isomerizations are important unimolecular reactions that result in the intramolecular rearrangement of atoms, and their rate parameters are of the same order of magnitude as other unimolecular reactions. Consequently, they can have significant impact on product distributions in high-temperature processes. A large number of different types of isomerization reactions seem to be possible, in which stable as well as radical species serve as reactants (Benson, 1976). Unfortunately, with the exception of *cis-trans* isomerizations, accurate kinetic information is scarce for many of these reactions. This is, in part, caused by experimental difficulties associated with the detection of isomers and with the presence of parallel reactions. However, with computational quantum mechanics theoretical estimations of barrier heights in isomerizations are now possible.

At present rate parameters for *cis-trans* isomerization reactions can be estimated by using the empirical model involving biradical transition states (Benson, 1976). That is, the transition state can be viewed as the —C·—C·— biradical, which rapidly rotates. Experimental rate parameters for a variety of *cis-trans* isomerization reactions are presented in Table XI. As seen from this table, the A factors for these reactions are consistent with a tight transition-state model. Although not directly evident from Table XI, activation energies of *cis–trans* isomerizations are directly related to the strengths of the pi bonds being broken. For example, pi-bond strengths decrease with the increasing substitution of the adjoining carbon atoms, and with increasing polarizability of the groups and atoms involved, and this is closely followed with the decrease in the activation energy for isomerization.

Atom migrations in molecules represent another class of isomerization reactions. Examples for the isomerizations of radical and stable species include

$$CH_3CH_2CH_2 \rightleftharpoons CH_3CHCH_3, \tag{71}$$

$$CHCH \rightleftharpoons CH_2C. \tag{72}$$

Rate parameters for these classes a isomerizations are extremely scarce. Our computational studies, however, suggest that for 1,2 hydrogen shift in alkanes, barrier heights should be about 25–40 kcal/mol, the same order of magnitude as the strain energy associated with the formation of three-centered rings. In Table XI some values reported in the literature are also presented (Dente and Ranzi, 1983).

Another class of isomerization involves cyclization–decyclization reactions. For example, cyclic hydrocarbons may decompose with the formation of unsaturated products. Again these reactions may be viewed to go through biradical transition states, as seen, for example, in

$$c\text{-}C_3H_6 \rightleftharpoons C_3H_6. \tag{73}$$

Rate parameters for some decyclization processes are also presented in Table XI. From the principles of detailed balancing, rate parameters for the reverse reactions, i.e., cyclizations, can be calculated.

It is important to recognize that the various isomerization reactions as written above can be broken down into sequences of simpler elementary steps. For example, the decyclization of $c\text{-}C_3H_6$ can be viewed to occur via the scission of a C–C bond, first leading to the formation of the $CH_2CH_2CH_2$ biradical intermediate, followed by 2–1 hydrogen shift in the second step, leading to the production of CH_3CHCH_2. However, since the lifetimes of biradicals are expected to be extremely short, decyclization reactions (as well as their reverse, cyclizations) can be represented by a single reaction for practical purposes.

At high temperatures and low pressures, the unimolecular reactions of interest may not be at their high-pressure limits, and observed rates may become influenced by rates of energy transfer. Under these conditions, the rate constant for unimolecular decomposition becomes pressure- (density)-dependent, and the canonical transition state theory would no longer be applicable. We shall discuss energy transfer limitations in detail later.

D. BIMOLECULAR REACTIONS

Bimolecular reactions may be classified into two major groups: direct metathesis and association reactions. The latter also are related to the reverse of the unimolecular reactions discussed above. However, as we shall see, significant differences would exist when there are energy transfer limitations. It is also convenient further to classify bimolecular reactions as radical–molecule, radical–radical, and molecule–molecule reactions. The application of TST to bimolecular reactions, described symbolically by

$$A + B \rightleftharpoons AB^\ddagger \rightarrow R + S, \tag{74}$$

leads to the following:

$$-d[A]/dt = (kT/h)(R'T)(T/T_0)^{\langle \Delta c_p^\ddagger/R \rangle}$$
$$\exp[2 - \langle \Delta c_p^\ddagger \rangle (T - T_0)/RT] \exp(\Delta S^\ddagger(T_0)/R) \exp(-E/RT)[A][B], \tag{75}$$

where R' is the gas constant in concentration units (82.05 atm-cm^3/mol-K), and the other symbols have the same meaning as described before. In this case, the experimental activation energy is related to enthalpy change by $E = \Delta H^\ddagger + 2RT$. The preceding expression also exhibits a complex (non-Arrhenius) temperature dependency, and reduces to the following familiar form when $\langle \Delta c_p^\ddagger \rangle$ is small:

$$-d[A]/dt = (kTe^2/h)(R'T) \exp(\Delta S^\ddagger/R) \exp(-E/RT)[A][B] \text{ mol/cm}^3\text{-s}. \tag{76}$$

That is, TST predicts a T^2 dependency for the A factors, and we shall adopt this in the remainder of this article. Clearly, when E is large, the exponential term will again dominate the temperature dependency of the rate expression, and the Arrhenius plots will be linear. However, for small activation energies, the T^2 term will contribute significantly to the temperature dependency of the rate expression, and this leads to curved Arrhenius plots which we shall discuss below.

It is also important to recognize that the A factors for all bimolecular reactions are limited by the collision frequency, Z_{A-B}, which is given by

$$Z_{A-B} = (\sigma_{A-B})^2 \sqrt{(8\pi kT/\mu_{A-B})} N_A \text{ mol/cm}^3\text{-s}, \tag{77}$$

where $\sigma_{A-B} = (\sigma_A + \sigma_B)/2$, $\mu_{A-B} = (m_A m_B)/(m_A + m_B)$, σ_A, σ_B, and m_A, m_B are the collision diameters and molecular masses of A and B, respectively, and N_A is Avogadro's number. Since the numerical value of Z_{A-B} is in the range $10^{13.5}$–$10^{14.5}$, and that for the group $(kTe^2/h)(R'T)$ is $10^{18.5}$–$10^{19.5}$ at process temperatures (500 < T < 1,500 K), the entropy must always decrease with the formation of the transition state in bimolecular reactions. This, of course, is expected, since the transition state AB‡ corresponds to a more orderly configuration of species A and B.

E. Direct Metathesis Reactions

Direct metathesis reactions are characterized by tight transition states and involve the transfer of atoms or radicals as a consequence of the close proximity of the two reactants, and they are the only type of reactions that are not subject to becoming energy-transfer–limited at high temperatures and low pressures. A typical energy diagram for metathesis reactions is shown in Fig. 8c. Examples of such reactions include

$$Cl + CH_4 \rightleftharpoons HCl + CH_3, \quad (78)$$

$$H + C_2H_6 \rightleftharpoons H_2 + C_2H_5, \quad (79)$$

$$CH_3 + C_2H_6 \rightleftharpoons CH_4 + C_2H_5. \quad (80)$$

Since radical and atom metathesis reactions generally have low activation energies, their rate coefficients are expected to exhibit non-Arrhenius behavior because of the increased importance of the T^2 term noted previously. In Fig. 9, rate coefficient data for H-atom abstractions from CH_4

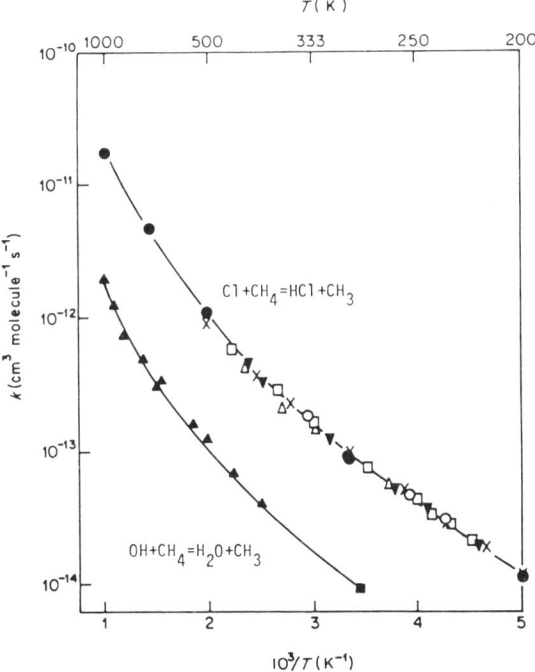

FIG. 9. Arrhenius plots for the reactions $Cl + CH_4 \rightleftharpoons HCl + CH_3$ (adapted from Heneghan et al., 1981), and $OH + CH_4 \rightleftharpoons H_2O + CH_3$. Solid lines represent TST-based rate coefficient fits to the experimental data (see text). Reprinted with permission from Tully, F. P., and Ravishankara, A. R., J. Phys. Chem. **84**, 3126, Copyright © 1980 American Chemical Society.

by Cl and OH are presented. These reactions are among the best-studied atom metathesis reactions, and as expected, their rates exhibit considerable non-Arrhenius behavior. TST-based rate coefficient fits to these data were found to be $5.18 \times 10^6 T^{2.11} \exp(-1,590/RT)$ cm^3/mol-s for the reaction Cl + CH$_4 \rightleftharpoons$ HCl + CH$_3$ (Heneghan et al., 1981), and $1.55 \times 10^6 T^2 \exp(-2,470/RT)$ cm^3/mol-s for the reaction OH + CH$_4 \rightleftharpoons$ H$_2$O + CH$_3$ (Tully and Ravishankara 1980). These fits correspond to the solid lines that are also shown in Figure 9.

Recently, transition state theory calculations were applied to a class of reactions involving OH radicals and haloalkanes, again to account systematically for the expected curvature in Arrhenius plots for these reactions (Cohen and Benson, 1987a). Subsequently, empirical relationships were also derived for the *a priori* determination of pre-exponential factors (A) and activation energies (E) based on an assumed $T^{1.5}$ dependency of the pre-exponential factor (Cohen and Benson, 1987b). This and related studies clearly illustrate the broad utility of transition state theory in the systematic development of detailed chemical kinetic mechanisms.

In Table XI, rate constants for a number of prototype metathesis reactions are presented. As is evident from this table, atom metatheses are characterized by A factors of the order $10^{9.0}$ cm^3/s-mol. This A value corresponds to an entropy decrease ($-\Delta S^\ddagger$) of about 20–30 cal/mol-K associated with the formation of the transition state (TS) and can be explained by considering the following:

$$\Delta S^\ddagger = S_{TS} - S_A(\text{atom}) - S_B(\text{molecule}). \tag{81}$$

We expect that $S_{TS} \approx S_B$(molecule) with minor corrections for differences in symmetry and spin; thus, $\Delta S^\ddagger \approx -S_A$(atom). Since the absolute entropies of atoms are in the range 25–35 cal/mol-K (Chase et al., 1985), $-\Delta S^\ddagger$ would be expected to be in this range without any corrections.

In contrast, polyatomic radical metathesis reactions have A factors that are slightly lower, typically in the range 10^6–10^7 cm^3/s-mol, suggesting entropy decreases in the range 30–40 cal/mol-K. This again is expected in view of the preceding discussion, since the absolute entropies of radicals are larger than those of atoms.

Several empirical rules also exist for the estimation of activation energies of metathesis reactions. A useful method involves the consideration of the following prototype reaction written in the *exothermic direction*:

$$A + BC \rightleftharpoons AB + C, \tag{82}$$

for which the activation energy E in kcal/mol is given by (Alfassi and Benson 1973)

$$E = F_A \times F_C, \tag{83}$$

where F_A and F_C are empirical constants for the transferring atoms and radicals that are also tabulated in Table XI. For example, the activation energy of $Cl + C_2H_6 \rightleftharpoons HCl + C_2H_5$ can be estimated to be $E = 0.57 \times 2.85 = 1.62$ kcal/mol, in which $A = Cl$ ($F_A = 0.57$) and $C = C_2H_5$ ($F_C = 2.85$). This value of E is very close to the experimental value of about 1.0 kcal/mol.

Polanyi relationships also represent another useful method for the estimation of activation energies in metathesis reactions (Laidler, 1987). These methods rely on the assumption that activation energies are linearly related to the heats of reaction:

$$E = a(-\Delta H_r) + b, \tag{84}$$

where a and b are empirical constants. These methods are reliable primarily for homolytic reactions. Polanyi relationships frequently fail when there is charge separation involved in the transition state, which would be the case when atoms or groups with different electronegativities are involved.

Semenov's empirical formulation represents a special case of a Polanyi relationship in which the activation energy for an exothermic hydrocarbon metathesis reaction is given by (Semenov, 1958)

$$E = 11.5 - 0.25(-\Delta H_r) \text{ kcal/mol}. \tag{85}$$

The application of this relation to the forward branch of the reaction $H + C_2H_6 \rightleftharpoons H_2 + C_2H_5$ results in the determination of $E = 10$ kcal/mol, which is very close to the experimental value of 9.7 kcal/mol. On the other hand, the application of the Semenov relationship to the reaction $Cl + C_2H_6 = HCl + C_2H_5$ yields an activation energy of 10 kcal/mol, which is substantially higher than the experimental value of about 1.0 kcal/mol. The failure of the Semenov relationship in the latter reaction is not surprising, because of the charged transition state that arises because of the involvement of the highly electronegative Cl atom with H.

Hirschfelder's rules represent another useful tool that can be used to estimate the activation energies of exothermic metathesis reactions (Laidler, 1987). According to these rules, the activation energy of an atom metathesis reaction is given by

$$E = 0.055 \times (BDE) \quad \text{(atoms)}, \tag{86}$$

where BDE represents the bond dissociation energy for the bond being broken. For radical metathesis, E is given by

$$E = 0.13 \times (BDE) \quad \text{(radicals)}. \tag{87}$$

The bond-energy bond-order (BEBO) method developed by Johnston and Parr (1963), in spite of its nonkinetic basis, represents a broadly applicable empirical approach to estimating activation energies of metathesis reactions.

This method, since its original proposal for H transfer reactions, has been modified to include other atoms and radicals as well (Mayer et al., 1967); however, it was recently shown to be inadequate to handle metal atom reactions (Rogowski et al., 1989). The fundamental assumption of the BEBO method is that the reaction occurs via an energy path along which the sum of bond orders for the bonds being broken and formed remains unity. The bond order (n) is related to bond length (r) via the following relationship (Pauling, 1960):

$$r = r_s - 0.26 \ln(n), \tag{88}$$

and bond energy (E) is related to bond order via

$$E = E_s(n)^p, \tag{89}$$

where p is an empirical constant, and s corresponds to single bonds between the atoms involved. To illustrate the utility of the BEBO method, consider the following schematic metathesis reaction:

$$A + XB \rightleftharpoons [A \underset{r_1}{-} X \underset{r_2}{-} B] \rightleftharpoons AX + B, \tag{90}$$

where r_1 and r_2 represent intergroup distances in the transition state. The electronic energy can then be written as the sum of the following pair interactions:

$$V(r_1, r_2) = -E_1(AX) - E_2(XB) + E_3(AB), \tag{91}$$

where E_1 and E_2 are binding interactions, and E_3 is antibinding (repulsion) energy. Using the earlier definitions, the values for E_1 and E_2 are given by

$$E_1 = E_{1,s}(n)^p, \tag{92}$$

$$E_2 = E_{2,s}(1-n)^p, \tag{93}$$

and the repulsion energy E_3 is given by the following "anti-Morse" function (Mayer et al., 1967):

$$E_3 = E_{3,s} \exp(-\beta \Delta r_3)[1 + 0.5 \exp(-\beta \Delta r_3)], \tag{94}$$

where β is the Morse parameter, and Δr_3 is the difference between the A–B distance ($r_1 + r_2$) in the transition state and the equilibrium bond distance AB. Since E_1, E_2, and E_3 are all functions of bond order, the $V(r_1, r_2)$ becomes only a function of n. The activation energy can then be estimated by systematically increasing n until $V(r_1, r_2)$ becomes a maximum. The activation energy for the reaction is then given by

$$E_{act} = V_{max}(r_1, r_2). \tag{95}$$

Computational details of the BEBO method are discussed in Johnston (1966) and Brown (1981). As is evident from the foregoing discussion, although the BEBO method represents a general method to estimate activation energies, it is strictly applicable to bimolecular metathesis reactions. In addition, in spite of its computational rigor, the BEBO method often does not lead to the determination of activation energies that are more accurate than the other empirical methods discussed earlier.

It is important to note that radical–radical or atom–radical metathesis reactions have no activation energy barriers, and their energy diagrams resemble that of Fig. 8a. Examples for such reactions include the following:

$$C_2H_3 + C_2H_5 \rightleftharpoons C_2H_4 + C_2H_4, \tag{96}$$

$$C_2H_5 + H \rightleftharpoons C_2H_4 + H_2. \tag{97}$$

Rate coefficient parameters for a number of radical–radical metathesis reactions are also presented in Table XI.

Although relatively unimportant compared to radical–molecule reactions under process conditions, molecule–molecule metathesis reactions also occur. Reverses of the preceding radical–radical metathesis reactions are examples for molecule–molecule metathesis. Additional examples are provided in Table XI. The activation energies of molecule–molecule reactions correspond to their heats of reaction, since their reverses, radical–radical metathesis reactions, have zero activation energies. The A factors for molecule–molecule reactions are slightly higher than other types of metathesis reactions; they are of the order $10^9 \, \text{cm}^3/\text{s-mol}$. However, their activation energies, which are in the range 30–60 kcal/mol, are also higher than those of radical metathesis reactions. Therefore, in the presence of radical reaction pathways, molecule–molecule reactions contribute only to a minor extent to the progress of the reaction. However, molecule–molecule reactions can be very important in the early stages of a process (i.e., during the induction period), since they contribute to the generation of free radicals. In this regard, molecule–molecule reactions are similar to unimolecular decomposition reactions.

F. Association Reactions

In the high-pressure limit conditions considered in this section, association reactions are, in principle, the reverse of the fission reactions discussed previously. That is, although the association process initially results in the formation of a "chemically activated" adduct—as a consequence of net energy released by the exothermic association process—this energy is rapidly dissipated into the surrounding medium by intermolecular collisions. Clearly, if rapid de-energization is not possible, the chemically activated adduct may decompose back to reactants, or rearrange and form new products. Consequently, when energy transfer limitations exist, association reactions cannot simply be treated as the reverse of unimolecular fission reactions. We shall examine these issues in the next section.

Association reactions can be further classified as simple and complex, similarly to the unimolecular decomposition reactions treated earlier. Simple association reactions involve the formation of a single bond, such as those observed in atom and radical recombinations, for example:

$$H + C_2H_5 \rightleftharpoons C_2H_6, \quad (98)$$

$$CH_3 + CH_2Cl \rightleftharpoons C_2H_5Cl. \quad (99)$$

The treatment of these simple associations directly follows that of the simple fission reactions discussed previously. For example, these reactions proceed via the formation of a loose transition state and without an activation energy barrier. The rates and rate parameters of simple associations can be determined either directly, by the application of bimolecular TST, or from their reverse, simple unimolecular fission reactions, through the use of the principle of microscopic reversibility.

Complex associations involve the concerted formation and breakage of multiple bonds; examples include the following:

$$H + C_2H_2 \rightleftharpoons C_2H_3, \quad (100)$$

$$CH_3 + O_2 \rightleftharpoons CH_3O_2, \quad (101)$$

$$O + C_2H_4 \rightleftharpoons C_2H_4O, \quad (102)$$

$$CH_3 + C_2H_2 \rightleftharpoons CH_3C_2H_2. \quad (103)$$

Again the analysis of these reactions can be undertaken either directly by the use of TST or by considering their reverse, the complex fission reactions discussed earlier. Since unimolecular reactions were treated in considerable detail before, we will not repeat related issues here.

Occasionally, the rates of bimolecular reactions are observed to exhibit negative temperature dependencies, i.e., their rates decrease with increasing temperature. This counterintuitive situation can be explained via the transition state theory for reactions with no activation energy barriers; that is, preexponential terms can exhibit negative temperature dependencies for polyatomic reactions as a consequence of partition function considerations (see, for example, Table 5.2 in Moore and Pearson, 1981). However, another plausible explanation involves the formation of a bound intermediate complex (Fontijn and Zellner, 1983; Mozurkewich and Benson, 1984). To illustrate this, consider the following reaction scheme:

$$A + B \underset{k_{-1}}{\overset{k_1}{\rightleftharpoons}} [AB] \overset{k_2}{\longrightarrow} P + Q, \quad (104)$$

where [AB] represents an adduct. The rate of disappearance of reactants can be shown to be

$$-d[A]/dt = k_1 k_2/(k_{-1} + k_2)[A][B]. \quad (105)$$

When $k_{-1} \gg k_2$, this relation reduces to the following:

$$-d[A]/dt = k_2 K_1 [A][B], \quad (106)$$

where $K_1 = k_1/k_{-1}$ is the equilibrium constant for the formation of the adduct. Substituting $k_2 = k_{2_0} \exp(-E_2/RT)$ and $K_1 = \exp(\Delta S_1/R) \exp(-\Delta H_1/RT)$ leads to

$$-d[A]/dt = k_{2_0} \exp(\Delta S_r/R) \exp(-(E_2 + \Delta H_1)/RT)[A][B]. \quad (107)$$

Because ΔH_1 for association reactions has to be negative, the observed activation energy for the reaction can be zero, i.e., $E_2 = -\Delta H_1$, or negative, i.e., $E_2 < -\Delta H_1$. However, when $E_2 > -\Delta H_1$, the overall reaction rate will always exhibit a normal temperature dependency, i.e., the rate will increase with increasing temperature. The energy diagram for these reactions can be represented by Fig. 8d, and in view of the preceding analysis it is clear that a negative temperature dependency would be expected when $E_2 < E_1$. In some cases the intermediate complex can be stabilized by intermolecular collisions, and this leads to pressure-dependent rate coefficients in a manner similar to those observed in unimolecular reactions. We shall discuss these situations in the section on energy-transfer–limited processes (Section X).

Examples for complex bimolecular reactions with rates exhibiting negative temperature dependencies include

$$HO_2 + NO \rightleftharpoons [HOONO] \rightarrow OH + NO_2, \quad (108)$$
$$C_2H_3 + O_2 \rightleftharpoons [C_2H_3O_2] \rightarrow CH_2O + CHO, \quad (109)$$
$$C_2Cl_3 + O_2 \rightleftharpoons [C_2Cl_3O_2] \rightarrow COCl_2 + COCl. \quad (110)$$

In Fig. 10, experimental data for these reactions are presented. Two-parameter (Arrhenius) rate coefficient expressions for these three reactions were determined to be $2.1 \times 10^{12} \exp(483/RT) \, cm^3/mol\text{-s}$ (Fontijn and Zellner, 1983), $3.97 \times 10^{12} \exp(250/RT) \, cm^3/mole\text{-s}$ (Slagle et al., 1984), and $1.2 \times 10^{12} \exp(830/RT) \, cm^3/mol\text{-s}$ (Russell et al., 1988b), respectively. These fits, indicated by solid lines, are also shown in Fig. 10. The rates for the latter two reactions were also determined to be pressure-independent, suggesting that under the experimental conditions studied, the lifetime of the adduct must be short relative to collision times.

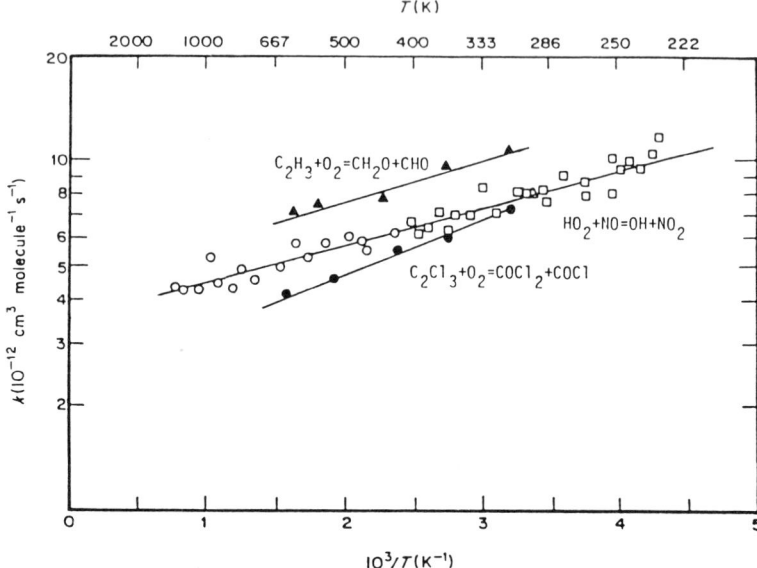

FIG. 10. Arrhenius plots for the reactions $HO_2 + NO \rightleftharpoons OH + NO_2$ (Howard, 1980), $C_2H_3 + O_2 \rightleftharpoons CH_2O + CHO$ (Slagle et al., 1984), and $C_2Cl_3 + O_2 \rightleftharpoons COCl_2 + COCl$ (Russell et al., 1988). Solid lines represent two-parameter (Arrhenius) rate coefficients fits to the experimental data (see text).

IX. Estimation of Rate Parameters by Quantum Mechanics

As we discussed earlier, according to the transition state theory (TST), the estimation of activation energies (E) and pre-exponential factors (A) of elementary reactions depends on our ability to determine the $\Delta S^\ddagger$ and $\Delta H^\ddagger$. In Fig. 11 the energy diagram for a prototype exothermic reaction is presented. In this diagram the reactant and product configurations are represented by the two minima, and the transition state corresponds to the saddle point. That is, the transition state would be stable to all geometric deformations except the reaction coordinate, which may be a bond distance, bond angle, or dihedral angle, depending on the nature of the reaction for simple molecules. However, when large molecules are involved, such as those observed in biochemical reactions, complex reaction coordinates are possible. That is, the minimum energy path between the reactants and the products may not be represented by a simple reaction coordinate, but may involve a concerted deformation of bonds, bond angles, and dihedral angles. This is an inherent difficulty associated with all methods that require the

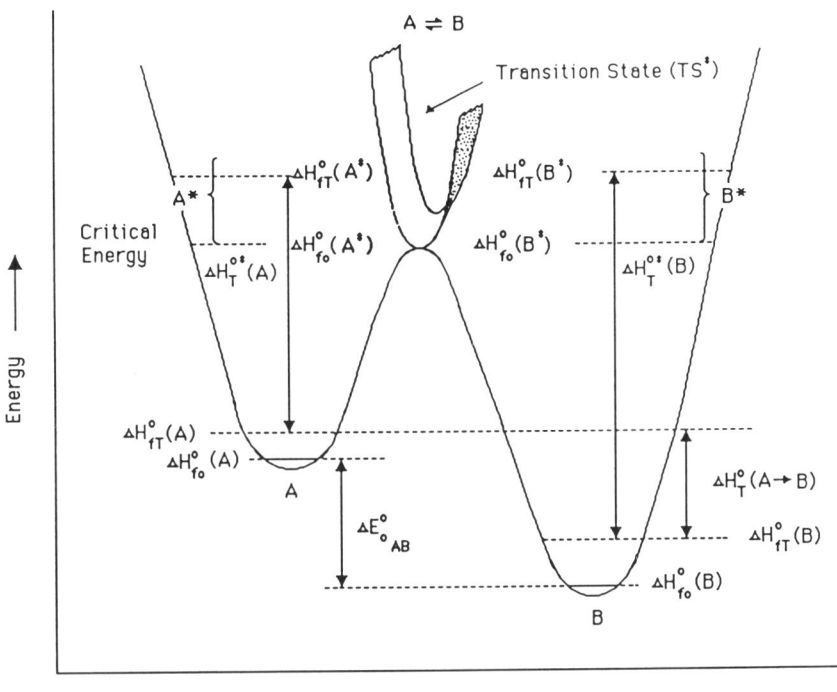

FIG. 11. Potential energy diagram for a hypothetical exothermic reaction, showing the transition state as the saddle point.

knowledge of the PES, and the numerical location of transition states (saddle points) is a subject of continuing research in this field.

We shall illustrate the application of these principles for the determination of rate parameters for unimolecular and bimolecular reactions using examples taken from processes that are of considerable interest to chemical engineering practice today: (1) the unimolecular decomposition of silane (SiH_4), which is an important reaction in the chemical vapor decomposition (CVD) of silicon in the manufacture of microelectronics (Till and Luxon, 1982); (2) hydrogen atom abstraction from methane (CH_4) by H and Cl radicals, which are important reactions in the conversion of methane (natural gas) into more valuable hydrocarbon products (Karra and Senkan, 1988b); and (3) addition of CH_3 and C_2H_3 radicals to C_2H_2, which are important in hydrocarbon pyrolysis and combustion, especially with regard to the formation of high–molecular-weight hydrocarbons, soot, and coke.

Two likely paths exist for the thermal unimolecular decomposition of silane. The first reaction path involves the scission of the SiH_3-H bond, with a stretched or loose transition state:

$$SiH_4 \rightleftharpoons SiH_3 + H \quad (\Delta H_r = 90.0 \text{ kcal/mol}). \tag{111}$$

For this reaction, the reaction coordinate is simply the Si–H bond distance. Since the reverse of this reaction is radical–radical recombination and should have no activation energy barrier, the E for forward reaction would be expected to be ΔH_r.

The other SiH_4 decomposition pathway involves H_2 elimination via a three-center cyclic transition state (Roenigk et al., 1987):

$$SiH_4 \rightleftharpoons SiH_2 + H_2 \quad (\Delta H_r = 60.0 \text{ kcal/mol}). \tag{112}$$

The reaction coordinate again is simple; it is the H–Si–H bond angle. In this case, however, the reverse reaction is a radical-molecule reaction, and we cannot make the a priori assumption that its activation energy would be zero. In fact, the literature is full of examples of radical–molecule reactions with large activation energies (Benson, 1976). As a result, we cannot also make the assumption that for the forward reaction $E = \Delta H_r$, as we did in the case of the Si–H bond fission reaction. At this point, we must resort to either quantum chemical calculations or experiments to resolve this issue.

In Figs. 12 and 13, the calculated potential energy diagrams for the two SiH_4 reactions are presented. These calculations were made by starting with the ground state SiH_4, and by fully optimizing the geometry of the molecule each time the reaction coordinate was changed and kept fixed at its new position. In the former case, the Si–H bond distance was increased as the reaction coordinate, and the geometry of the stretched SiH_3-H was optimized to determine the minimum energy configuration. These calculations were made by using the AM1 quantum mechanical molecular model developed recently (Dewar et al., 1985).

As seen from Fig. 12, quantum chemical calculations indeed support the fact that the activation energy for the SiH_3-H bond scission reaction corresponds to the bond dissociation energy or the heat of reaction, i.e., $\Delta H_r \approx E \approx 90$ kcal/mol. In addition, at the transition state, the Si–H bond distance is calculated to be about 2.5 Å, a clear indication of the looseness and the increase in entropy associated with the formation of the transition state.

In Fig. 13, the energy diagram for H_2 elimination from SiH_4 is presented. In this case the activation energy also seems to be closely related to the heat of reaction, i.e., $\Delta H_r \approx E \approx 60$ kcal/mol. This is a surprising and important result because it indicates that the reverse reaction, i.e., $SiH_2 + H_2 = SiH_4$, is expected to proceed with no activation energy barrier. In addition, this result suggests the possibility that SiH_2 may insert into double bonds with little or no activation energy barrier. Clearly, this is an important issue in mechanism

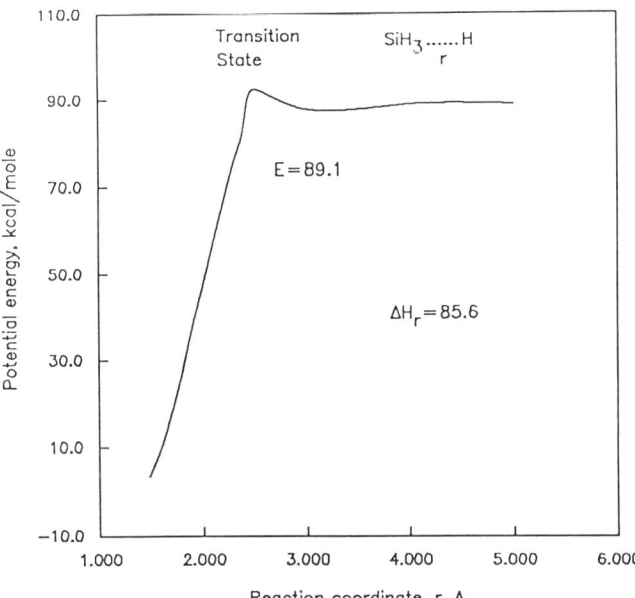

FIG. 12. Potential energy diagram for the unimolecular decomposition of SiH_4 to SiH_3 and H, showing the activation energy barrier E. The sketch above is a geometrical representation of the events (1 to r) along the reaction coordinate, which is the Si–H distance.

development for silane decomposition. The analysis of the transition state in this case indicates that interatomic distances remain very close to bonding distances, indicative of a tight transition state.

Following the determination of the geometry and the thermochemistry of transition states, the rate parameters for the two silane decomposition pathways can be obtained directly by the TST formulation presented earlier. These calculations have led to unimolecular rate constant expressions $10^{16.2} \exp(-91000/RT) \, \mathrm{s}^{-1}$, and $10^{13.2} \exp(-62000/RT) \, \mathrm{s}^{-1}$ for Si–H bond scission and H_2 elimination reactions, respectively. These results clearly

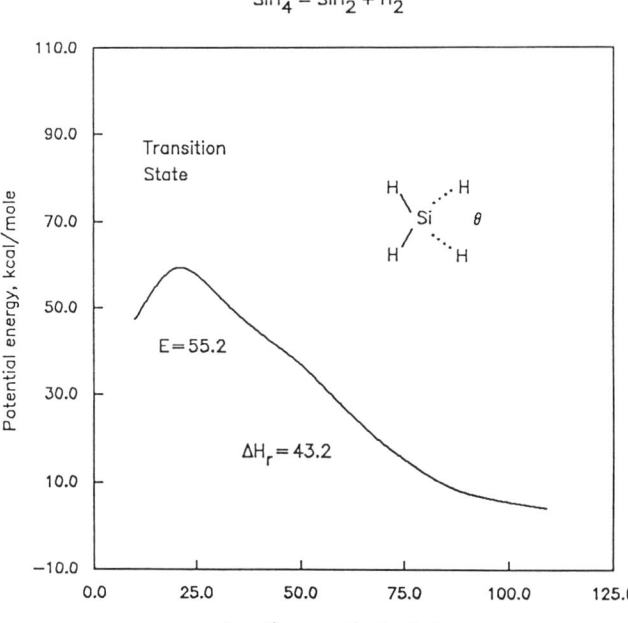

FIG. 13. Potential energy diagram for the unimolecular decomposition of SiH_4 to SiH_2 and H_2, showing the activation energy barrier E. The sketch above is a geometrical representation of the events (1 to r) along the reaction coordinate, which is the H–Si–H angle.

suggest that under conditions of practical interest, i.e., $500 < T < 1{,}500\,K$, the H_2 elimination reaction would be expected to dominate initial silane decomposition. We shall further examine silane decomposition reactions when we discuss energy transfer limitations.

Hydrogen abstraction from methane by Cl and H occurs via the following bimolecular metathesis reactions:

$$CH_4 + Cl \rightleftharpoons CH_3 + HCl \quad (\Delta H_r = -1.7\,\text{kcal/mol}), \quad (113)$$

$$CH_4 + H \rightleftharpoons CH_3 + H_2 \quad (\Delta H_r = -0.6\,\text{kcal/mol}). \quad (114)$$

Both reactions have been widely studied and have well-established activation

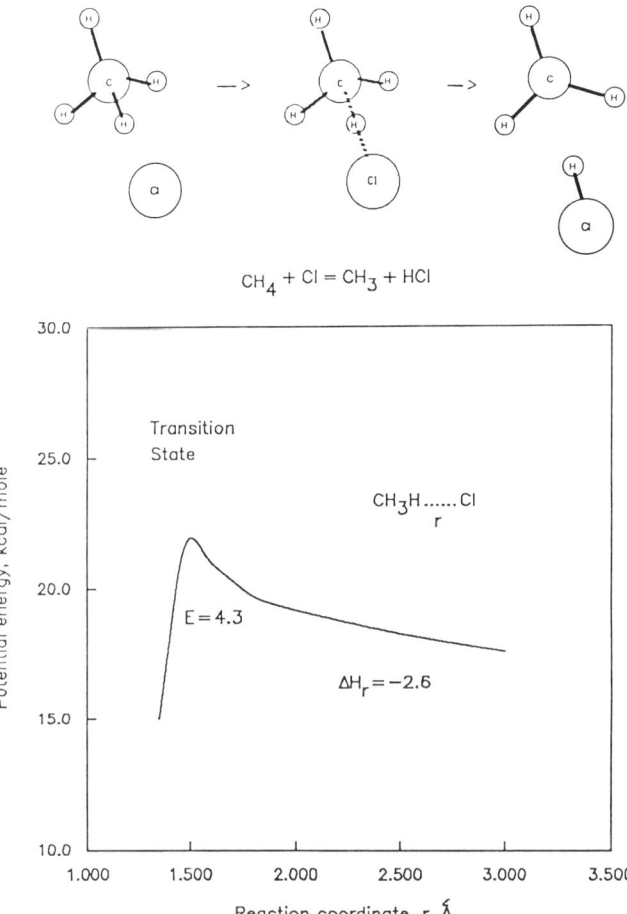

FIG. 14. Potential energy diagram for the bimolecular metathesis reaction $CH_4 + Cl = CH_3 + HCl$, showing the activation energy barrier E. The sketch above is a geometrical representation of the events (1 to r) along the reaction coordinate, which is the H–Cl distance.

energies. Thus, the comparison of model predictions with the experimental data is feasible.

The reaction coordinate for these reactions can be viewed as the interatomic distance (r) between one of the hydrogen atoms in CH_4 and the approaching radical. The calculated potential energy diagrams using the MNDO-PM3 formalism, are presented in Figs. 14 and 15, together with the ball-and-stick models for the reactions. As is evident from these figures, both reactions exhibit a pronounced activation energy barrier, in spite of their

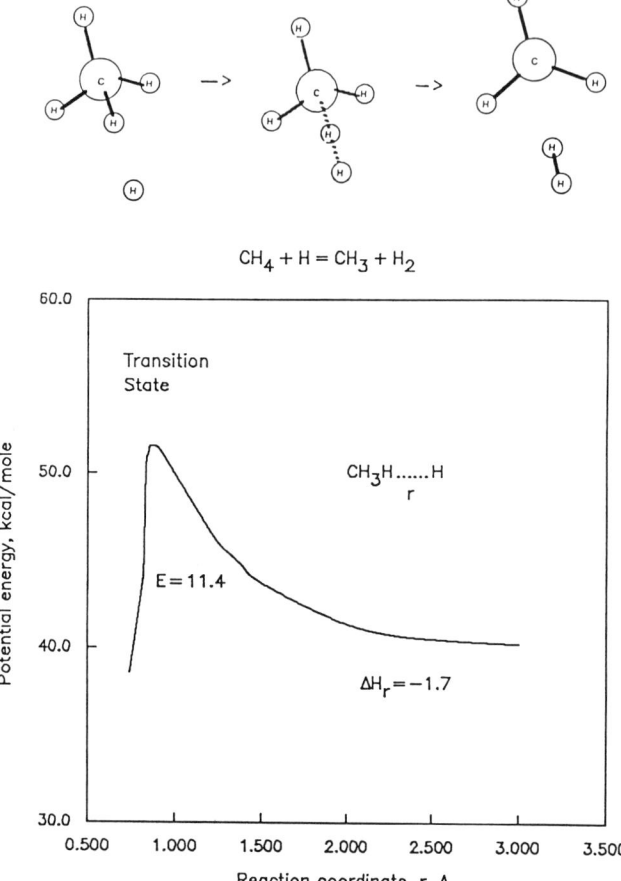

FIG. 15. Potential energy diagram for the bimolecular metathesis reaction $CH_4 + H = CH_3 + H_2$, showing the activation energy barrier E. The sketch above is a geometrical representation of the events (1 to r) along the reaction coordinate, which is the H–H distance.

exothermicities. From Fig. 14, the activation energy for hydrogen abstraction by Cl can be determined to be about 3–4 kcal/mol, which compares favorably with the available experimental value of 3.0 kcal/mol (Kerr and Drew, 1987).

Similarly, from Fig. 15, the activation energy for hydrogen abstraction by H can be determined to be about 10–12 kcal/mol. This value again compares very well with the available experimental value of 12.5 kcal/mol (Kerr and Drew, 1987). As is also evident from Figs. 14 and 15, both reactions proceed via tight and early transition states, i.e., the interatomic distances remain within bonding distances and the geometry of the transition state exhibits reactantlike features.

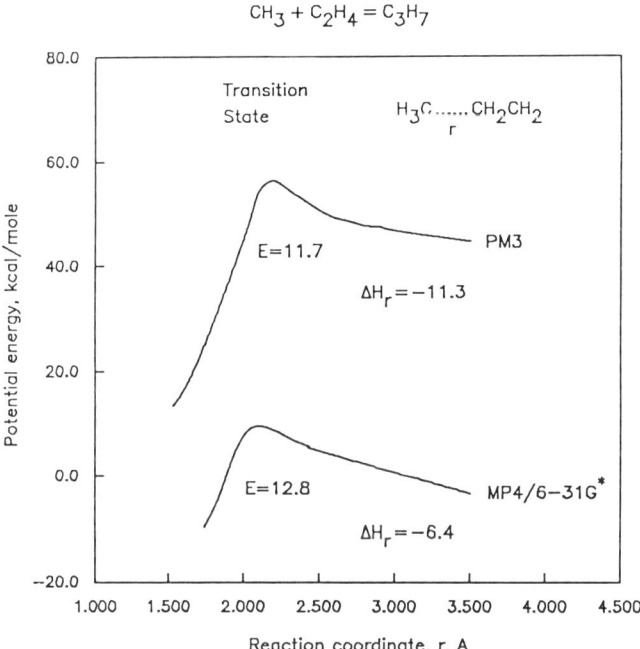

FIG. 16. Potential energy diagram for the addition of CH_3 to C_2H_4 both by the semiempirical MNDO formalism and by *ab initio* calculations (from Gonzales *et al.*, 1989). The sketch above is a geometrical representation of the events (1 to r) along the reaction coordinate, which is the $H_3C-CH_2CH_2$ distance.

The addition of CH_3 to C_2H_4,

$$CH_3 + C_2H_4 \rightleftharpoons CH_3CH_2CH_2, \tag{115}$$

is an important molecular-weight growth reaction, and thus has been studied both experimentally (Kerr and Drew, 1987) and theoretically (Hoyland, 1971; Dewar and Olivella, 1978; Arnaud *et al.*, 1985; Gonzales *et al.*, 1989). We have undertaken a semiempirical study, using the MNDO-PM3 formalism (Stewart, 1989a) and compared our results with those of *ab initio* calculations in Fig. 16 (Gonzales *et al.*, 1989). As seen from Fig. 16, although relative energies are different between the two sets of calculations, both methods predict barrier heights that are in the range 11–13 kcal/mol. Both of the theoretical determinations are higher than the experimental value of 8 kcal/mol (Kerr and Drew 1987), and suggests the need for additional measurements. Both our semiempirical calculations and *ab initio* ones suggest that this reaction proceeds via an early and tight transition state, which is also shown in Fig. 16.

The addition of C_2H_3 to C_2H_2,

$$C_2H_3 + C_2H_2 \rightleftharpoons C_4H_5, \qquad (116)$$

is believed to be important in the process of soot formation in flames (Frenklach *et al.*, 1986), and no experimental data exist for its rate parameters. In Fig. 17 the energy diagram, determined again by the MNDO-PM3 formalism, is presented together with a ball-and-stick diagram for reaction (116). These calculations suggest an activation energy of about 5.3 kcal/mol, which is a reasonably low value that is consistent with the highly sooting tendency of acetylene compared to ethylene. As seen in Fig. 17, reaction (116) proceeds via a tight and late transition state, i.e., the geometry of the transition state resembles the product, in contrast with the early transition state associated with addition of CH_3 to ethylene.

As is evident from these examples, computational quantum mechanics, semiempirical and *ab initio* methods alike, represent important new tools for the estimation of rate parameters from first principles. Our ability to estimate activation energies is particularly significant because until the advent of these techniques, no fundamentally based methods were available for the determination of this important rate parameter. It must be recognized, however, that these theoretical approaches still are at their early stages of development; that is to say, computational quantum chemical methods should only be used with considerable care and in conjunction with conventional methods of estimation discussed earlier in this article, as well with experiments.

X. Energy-Transfer–Limited Processes

Many association reactions, as well as their reverse unimolecular decompositions, exhibit rate parameters that depend both on temperature and pressure, i.e., density, at process conditions. This is particularly the case for molecules with fewer than 10 atoms, because these small species do not have enough vibrational and rotational degrees of freedom to retain the energy imparted to or liberated within the species. Under these conditions, energy transfer rates affect product distributions. Consequently, the treatment of association reactions, in general, would be different than that of the fission reactions.

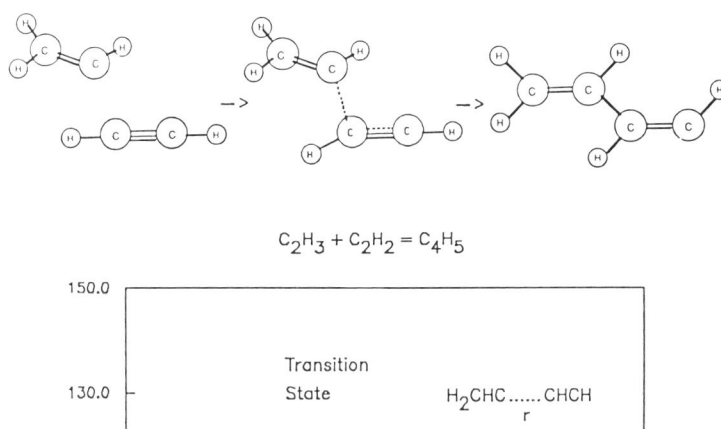

FIG. 17. Potential energy diagram for the addition of C_2H_3 to C_2H_2 using the semiempirical MNDO-PM3 formalism. The sketch above is a geometrical representation of the events (1 to r) along the reaction coordinate, which is the H_2CHC–CHCH distance.

Energy transfer limitations have long been recognized to affect the rates and mechanisms of fission and association reactions (Robinson and Holbrook, 1972; Laidler, 1987). In addition, it is increasingly being recognized that many exothermic bimolecular reactions can exhibit pressure-(density)-dependent rate parameters if they proceed via the formation of a bound intermediate. When energy transfer limitations exist, the rate coefficients exhibit non-Arrhenius temperature dependencies—i.e., the plots of $\ln(k)$ as a function of $1/T$ are curved.

The importance of energy transfer limitations can be illustrated best by considering unimolecular decomposition reactions, analyzed first by Lindemann and Hinshelwood (L–H) (Robinson and Holbrook, 1972). The following is a modified version of the L–H analysis:

$$A + M \underset{\beta k_d}{\overset{\beta k_e}{\rightleftharpoons}} A^* + M, \qquad (117)$$

$$A^* \xrightarrow{k_1} P + Q \quad \text{(product set 1)}, \qquad (118)$$

$$ \xrightarrow{k_2} R + S \quad \text{(Product set 2)}, \qquad (119)$$

....

The presence of additional product channels can also be considered. In this analysis M represents any collision partner ([M] = P/RT), the parameters k_e and k_d are the rate coefficients for energization and de-energization of A as a consequence of intermolecular collisions, and β represents the efficiency of collisional energy transfer given by (Gardiner and Troe, 1984)

$$\beta/(1 - \beta^{0.5}) = \langle \Delta E \rangle / (F_E RT), \qquad (120)$$

where $\langle \Delta E \rangle$ is the average energy transferred per collision per mole of gas, F_E is an empirical fitting constant that is in the range 1–3, R is the gas constant, and T is the absolute temperature. The collision efficiency β depends on nature of the gaseous environment through the use of $\langle \Delta E \rangle$, which has been determined for a number of gaseous species shown in Table XII (Gardiner

TABLE XII
AVERAGE COLLISIONAL ENERGIES TRANSFERRED FOR VARIOUS GASES (FROM GARDINER AND TROE, 1984)

Species	$\langle \Delta E \rangle$, kcal/mol
He	0.43
Ar	0.89
Xe	1.53
H_2	0.62
N_2	0.69
O_2	1.1
CO_2	2.4
N_2O	1.8
CH_4	2.1
CF_4	3.1
SF_6	2.3
C_2H_6	4.1
C_3H_8	4.3
C_4H_{10}	7.2
C_5H_{12}	4.3

and Troe, 1984). In general, $\langle \Delta E \rangle$ increases with increasing complexity and molecular weight of the species. However, the values shown in Table XII should not be treated as firm numbers, because of the large uncertainties present in the methods used to acquire them (Gardiner and Troe, 1984).

As a first approximation, k_e and k_d are the bimolecular collision rates, that is,

$$k_e = Z_{A-M} = (\sigma_{A-M})^2 \sqrt{(8\pi kT/\mu_{A-M})}, \qquad (121)$$

$$k_d = Z_{A^*-M} = (\sigma_{A^*-M})^2 \sqrt{(8\pi kT/\mu_{A^*-M})}, \qquad (122)$$

where the symbols have their previously defined meaning. The parameters k_1, k_2... are the rate coefficients for each of the decomposition channels open to the thermally activated A^*. By assuming that the fraction of energized A at any time is small, i.e., $[A^*]/[A] \ll 1$, the total rate of dissociation of A can be expressed as

$$-d[A]/dt = k_e \beta [M][A] \{k_1 + k_2 + \cdots\}/\{k_1 + k_2 + \cdots + k_d \beta [M]\}. \qquad (123)$$

In the high-pressure limit, i.e., $k_d \beta [M] \gg k_1 + k_2 + \cdots$, the rate of decomposition of A will be

$$-d[A]/dt = (k_e/k_d)\{k_1 + k_2 + \cdots\}[A] = k_\infty [A]. \qquad (124)$$

That is, the high-pressure limit dissociation rate coefficient of A, k_∞, is independent of the density ($[M]$) of the system. Under these conditions, $[A^*]$ is related to the thermal energy distribution function by

$$[A^*] = K(E, T)[A] = k_e/k_d [A], \qquad (125)$$

where $K(E, T)$ is given by (Robinson and Holbrook, 1972)

$$K(E, T) = (E/RT)^{s-1} \exp(-E/RT)/[(s-1)!]. \qquad (126)$$

In this expression, s is the number of internal harmonic oscillators, which is $3N - 5$ for linear molecules and $3N - 6$ for nonlinear molecules, and N is the number of atoms in the species. These are the precise conditions under which the high-pressure rate parameters discussed in the previous sections are applicable.

In the low-pressure limit, i.e., $k_d \beta [M] \ll k_1 + k_2 + \cdots$, the rate expression becomes

$$-d[A]/dt = k_e \beta [M][A] = k_0 \beta [M][A], \qquad (127)$$

where k_o is the low-pressure rate constant, which we also denote as k_e. As evident from the above expression, in the low-pressure limit the rate of decomposition of A becomes directly proportional to the rate of energization via intermolecular collisions. The decomposition of A between these two limiting situations is referred to as the "fall-off" regime, where the rate constant is given by

$$k_{L-H}/k_\infty = 1/\{1 + k_\infty/(k_0\beta[M])\} = X/(1 + X), \tag{128}$$

where the subscript L–H reemphasizes the Lindemann–Hinshelwood analysis used in this derivation. In the preceding expression, $X = k_0\beta[M]/k_\infty = [M]/[M_c]$, in which $[M_c]$ corresponds to the center of fall-off, which can be determined by setting $k_0\beta[M_c] = k_\infty$.

The rates of formation of products P and Q, and R and S, respectively, are also given by

$$d[P]/dt = d[Q]/dt = k_e\beta[M][A]k_1/\{k_1 + k_2 + \cdots + k_d[M]\}, \tag{129}$$

$$d[R]/dt = d[S]/dt = k_e\beta[M][A]k_2/\{k_1 + k_2 + \cdots + k_d\beta[M]\}\ldots \tag{130}$$

In spite of the proper qualitative features of the Lindemann–Hinshelwood model, it does not correctly predict the much broader experimental fall-off behavior; this is shown in Fig. 18, in which $\log(k/k_\infty)$ is plotted as a function of $\log(M = P/RT/M_c = P_c/RT)$. As evident from this figure, the actual rate at the center of fall-off (i.e., at P_c) is depressed relative to the L–H model; consequently, the transition of rate from low- to high-pressure limit occurs more gradually.

Powerful formalisms such as the Rice–Rampsperger–Kassel–Marcus (RRKM) method exist to analyze simple energy-transfer–limited unimolecular reactions in detail (see, for example, Robinson and Holbrook,

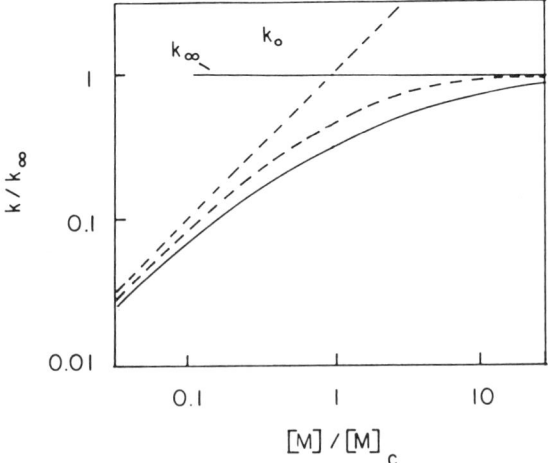

FIG. 18. Comparison of the Lindemann–Hinshelwood model (dashed line) with experimental data (solid line).

1972; Laidler, 1987). However, their use in the development of detailed chemical kinetic models is seldom justified because of the complexities involved in the calculations, and the need accurately to know data that are frequently unknown or difficult to estimate. In addition, such calculations may be unnecessary if the associated reactions subsequently are determined to be unimportant by sensitivity analysis.

Recently, Troe developed an empirical model to resolve the discrepancy between the L–H model and the actual fall-off, in which a fall-off broadening factor $F(X)$ was introduced (Gardiner and Troe, 1984):

$$k_{\text{actual}} = k_{\text{L-H}}F(X), \tag{131}$$

and $F(X)$ was empirically determined to be

$$\log(F(X)) = \log(F_c)/[1 + (\log(X))^2]. \tag{132}$$

In the preceding expression, $\log(F_c)$ is related to the depression of the fall-off curve at the center relative to the L–H expression in a $\log(k/k_\infty)$ vs. $\log(X/(1 + X))$ plot. The values for F_c can then be related to the properties of specific species and reaction and temperature using methods discussed in Gardiner and Troe (1984). In Fig. 19, values of F_c for a variety of hydrocarbon decompositions are presented. As evident from this figure, in the limit of zero or infinite temperatures and pressures, all reactions exhibit Lindemann–Hinshelwood behavior and F_c approaches unity. From this figure, it is clear that L–H analysis generally does an adequate job in

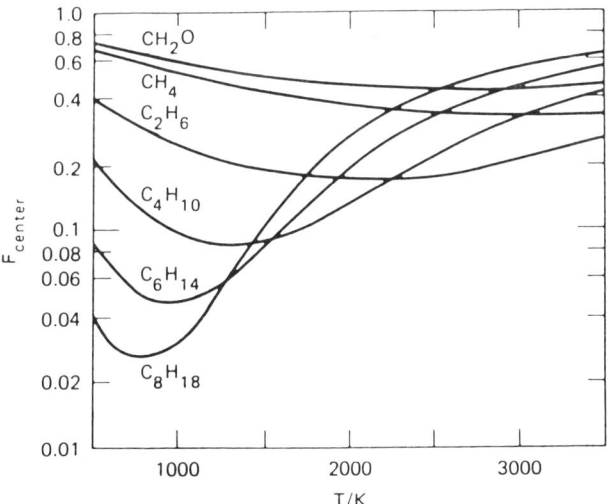

FIG. 19. The F_c values as a function of temperatures for a variety of hydrocarbons. (From Golden and Larson, 1984.)

describing the fall-off behavior of small molecule reactions under a broad range of temperatures, i.e., F_cs are close to unity. This is because small molecule reactions are closer to their low-pressure limits at process conditions of about 1 atm pressure and temperatures in the range 500–1,500 K. Clearly, for larger molecules the L–H analysis does a poor job at process conditions, as evident by very small F_cs consistent with the experimentally observed depression of the fall-off curve.

To illustrate these issues better, the pressure at the center of fall-off (P_c) is presented in Fig. 20. As seen from this figure, the unimolecular decompositions of small molecules are at their low-pressure limits at atmospheric pressure, and at process temperatures, $k_{\text{actual}} = k_0 \beta [\text{M}]$. Decompositions of larger molecules, on the other hand, are closer to their high-pressure limits. It is important to recognize that the unimolecular decompositions of hydrocarbons from CH_4 to C_3H_8 exhibit differing degrees of fall-off under process conditions, and this must be properly accounted for in the development of accurate detailed chemical kinetic models.

Troe's analysis summarized above requires the knowledge of both low- and high-pressure rate constants, in addition to an empirically determined F_c to describe the actual fall-off behavior. We already discussed methods for the estimation of high-pressure rate parameters. The low-pressure rate parameters can be estimated by recognizing the fact that k_0 represents pure energy transfer limitations, and thus can be determined from rate of collisional energization of A and from the thermal energy distribution function $K(E, T)$:

$$k_0 = k_e = (\sigma_{A-M})^2 \sqrt{(8\pi kT/\mu_{A-M})}(E/RT)^{s-1} \exp(-E/RT)/[(s-1)!]. \tag{133}$$

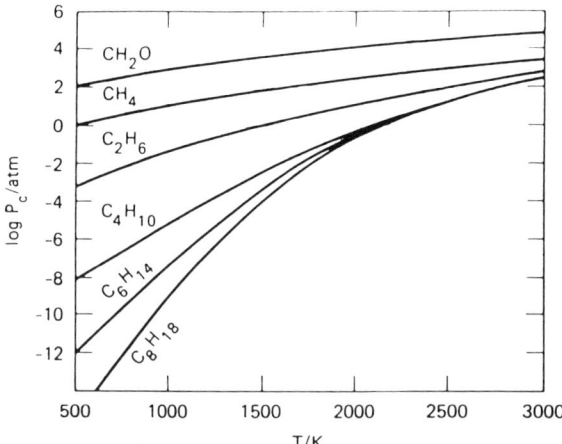

FIG. 20. Pressure at the center of fall-off (P_c) as a function of temperature for a variety of hydrocarbons. (From Golden and Larson, 1984.)

That is, the model assumes that all the energized A (i.e., A*) react, and the latter terms in the preceding expression represent the fraction [A*]/[A]. Benson (1960) has also shown that $E_0 = E_\infty - (S - \frac{3}{2})RT$. That is, the low pressure limit activation energy should be less than the high pressure limit one.

Although this method is rigorous, and its application straightforward, the approach nevertheless is tedious to apply in detailed chemical kinetic modeling because of the need to estimate both the high- and low-pressure rate parameters for each pressure-dependent elementary reaction. Consequently, we shall discuss methods that are less rigorous but equally accurate and easier to implement.

Kassel's unimolecular quantum-RRK (QRRK) method rectifies many of the application-related difficulties associated with the rigorous RRKM method and is relatively easier to apply than Troe's method. In this approach, the storage of excess energy is assumed to be quantized vibrational energy ($n = E/h\nu$), and it is assumed that the $3N - 6$ vibrations of a (non-linear) molecule can be represented by a single frequency, which may either be the geometric (ν_g) or the arithmetic (ν_a) mean frequency. Data to evaluate mean vibrational frequencies usually are available (Chase et al., 1985; Shimanouchi, 1972). Additional data needed in the QRRK analysis include the high-pressure limit rate parameters, i.e., A and $E_0 (m = E_0/h\nu)$, which generally are available or estimable using the methods discussed earlier. The QRRK representation of the rate of formation of products, e.g., P and Q, can be then shown to be (Robinson and Holbrook, 1971)

$$d[P]/dt = \sum_{\substack{E=E_{01} \\ (n=m_1)}}^{\infty} k_1(E)\{k_d\beta[M]K(E,T)\}/\{k_1(E)+k_2(E)+ \cdots k_d\beta[M]\}, \quad (134)$$

$$d[R]/dt = \sum_{\substack{E=E_{02} \\ (n=m_2)}}^{\infty} k_2(E)\{k_d\beta[M]K(E,T)\}/\{k_1(E)+k_2(E)+ \cdots k_d\beta[M]\}, \quad (135)$$

where k_1 and k_2 are the quantized versions of the high-pressure rate constants, which are given by

$$k_1(E) = A_1 n!(n - m_1 + s - 1)!/\{(n - m_1)!(n + s - 1)!\}, \quad (136)$$

$$k_2(E) = A_2 n!(n - m_2 + s - 1)!/\{(n - m_2)!(n + s - 1)!\}, \quad (137)$$

and $K(E, T)$ represents the quantized thermal energy distribution function,

$$K(E, T) = \exp(-nh/kT)(1 - \exp(-h/kT))^s \{(n + s - 1)!/n!(s - 1)!\}. \quad (138)$$

To illustrate the application of the QRRK method, consider the unimolecular decomposition of SiH_4 we considered previously. The parameters

TABLE XIII
QRRK PARAMETERS FOR THE UNIMOLECULAR DECOMPOSITION OF SiH_4

Reaction	High-Pressure Rate Parameters[a]	
	A, 1/s	E, kcal/mol
$SiH_4 \rightleftharpoons SiH_3 + H$	1.58×10^{16}	91
$SiH_4 \rightleftharpoons SiH_2 + H_2$	1.58×10^{13}	62

SiH_4 Properties	
v_g, 1/s	1,437
σ, Å	4.08
ε/k, K	207.6

Bath Gas Properties	
Name	Helium
σ, Å	2.57
ε/k, K	10.8
$\langle \Delta E \rangle$, kcal/mol	0.43

[a]Estimated using methods discussed in this paper.

necessary to undertake QRRK analysis are presented in Table XIII. The high-pressure rate parameters for the reactions were estimated using the methods outlined previously; the geometric mean frequency v_g was obtained from Shimanouchi (1972); the collisional energy transfer capacity of the bath gas was obtained from Gardiner and Troe (1984); and the Lennard–Jones parameters were obtained from Hirschfelder et al. (1954). From Fig. 21, it is clear that under conditions of practical interest, i.e., temperatures in the range 500–1,500 K, the associated reactions are in the fall-off regime, and that SiH_4 decomposition occurs primarily via H_2 elimination, and not by simple Si–H bond fission. It is important to note that this is in complete contrast with the decomposition of CH_4. Methane decomposes primarily by C–H bond fission, and not by the H_2 elimination process.

The QRRK approach illustrated above also constitutes the basis to analyze the behavior of the reverse, i.e., association, reactions that proceed through "chemically activated" transition states. Recently Dean (1985) reformulated the unimolecular quantum-RRK method of Kassel and devised a practical method for the proper description of the fall-off behavior of bimolecular reactions, including reactions when multiple product channels are present. The method developed was shown to describe the behavior of a large variety of bimolecular reactions with considerable success (Dean and Westmoreland, 1987; Westmoreland et al., 1986).

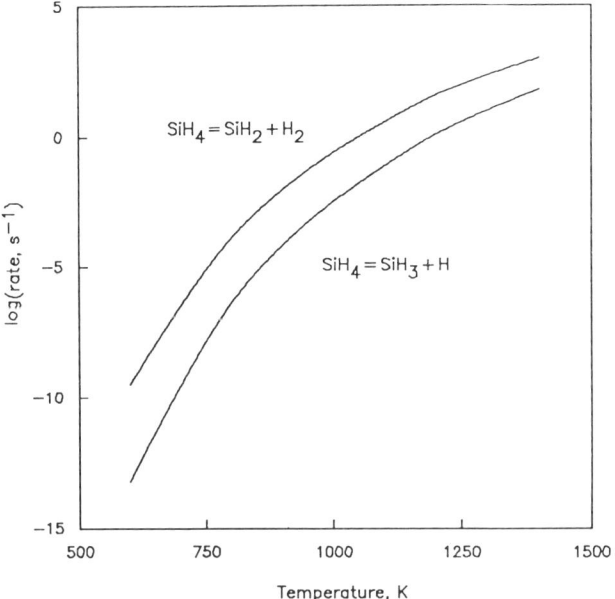

Fig. 21. Unimolecular QRRK predictions for the rate of decomposition of SiH_4.

To discuss the major features of the bimolecular QRRK, consider the following reaction mechanism:

$$P + Q \underset{k_{-1}}{\overset{k_1}{\rightleftharpoons}} A^*, \tag{139}$$

$$A^* + M \xrightarrow{\beta k_d} A + M, \tag{140}$$

$$A^* \xrightarrow{k_2} R + S, \tag{141}$$

where k_1 is the high-pressure recombination rate of P and Q, and k_{-1} and k_2 are energy-dependent rate parameters for decomposition of the chemically activated adduct (A*) back to reactants and to new products, respectively. In this case, the energy distribution function of the excited species A* will be given by $f(E, T)$, the chemical activation energy distribution function. Since $f(E, T)$ would be different from the thermal energy distribution function $k(E, T)$ when fall-off conditions exist, association reactions, in general, are not the reverse of fission reactions. The chemical activation energy distribution function can be shown to be related to $K(E, T)$ via the following expression (Robinson and Holbrook, 1972):

$$f(E, T) = k_{-1}(E)K(E, T) \bigg/ \sum_{E_{\min}}^{\infty} k_{-1}K(E, T). \qquad (142)$$

The rate of formation of products can be shown to be

$$d[R]/dt = k_1[P][Q] \sum_{E_{\min}}^{\infty} [k_2(E)f(E, T)/\{k_{-1}(E)+k_2(E)+\beta k_d[M]\}], \qquad (143)$$

$$d[A]/dt = \beta k_d[M]k_1[P][Q] \sum_{E_{\min}}^{\infty} [f(E, T)/\{k_{-1}(E)+k_2(E)+\beta k_d[M]\}]. \qquad (144)$$

The corresponding quantized expressions for k_1 and k_2 are

$$k_{-1}(E) = A_{-1}n!(n - m_{-1} + s - 1)!/\{(n - m_{-1})!(n + s - 1)!\}, \qquad (145)$$

$$k_2(E) = A_2 n!(n - m_2 + s - 1)!/\{(n - m_2)!(n + s - 1)!\}, \qquad (146)$$

where $m_{-1} = E_{-1}/h\nu$ and $m_2 = E_2/h\nu$.

To illustrate the utility of the bimolecular QRRK theory, consider the recombination of CH_2Cl and CH_2Cl radicals at temperatures in the range 800–1,500°C. This recombination process is important in the chlorine-catalyzed oxidative pyrolytic (CCOP) conversion of methane into more valuable C_2 products, and it has been studied recently by Karra and Senkan (1988a). The following composite reaction mechanism represents the complex process:

$$CH_2Cl + CH_2Cl \underset{k_{-1}}{\overset{k_1}{\rightleftharpoons}} [1,2 - C_2H_4Cl_2]^*, \qquad (147)$$

$$[1,2 - C_2H_4Cl_2]^* \overset{\beta k_d}{\rightleftharpoons} 1,2 - C_2H_4Cl_2, \qquad (148)$$

$$[1,2 - C_2H_4Cl_1]^* \overset{k_2}{\rightleftharpoons} C_2H_3Cl + HCl, \qquad (149)$$

$$[1,2 - C_2H_4Cl_2]^* \overset{k_3}{\rightleftharpoons} C_2H_4Cl + Cl. \qquad (150)$$

Other reactions, such as C–H bond scissions and H_2 eliminations, although possible, have been shown to be unimportant under the conditions studied. In Table XIV, the parameters needed for the QRRK analysis of the recombination of CH_2Cl radicals are presented. The methods and sources used to obtain these data are the same as those noted in the discussion of the unimolecular QRRK method. In Fig. 22, the apparent rate coefficients for

TABLE XIV
QRRK Parameters for the Recombination of CH_2Cl and CH_2Cl

Reaction	High-Pressure Rate Parameters[a]	
	A, 1/s	E, kcal/mol
$CH_2Cl + CH_2Cl \rightleftharpoons 1,2\text{-}C_2H_4Cl_2$	1.00×10^{13}	0.0
$1,2\text{-}C_2H_4Cl_2 \rightleftharpoons CH_2Cl + CH_2Cl$	8.68×10^{15}	87
$1,2\text{-}C_2H_4Cl_2 \rightleftharpoons C_2H_3Cl + HCl$	3.98×10^{13}	58
$1,2\text{-}C_2H_4Cl_2 \rightleftharpoons C_2H_4Cl + Cl$	1.26×10^{16}	81

$1,2\text{-}C_2H_4Cl_2$ Properties	
v_g	1,018 wavenumbers
σ, A	5.039
ε/R, K	423.20

Bath Gas Properties	
Name	CH_4
σ, Å	3.33
ε/k, K	136.5
$\langle \Delta E \rangle$, kcal/mol	2.1

[a]Estimated using methods discussed in this paper.

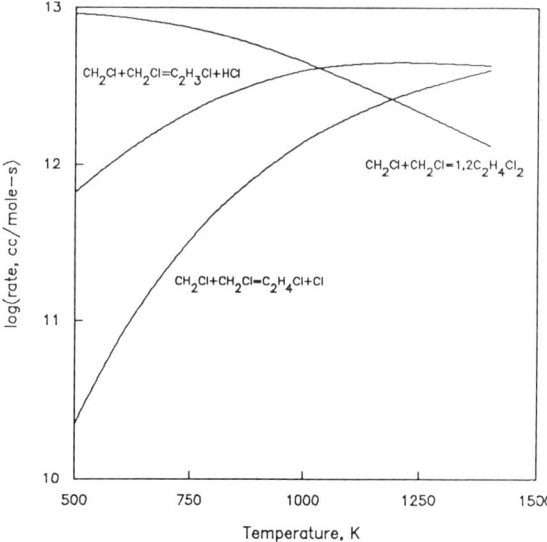

FIG. 22. Bimolecular QRRK predictions for the rates of recombination of CH_2Cl—CH_2Cl.

each of the rate parameters are presented based on the composite analysis of the preceding reactions. From this figure, it is clear that the energy transfer limitations are expected to exist over a broad range of conditions, and that all of the reaction paths considered are of equal importance at temperatures of the order 1,300 K.

In summary, energy transfer limitations must be considered, in principle, in all detailed chemical kinetic modeling, and especially in modeling the reactions of molecules containing less than about 10 to 12 atoms. It is important, however, to note that although powerful formalisms exist to account properly for energy transfer limitations, their practical utility depends on the availability of accurate "high-pressure" reaction rate parameters. That is, a highly sophisticated energy transfer analysis is of little quantitative value if high-pressure rate parameters were uncertain.

XI. Heterogeneous Reactions

Although the major objective of this paper has been to illustrate the concepts involved in the development of detailed chemical kinetic mechanisms for reactions taking place in the gas phase, a very short introduction is provided here to illustrate the application of the foregoing concepts to heterogeneous systems.

The philosophy used to develop detailed chemical kinetic mechanisms for gas-phase reactions can, in principle, be extended to treat heterogeneous reactions, provided diffusion is also considered in the final analysis. Clearly, the problem in heterogeneous catalysis is considerably more complex because of the close proximity of a large number of atoms and their collective effect on reaction kinetics and mechanisms, and the inevitable variation of catalyst structure with time — for example, as a result of sintering and poisoning.

In the absence of transport limitations, the processes of adsorption, surface diffusion, surface reaction, and desorption can be treated via the transition state theory (Baetzold and Somorjai, 1976; Zhdanov et al., 1988). For example, the application of the TST to a single site adsorption process,

$$A + [\ ] \rightleftharpoons [A]^\ddagger \rightarrow [A], \tag{151}$$

can be shown to lead to the following relation for the rate of adsorption:

$$r_{as} = (kT/h)(R'T)(T/T_0)^{\langle \Delta c_p^\ddagger/R \rangle}$$
$$\times \exp[2 - \langle \Delta c_p^\ddagger \rangle (T - T_0)/RT] \exp(\Delta S^\ddagger(T_0)/R) \exp(-E_a/RT)[A]C_v \tag{152}$$

in mol/cm²-s units, where E_a is the activation energy for adsorption, C_v is the surface concentration of vacant and available sites in mol/cm², and the other terms have their previously defined meanings. Again, a complex temperature dependency for the rate of adsorption is predicted by the transition state theory. When $\langle \Delta c_p^\ddagger \rangle$ is small, the expression for adsorption rate reduces to the following form:

$$r_{as} = (kTe^2/h)(R'T)\exp(\Delta S^\ddagger/R)\exp(-E_a/RT)[A]C_v \text{ mol/cm}^2\text{-s}. \quad (153)$$

For non–site-specific physical adsorption processes, C_v will correspond to the monolayer capacity of the entire surface, C_M, which is of the order 10^{-9} mol/cm² for many species. However, if selective adsorption is involved, C_v can be considerably less than C_M. At process temperatures in the range 300–1,000 K, the term $(kTe^2/h)(R'T)$ would be of the order 10^{18} cm³/mol-s. For the case of $C_v = C_M = 10^{-9}$ mol/cm², the r_{as} corresponds to the maximum level permitted by the TST, that is,

$$r_{as} = 10^9 \exp(\Delta S^\ddagger)\exp(-E_a/RT)[A] \text{ mol/cm}^2\text{-s}. \quad (154)$$

It should be recognized that the rates of adsorption are also limited by the wall collision frequency, as given by

$$r_{as}(\text{wall}) = s_0(RT/2\pi M_A)^{0.5}\exp(-E_a/RT)[A], \quad (155)$$

where s_0 is the sticking coefficient. The maximum numerical value of the pre-exponential term in Eq. (155) can be shown to be limited to about 10^4 at process conditions, i.e., 300–1,000 K and $M_A \approx 30$, and by assuming s_0 to be unity. Even under these limiting conditions it can be seen that $\Delta S^\ddagger$ must always decrease with the formation of the transition state for adsorption, which of course is well known to be the case. An estimate for entropy change associated with any adsorption process can be made by

$$-\Delta S_a = \Delta H_a/T, \quad (156)$$

where ΔH_a is the heat of adsorption. The ΔH_a would be approximately the latent heat of condensation for the case of physical adsorption, and the resulting entropy change corresponds to the loss of one translational degree of freedom of the adsorbed species. However, in chemisorption processes, the ΔH_a can be significantly large, i.e., of the order of bond dissociation energies (Toyoshima and Somorjai, 1979). Under these circumstances, a large decrease in entropy associated with the formation of the transition state, and thus a large decrease in the pre-exponential term of the adsorption rate constant, would be expected; this is indeed the case (Baetzold and Somorjai, 1976; Boudart and Djega-Mariadassou, 1984; Zhdanov et al., 1988).

Adsorption processes are almost always exothermic, and frequently have little activation energy barriers, i.e., $E_a \approx 0.0$. However, for certain complex chemisorptions, e.g., processes that involve the rearrangement of chemical bonds, the adsorption processes have been noted to exhibit an activation energy barrier (Boudart and Djega-Mariadassou, 1984).

Desorption processes, on the other hand, are always endothermic, and thus have activation energy barriers, E_d, that are at least equal to the heats of adsorption. The application of TST to a simple desorption process,

$$[A] \rightleftharpoons [A]^{\ddagger} \rightarrow A + [\ ], \qquad (157)$$

leads to the following relation when $\langle \Delta c_p^{\ddagger} \rangle$ is small:

$$r_d = (kTe/h)\exp(\Delta S^{\ddagger}/R)\exp(-E_d/RT)C_A \text{ mol/cm}^2\text{-s}. \qquad (158)$$

where E_d is the activation energy for desorption. The (kTe/h) term is again of the order 10^{13}/s at process temperatures. As in the case of unimolecular gas-phase reactions, the nature of the transition state establishes the order of magnitude of the pre-exponential term. For a "tight" transition state, $\Delta S^{\ddagger}$ will be very small; thus, the pre-exponential factor for the desorption process will be of the order 10^{13}/s. Clearly, larger pre-exponential factors may be possible for desorptions that proceed through a "loose" transition state. Useful compilations exist summarizing the ranges of pre-exponential factors expected in heterogeneous catalysis (Baetzold and Somorjai, 1976; Zhdanov et al., 1988). However, the a priori determination of activation energies remains a challenging task.

In heterogeneous catalysis, the catalyst often exists in clusters spread over a porous carrier. Experimentally, it is well established that reactivity and selectivity of heterogeneous reactions change enormously with cluster size. Thus, theoretical studies on clusters are particularly important to establish a basis for the determination of their optimal size and geometry. Cluster models are also important for studying the chemistry and reactivity of perfect crystal faces and the associated adsorption and desorption processes in heterogeneous catalysis (Bauschlicher et al., 1987).

The application of quantum mechanics to heterogeneous reactions is a research area under rapid growth. Although there is a significant overlap with the gas-phase reactions discussed earlier in this article, the application of quantum mechanics to heterogeneous processes is considerably more difficult because of the need to account for a large number of atoms to describe properly the physical and chemical features of condensed phases. This especially becomes a problem when transition metals are involved, because of the presence of a large number of valence electrons (Veillard, 1986). As noted previously, since the computational resources needed to solve Schrödinger's equation numerically increase approximately with the fourth power of the number of atoms and electrons present in the system, the application of *ab initio* methods, even to simple transition metals, becomes an enormously time-consuming task. Consequently, such methods have been applied only to small metal clusters (see, for example, Bauschlicher et al., 1987, and Almlof and Luthi, 1987). When a large number of metal atoms are involved, the method of approach inevitably must include some degree of empiricism. Semi-empirical methods, however, may be adequate for the required tasks, as they can be used to identify promising metal cluster sizes and geometries that subsequently can be studied using more rigorous ab initio methods.

A review of research on the application of quantum mechanics to heterogeneous reactions is outside the scope of this article. Thus, the readers are referred to sources such as Surface Science, Vol. 156, which contains the proceedings of the Third International Meeting on Small Particles and Inorganic Clusters, Berlin, West Germany, July 9–13, 1984, and Supercomputer Research in Chemistry and Chemical Engineering, ACS Symposium Series No. 353, 1987.

In summary, although the application of detailed chemical kinetic modeling to heterogeneous reactions is possible, the effort needed is considerably more involved than in the gas-phase reactions. The thermochemistry of surfaces, clusters, and adsorbed species can be determined in a manner analogous to those associated with the gas-phase species. Similarly, rate parameters of heterogeneous elementary reactions can be estimated, via the application of the transition state theory, by determining the thermochemistry of saddle points on potential energy surfaces.

XII. An Example: Detailed Chemical Kinetic Modeling of the Oxidation and Pyrolysis of CH_3Cl/CH_4

In this section we discuss the development, verification, and predictive utility of a detailed chemical kinetic mechanism for the oxidation and pyrolysis of CH_3Cl/CH_4. Understanding the chemical kinetics of high-temperature oxidation and pyrolysis of CH_3Cl/CH_4 is important for a number of reasons. First, combustion is increasingly being used to destroy hazardous chemical wastes, which frequently contain chlorinated hydrocarbons (Oppelt, 1987), and CH_3Cl is the simplest chlorinated hydrocarbon, whose combustion mechanism must be better understood for the improved design and operation of incinerators. Second, understanding the effects of chlorine on hydrocarbon pyrolysis and oxidative pyrolysis is important for better exploitation of high-temperature reactions as a manufacturing process to convert methane (natural gas) into more useful products via chlorine catalysis (Senkan, 1987).

An elementary reaction set describing the high-temperature reactions of CH_3Cl and CH_4 is presented in Table XV, together with the rate parameters

TABLE XV
CHEMICAL KINETIC MECHANISM FOR THE OXIDATION AND PYROLYSIS OF CH_3Cl/CH_4 ($k = AT^n \exp(-E/RT)$, IN cm, cal, s, mol units)

No.	Reaction	A	n	E	Reference[a]
	C_1 Kinetics				
1	$CH_3Cl \rightleftharpoons CH_3 + Cl$	3.42E32	−5.93	99,365.	a, ac
2	$CH_4 \rightleftharpoons CH_3 + H$	1.10E33	−5.9	105,150.	b, ac
3	$CH_4 + H \rightleftharpoons CH_3 + H_2$	5.50E07	1.97	11,207.	c
4	$CH_4 + Cl \rightleftharpoons CH_3 + HCl$	5.16E06	2.11	1,580.	d
5	$CH_4 + O_2 \rightleftharpoons CH_3 + HO_2$	7.94E13	0.	55,887.	e
6	$CH_4 + HO_2 \rightleftharpoons CH_3 + H_2O_2$	2.00E13	0.	18,000.	e
7	$CH_4 + O \rightleftharpoons CH_3 + OH$	1.20E07	2.1	7,624.	f
8	$CH_4 + OH \rightleftharpoons CH_3 + H_2O$	1.60E06	2.1	2,462.	f
9	$CH_4 + ClO \rightleftharpoons CH_3 + HOCl$	1.00E12	0.	7,500.	est.
10	$CH_3Cl + Cl \rightleftharpoons CH_2Cl + HCl$	3.16E13	0.	3,300.	g
11	$CH_3Cl + H \rightleftharpoons CH_3 + HCl$	7.00E13	0.	5,000.	h
12	$CH_3Cl + H \rightleftharpoons CH_2Cl + H_2$	3.00E13	0.	11,000.	est.
13	$CH_3Cl + CH_3 \rightleftharpoons CH_2Cl + CH_4$	1.00E12	0.	9,400.	g
14	$CH_3Cl + O_2 \rightleftharpoons CH_2Cl + HO_2$	6.31E13	0.	54,000.	est.
15	$CH_3Cl + HO_2 \rightleftharpoons CH_2Cl + H_2O_2$	3.00E13	0.	16,000.	est.
16	$CH_3Cl + O \rightleftharpoons CH_2Cl + OH$	1.30E13	0.	6,900.	g
17	$CH_3Cl + OH \rightleftharpoons CH_2Cl + H_2O$	2.54E06	1.97	1,190.	ad
18	$CH_3Cl + ClO \rightleftharpoons CH_2Cl + HOCl$	2.00E12	0.	12,000.	est.
19	$CH_3 + O_2 \rightleftharpoons CH_3O + O$	1.50E13	0.	28,681.	k
20	$CH_3 + O_2 \rightleftharpoons CH_2O + OH$	5.34E13	0.	34,574.	k
21	$CH_3 + O \rightleftharpoons CH_2O + H$	1.26E14	0.	2,000.	f
22	$CH_3 + OH \rightleftharpoons CH_3O + H$	4.52E14	0.	15,500.	l
23	$CH_3 + OH \rightleftharpoons CH_2O + H_2$	4.00E12	0.	0.	m
24	$CH_3 + ClO \rightleftharpoons CH_2O + HCl$	6.31E12	0.	0.	est.
25	$CH_3 + HO_2 \rightleftharpoons CH_3O + OH$	2.00E13	0.	0.	e
26	$CH_3 + CH_2O \rightleftharpoons CH_4 + CHO$	1.00E10	0.5	6,000.	e
27	$CH_2Cl + O_2 \rightleftharpoons CH_2O + Cl + O$	1.50E13	0.	30,300.	est.
28	$CH_2Cl + O_2 \rightleftharpoons CHClO + OH$	4.00E13	0.	34,000.	est.
29	$CH_2Cl + O \rightleftharpoons CH_2O + Cl$	1.00E14	0.	1,000.	est.
30	$CH_2Cl + OH \rightleftharpoons CH_2O + HCl$	6.31E12	0.	0.	est.
31	$CH_2Cl + OH — CH_2O + H + Cl$	5.00E14	0.	15,000.	est.
32	$CH_2Cl + ClO \rightleftharpoons CHClO + HCl$	6.31E12	0.	0.	est.
33	$CH_2Cl + HO_2 — CH_2O + OH + Cl$	1.00E13	0.	0.	est.
34	$CH_2Cl + CH_2O \rightleftharpoons CH_3Cl + CHO$	3.16E11	0.	5,000.	est.
35	$CH_3O + O_2 \rightleftharpoons CH_2O + HO_2$	1.00E13	0.	7,170.	e
36	$CH_3O + M \rightleftharpoons CH_2O + H + M$	2.00E14	0.	19,870.	f
37	$CH_2O + M \rightleftharpoons CHO + H + M$	5.00E16	0.	76,480.	f
38	$CH_2O + H = CHO + H_2$	2.50E09	1.27	11,000.	k
39	$CH_2O + O \rightleftharpoons CHO + OH$	1.70E06	2.32	6,200.	k
40	$CH_2O + OH \rightleftharpoons CHO + H_2O$	6.90E04	2.65	−8,000.	k
41	$CH_2O + Cl \rightleftharpoons CHO + HCl$	5.00E13	0.	500.	n
42	$CH_2O + HO_2 \rightleftharpoons CHO + H_2O_2$	1.00E12	0.	7,888.	e
43	$CHClO + M \rightleftharpoons CO + HCl + M$	1.00E17	0.	40,000.	est.
44	$CHClO + H \rightleftharpoons CHO + HCl$	2.00E13	0.	4,500.	est.

TABLE XV Continued

No.	Reaction	A	n	E	Reference[a]
45	$CHClO + O \longrightarrow CO + Cl + OH$	1.00E13	0.	1,000.	est.
46	$CHClO + OH \longrightarrow CO + H_2O$	1.00E13	0.	2,000.	est.
47	$CHCLO + Cl \longrightarrow CO + Cl + HCl$	1.00E13	0.	1,000.	est.
48	$CH_2 + O_2 \rightleftharpoons CO_2 + H + H$	1.30E13	0.	0.	m
48	$CH_2 + O_2 \rightleftharpoons CH_2O + O$	5.00E13	0.	9,000.	m
49	$CH_2 + O_2 \rightleftharpoons CHO + OH$	1.00E14	0.	3,700.	m
50	$CH_2 + O \rightleftharpoons CO + H + H$	8.00E13	0.	0.	m
	C_2 Kinetics				
51	$CH_3 + CH_3 \rightleftharpoons C_2H_5 + H$	7.80E11	0.	13,039	j
52	$CH_3 + CH_3 \rightleftharpoons C_2H_4 + H_2$	1.00E16	0.	31,792.	k
53	$CH_3 + CH_2Cl \rightleftharpoons C_2H_5Cl$	2.98E35	−6.87	7,694.	l
54	$CH_3 + CH_2Cl \rightleftharpoons C_2H_4 + HCl$	1.48E21	−2.19	5,207.	i
55	$CH_3 + CH_2Cl \rightleftharpoons C_2H_5 + Cl$	3.21E10	1.027	4,696.	i
56	$CH_2Cl + CH_2Cl \rightleftharpoons C_2H_4Cl_2$	1.10E36	−7.2	8,600.	i
57	$CH_2Cl + CH_2Cl \rightleftharpoons C_2H_3Cl + HCl$	1.31E24	−3.25	8,172.	i
58	$CH_2Cl + CH_2Cl \rightleftharpoons C_2H_4Cl + Cl$	1.91E17	−1.013	9,655.	i
59	$CH_2 + CH_3 \longrightarrow C_2H_4 + H$	4.00E13	0.	0.	aa
60	$CH_2 + CH_2Cl \longrightarrow C_2H_4 + Cl$	5.00E13	0.	0.	est.
61	$CH_2 + CH_2 \longrightarrow C_2H_2 + H + H$	1.00E14	0.	0.	ae
62	$CH_2 + C_2H_2O \rightleftharpoons C_2H_4CO$	1.00E12	0.	0.	ae
63	$CH_2 + C_2HO \rightleftharpoons C_2H_3 + CO$	3.00E13	0.	0.	ae
64	$C_2H_6 \rightleftharpoons CH_3 + CH_3$	2.00E32	−5.	92,225.	b
65	$C_2H_6 + H \rightleftharpoons C_2H_5 + H_2$	5.40E02	3.5	5,210.	f
66	$C_2H_6 + Cl \rightleftharpoons C_2H_5 + HCl$	4.64E13	0.	179.	o
67	$C_2H_6 + CH_3 \rightleftharpoons C_2H_5 + CH_4$	0.55E00	4.0	8,294.	f
68	$C_2H_6 + CH_2Cl \rightleftharpoons C_2H_5 + CH_3Cl$	1.00E12	0.	8,500.	est.
69	$C_2H_6 + O \rightleftharpoons C_2H_5 + OH$	3.00E07	2.0	5,115.	f
70	$C_2H_6 + OH \rightleftharpoons C_2H_5 + H_2O$	6.30E06	2.0	645.	f
71	$C_2H_5Cl \rightleftharpoons C_2H_4 + HCl$	1.11E14	−0.083	57,790.	p
72	$C_2H_5Cl + H \rightleftharpoons C_2H_5 + HCl$	6.31E13	0.	8,600.	s
73	$C_2H_5Cl + H \rightleftharpoons C_2H_4Cl + H_2$	5.00E13	0.	10,000.	est.
74	$C_2H_5Cl + Cl \rightleftharpoons C_2H_4Cl + HCl$	1.41E13	0.	616.	o
75	$C_2H_5Cl + CH_3 \rightleftharpoons C_2H_4Cl + CH_4$	1.00E12	0.	8,500.	est.
76	$C_2H_5Cl + CH_2Cl \rightleftharpoons C_2H_4Cl + CH_2Cl$	3.16E12	0.	9,000.	est.
77	$C_2H_5Cl + O \rightleftharpoons C_2H_4Cl + OH$	7.76E13	0.	6,600.	est.
78	$C_2H_5Cl + OH \rightleftharpoons C_2H_4Cl + H_2O$	5.00E13	0.	4,000.	est.
79	$C_2H_4Cl_2 \rightleftharpoons C_2H_3Cl + HCl$	6.61E13	−0.084	58,000.	p
80	$C_2H_4Cl_2 + H \rightleftharpoons C_2H_4Cl + HCl$	6.31E13	0.	8,400.	est.
81	$C_2H_4Cl_2 + H \rightleftharpoons C_2H_3Cl_2 + H_2$	5.00E13	0.	10,000.	est.
82	$C_2H_4Cl_2 + Cl \rightleftharpoons C_2H_3Cl_2 + HCl$	2.51E13	0.	3,100.	o
83	$C_2H_4Cl_2 + CH_3 \rightleftharpoons C_2H_3Cl_2 + CH_4$	1.00E12	0.	8,500.	est.
84	$C_2H_4Cl_2 + CH_2Cl \rightleftharpoons C_2H_3Cl_2 + CH_3Cl$	3.16E12	0.	9,000.	est.
85	$C_2H_4Cl_2 + O \rightleftharpoons C_2H_3Cl_2 + OH$	5.00E13	0.	7,000.	est.
86	$C_2H_4Cl_2 + OH \rightleftharpoons C_2H_3Cl_2 + H_2O$	3.98E13	0.	4,000.	est.
87	$C_2H_5 \rightleftharpoons C_2H_4 + H$	7.00E25	−4.1	42,984	b
88	$C_2H_5 + H \rightleftharpoons C_2H_4 + H_2$	1.90E12	0.	0.	r

TABLE XV Continued

No.	Reaction	A	n	E	Reference[a]
89	$C_2H_5 + Cl \rightleftharpoons C_2H_4 + HCl$	2.00E12	0.	0.	est.
90	$C_2H_5 + O_2 \rightleftharpoons C_2H_4 + HO_2$	2.00E12	0.	5,000.	b
91	$C_2H_4Cl \rightleftharpoons C_2H_4 + Cl$	1.05E20	−2.36	22,000.	est.
92	$C_2H_4Cl + H \rightleftharpoons C_2H_4 + HCl$	3.16E12	0.	0.	est., ac
93	$C_2H_4Cl + Cl \rightleftharpoons C_2H_3Cl + HCl$	1.00E13	0.	3,000.	s
94	$C_2H_3Cl_2 \rightleftharpoons C_2H_3Cl + Cl$	4.95E20	−2.35	20,000.	t
95	$C_2H_3Cl_2 + H \rightleftharpoons C_2H_3Cl + HCl$	1.00E13	0.	1,000.	est.
96	$C_2H_4 + M \rightleftharpoons C_2H_3 + H + M$	3.10E17	0.	98,160.	r
97	$C_2H_4 + M \rightleftharpoons C_2H_2 + H_2 + M$	3.00E17	0.	79,800	r
98	$C_2H_4 + Cl \rightleftharpoons C_2H_3 + HCl$	1.00E14	0.	7,000.	u
99	$C_2H_4 + H \rightleftharpoons C_2H_3H_2$	7.00E14	0.	14,500.	r
100	$C_2H_4 + CH_3 \rightleftharpoons C_2H_3 + CH_4$	3.97E11	0.	7,988.	f
101	$C_2H_4 + CH_2Cl \rightleftharpoons C_2H_3 + CH_3Cl$	2.00E12	0.	12,000.	est.
102	$C_2H_4 + O \rightleftharpoons CH_3 + CHO$	1.60E09	1.2	741.	v
103	$C_2H_4 + OH \rightleftharpoons C_2H_3 + H_2O$	3.50E13	0.	3,012.	v
104	$C_2H_4 + OH \rightleftharpoons CH_3 + CH_2O$	1.30E12	0.	−765.	f
105	$C_2H_4 + ClO \rightleftharpoons CH_2Cl + CH_2O$	5.00E12	0.	0.	est.
106	$C_2H_3Cl \rightleftharpoons C_2H_2 + HCl$	2.75E17	−1.3	69,312.	w, ac
107	$C_2H_3Cl + H \rightleftharpoons C_2H_3 + HCl$	1.00E14	0.	4,500.	g
108	$C_2H_3Cl + H \rightleftharpoons C_2H_2Cl + H_2$	1.00E14	0.	10,000.	est.
109	$C_2H_3Cl + Cl \rightleftharpoons C_2H_2Cl + HCl$	1.00E14	0.	5,000.	est.
110	$C_2H_3Cl + CH_3 \rightleftharpoons C_2H_2Cl + CH_4$	1.00E12	0.	11,000.	est.
111	$C_2H_3Cl + CH_2Cl \rightleftharpoons C_2H_2Cl + CH_3Cl$	1.00E12	0.	12,000.	est.
112	$C_2H_3Cl + O \rightleftharpoons CHClO + CH_2$	5.24E11	0.	0.	est.
113	$C_2H_3Cl + OH \rightleftharpoons CH_3 + CHClO$	5.00E12	0.	0.	est.
114	$C_2H_3Cl + OH \rightleftharpoons C_2H_2Cl + H_2O$	5.00E13	0.	3,000.	est.
115	$C_2H_3Cl + ClO \rightleftharpoons CH_2Cl + CHClO$	5.00E12	0.	0.	est.
116	$C_2H_3 \rightleftharpoons C_2H_2 + H$	9.30E22	−3.7	37,255.	b
117	$C_2H_3 + H \rightleftharpoons C_2H_2 + H_2$	1.00E13	0.	0.	r
118	$C_2H_3 + Cl \rightleftharpoons C_2H_2 + HCl$	1.00E13	0.	0.	est.
119	$C_2H_3 + O_2 \rightleftharpoons C_2H_2 + HO_2$	1.60E13	0.	10,400.	f
120	$C_2H_3 + O_2 \rightleftharpoons CH_2O + CHO$	4.00E12	0.	0.	x
121	$C_2H_3 + O \rightleftharpoons C_2H_2O + H$	3.00E13	0.	0.	m
122	$C_2H_2 + Cl \rightleftharpoons C_2H_2Cl$	5.13E18	−2.74	−1,779.	y, ac
123	$C_2H_2Cl + O_2 \rightleftharpoons CHClO + CHO$	2.00E12	0.	0.	est.
124	$C_2H_2Cl + O \rightleftharpoons C_2H_2O + Cl$	3.00E13	0.	0.	est.
125	$C_2H_2 + O \rightleftharpoons CH_2 + CO$	4.10E08	1.5	1,697.	z
126	$C_2H_2 + O \rightleftharpoons C_2HO + H$	4.00E14	0.	10,660.	z
127	$C_2H_2 + OH \rightleftharpoons C_2H_2O + H$	3.00E12	0.	1,100.	f
128	$C_2H_2 + ClO \rightleftharpoons C_2H_2O + Cl$	3.00E12	0.	0.	est.
129	$C_2H_2O + H \rightleftharpoons CH_3 + CO$	2.00E13	0.	0.	ae
130	$C_2H_2O + Cl \rightleftharpoons CH_2Cl + CO$	5.00E13	0.	3,000.	est.
131	$C_2H_2O + H \rightleftharpoons C_2HO + H_2$	3.00E13	0.	1,434.	m
132	$C_2H_2O + Cl \rightleftharpoons C_2HO + HCl$	5.00E13	0.	1,500.	est.
133	$C_2H_2O + M \rightleftharpoons CH_2 + CO + M$	2.30E15	0.	57,600.	ae
134	$C_2H_2O + O \rightleftharpoons CHO + CHO$	2.00E13	0.	2,300.	m
135	$C_2H_2O + O \rightleftharpoons CH_2O + CO$	2.00E13	0.	0.	m

TABLE XV Continued

No.	Reaction	A	n	E	Reference[a]
136	$C_2H_2O + OH \rightleftharpoons CH_2O + CHO$	1.00E13	0.	0.	f
137	$C_2H_2O + OH \rightleftharpoons C_2HO + H_2O$	1.00E13	0.	2,630.	m
138	$C_2HO + H \rightleftharpoons CH_2 + CO$	1.50E14	0.	0.	z
139	$C_2HO + O \rightarrow CO + CO + H$	1.00E14	0.	0.	m
140	$C_2HO + O_2 \rightarrow CO + CO + OH$	1.46E12	0.	2,500.	m
	CO and H_2 Kinetics				
141	$CHO + M \rightleftharpoons CO + H + M$	7.10E14	0.	16,802.	ab
142	$CHO + H \rightleftharpoons CO + H_2$	2.00E14	0.	0.	ab
143	$CHO + OH \rightleftharpoons CO + H_2O$	5.00E13	0.	0.	ab
144	$CHO + O \rightleftharpoons CO + OH$	3.00E13	0.	0.	ab
145	$CHO + O_2 \rightleftharpoons CO + HO_2$	3.00E12	0.	0.	ab
146	$CHO + Cl \rightleftharpoons CO + HCl$	1.00E14	0.	0.	ab
147	$CHO + CHO \rightleftharpoons CH_2O + CO$	2.00E13	0.	0.	ab
148	$CO + O_2 \rightleftharpoons CO_2 + O$	5.00E13	0.	63,169.	ab
149	$CO + OH \rightleftharpoons CO_2 + H$	4.40E06	1.5	−740.	ab
150	$CO + O + M \rightleftharpoons CO_2 + M$	5.30E13	0.	−4,538.	ab
151	$CO + HO_2 \rightleftharpoons CO_2 + OH$	1.50E14	0.	23,573.	ab
152	$CO + ClO \rightleftharpoons CO_2 + Cl$	1.00E13	0.	1,000.	ab
153	$H + O_2 \rightleftharpoons O + OH$	1.20E17	−0.91	16,504.	ab
154	$O + H_2 \rightleftharpoons H + OH$	1.50E07	2.0	7,547.	ab
155	$H_2 + OH \rightleftharpoons H_2O + H$	1.00E08	1.6	3,296.	ab
156	$O + H_2O \rightleftharpoons OH + OH$	1.50E10	1.14	17,244.	ab
157	$H + H + M \rightleftharpoons H_2 + M$	6.40E17	−1.0	0.	ab
158	$O + O + M \rightleftharpoons O_2 + M$	1.00E17	−1.0	0.	ab
159	$O + H + M \rightleftharpoons OH + M$	3.00E19	−1.0	0.	ab
160	$H + OH + M \rightleftharpoons H_2O + M$	1.41E23	−2.0	0.	ab
161	$H + O_2 + M \rightleftharpoons HO_2 + M$	7.00E17	−0.8	0.	ab
162	$H + HO_2 \rightleftharpoons H_2 + O_2$	2.50E13	0.	693.	ab
163	$H + HO_2 \rightleftharpoons OH + OH$	1.50E14	0.	1,003.	ab
164	$OH + HO_2 \rightleftharpoons H_2O + O_2$	2.00E13	0.	0.	ab
165	$O + HO_2 \rightleftharpoons O_2 + OH$	2.00E13	0.	0.	ab
166	$O + OH + M \rightleftharpoons HO_2 + M$	1.00E17	0.	0.	ab
167	$HO_2 + HO_2 \rightleftharpoons H_2O_2 + O_2$	2.00E12	0.	0.	ab
168	$H + H_2O_2 \rightleftharpoons H_2O + OH$	1.00E13	0.	3,583.	ab
169	$H_2O_2 + M \rightleftharpoons OH + OH + M$	1.20E17	0.	45,379.	ab
170	$H_2O_2 + Cl \rightleftharpoons HO_2 + HCl$	1.26E13	0.	2,000.	ab
171	$HCl + M \rightleftharpoons H + Cl + M$	4.36E13	0.	81,760.	ab
172	$HCL + H \rightleftharpoons H_2 + Cl$	7.94E12	0.	3,400.	ab
173	$O + HCl \rightleftharpoons OH + Cl$	3.16E13	0.	6,700.	ab
174	$OH + HCl \rightleftharpoons Cl + H_2O$	1.58E13	0.	1,000.	ab
175	$O + ClO \rightleftharpoons Cl + O_2$	9.70E12	0.	507.	ab
176	$Cl + HO_2 \rightleftharpoons HCl + O_2$	1.08E13	0.	−338.	ab
177	$Cl + HO_2 \rightleftharpoons OH + ClO$	2.47E13	0.	894.	ab
178	$ClO + HO_2 \rightleftharpoons HOCl + O_2$	3.55E11	0.	1,410.	ab
179	$ClO + H_2 \rightleftharpoons HOCl + H$	1.00E13	0.	13,500.	ab
180	$H + HOCl \rightleftharpoons HCl + OH$	1.00E13	0.	1,000.	ab

TABLE XV Continued

No.	Reaction	A	n	E	Reference[a]
181	Cl+HOCl=HCl+ClO	1.00E13	0.	2,000.	ab
182	O+HOCl=OH+ClO	5.00E13	0.	1,500.	ab
183	OH+HOCl=H$_2$O+ClO	1.80E12	0.	3,000.	ab
184	HOCl+M=OH+Cl+M	1.00E18	0.	55,000.	ab

a. Kondo, O., Suito, K., and Mukarami, I. (1980). The thermal unimolecular decomposition of methyl chloride behind shock waves. *Bull. Chem. Soc. Japan* **53**, 2133.

b. Harris, S. J., Weiner, A. M., Blint, R. J., and Goldsmith, J. E. M. (1988). Concentration profiles in rich and sooting ethylene flames. Paper presented at the *Twenty First Symposium (International) on Combustion*, p. 1033. The Combustion Institute, Pittsburgh.

c. Schatz, G. C., Wagner, A. F., and Dunning, T. H., Jr. (1984). A theoretical study of deuterium isotope effects in the reactions H$_2$+CH, and H+CH$_2$. *J. Phys. Chem.* **88**, 221.

d. Heneghan, S. P., Knoot, P. A., and Benson, S. W. (1981). The temperature coefficient of the rates in systems Cl+CH$_4$=CH$_4$+HCl, thermochemistry of methyl radical. *Int. J. Chem. Kin.* **43**, 677.

e. Westbrook, C. K., and Dryer, F. A. (1984). Chemical kinetic modeling of hydrocarbon combustion. *Prog. Energy Combust. Sci.* **10**, 1.

f. Warnatz, J. (1984). Rate coefficients in the CHO system, in *Combustion Chemistry* (W. C. Gardiner, Jr., Ed.). Springer-Verlag, New York.

g. *CRC Handbook of Bimolecular and Termolecular Gas Reactions: Volume 1.* J. A. Kerr and S. J. Moss, Eds., CRC Press, Inc. (1981).

h. Westenberg, A. A., and de Haas, N. (1975). Rates of H+CH$_2$X reactions. *J. Chem. Phys.* **62**, 3321.

i. Karra, S. B., and Senkan, S. M. (1988). Analysis of the chemically activated CH$_2$CHCH$_2$Cl and CH$_2$/CH$_2$Cl recombination reactions at clevated temperatures using the QRRK method. *Ind. Eng. Chem. Research* (in press).

j. Keifer, J. H., and Budach, K. A. (1984). The very-high temperature pyrolysis of ethane: Evidence against high rates for dissociative recombination reactions of methyl radicals. *Int. J. Chem. Kinetics* **16**, p. 679.

k. Hsu, D. S. Y., Shaub, W. M., Creamer, T., Gutman, D., and Lin, M. C. (1983). Kinetic modeling of CO production from the reaction of CH$_2$ with O$_2$ in shock waves. *Ber. Bunsenges. Phys. Chem.* **87**, 909.

l. Paezko, G., Lefdal, P. M., and Peters, N. (1988). Reduced reaction schemes for methane, methanol, and propane flames. Paper presented at the *Twenty First Symposium (International) on Combustion*, p. 739. The Combustion Institute, Pittsburgh.

m. Miller, J. A., Mitchell, R. E., Smooke, M. D., and Kee, R. J. (1982). Toward a comprehensive chemical mechanism for the oxidation of acetylene. *Nineteenth Symposium (International) on Combustion*, p. 181. The Combustion Institute, Pittsburgh.

n. DeMore, W. B., Molina, M. J., Watson, R. T., Golden, D. M., Hampson, R. F., Kurylo, M. J., Howard, C. J., and Ravishankara, A. R. (1985). *Chemical Kinetics and Photochemical Data for Use in Stratospheric Modeling.* Evaluation No. 6. JPL Publication 85-37.

o. Wine, P. H., and Semmes, D. H. (1983). Kinetics of Cl($^2P_1$) Reactions with the Chloroethanes CH$_1$CH$_2$Cl, CH$_1$CHCl$_2$, CH$_1$ClCH$_2$Cl and CH$_2$CHCl$_1$. *J. Phys. Chem.* **87**, 3572.

p. Hasslef, J. C., Setser, D. W., and Johnson, R. L. (1966). Chemical activation and non-equilibrium unimolecular reactions of C$_2$H$_2$Cl and 1.2-C$_2$H$_2$Cl$_2$ molecules. *J. Chem. Phys.* **45**, 3231.

r. Olson, D. B., and Gardiner, Jr., W. C. (1978). Combustion of methane in fuel-rich mixtures. *Combust. Flame* **32**, 151.

DETAILED CHEMICAL KINETIC MODELING   181

s. Kondrattiev, V. N. (1972). *Rate Constants of Gas Phase Reactions.* COM-72-10014, National Bureau of Standards, Washington.

t. Ashmore, P. G., Owen, A. J., and Robinson, P. J. (1982). Chlorine-catalyzed pyrolysis of 1,2-dichloroethane. Part 2—Unimolecular decomposition of 1,2-dichloroethyl radical and its reverse reaction. *J. Chem. Soc. Faraday Trans. I.* **78**, 667.

u. Weissman, M., and Benson, S. W. (1984). Pyrolysis of methyl chloride, a pathway in the chlorine-catalyzed polymerization of methane. *Int. J. Chem. Kin.* **16**, 307.

v. Hennessey, R. J., Robinson, C., and Smith, D. B. (1988). A comparative study of methane and ethane flames. *Twenty First Symposium (International) on Combustion*, p. 761. The Combustion Institute, Pittsburgh.

w. Zabel, F. (1977). Thermal gas-phase decomposition of chloroethylenes, II. Vinyl chloride. *Int. J. Chem. Kinetics*, **9**, 651.

x. Park, J. Y., Heaven, M. C., and Gutman, D. (1984). Kinetics and mechanism of the reactions of vinyl radical with molecular oxygen. *Chem. Phys. Lett.* **104**, 469.

y. Brunning, J., and Stief, L. J. (1985). Pressure dependence of the absolute rate constant for the reaction $Cl + C_2H_2$ from 210-361 K. *J. Chem. Phys.* **83**(3), 1005.

z. Frank, P., Bhaskaran, K. A., and Just, Th., Acetylene oxidation: The reaction $C_2H_2 + O$ at high temperatures (1988). *Twenty First Symposium (International) on Combustion*, p. 885. The Combustion Institute, Pittsburgh.

aa. Roth, P., and Just, Th. (1984). Kinetics of the high temperature, low concentration CH_2 oxidation verified by H and O atom measurements. *Twentieth Symposium (International) on Combustion*, p. 807. The Combustion Institute, Pittsburgh.

ab. Chang, W. D., Karra, S. B., and Senkan, S. M. (1987). A computational study of chlorine inhibition of CO flames. *Combust. Flame* **69**, 113.

ac. Rate parameters take into account fall-off behavior at 1 atm, and were determined using the quantum RRK method (Robinson and Holbrook 1972) or Troe's formalism (Gardiner and Troe, 1984).

ad. Jeong, K.-M., and Kaufman, F. Kinetics of the reaction of hydroxyl radicals with methane and with nine Cl- and F-substituted methanes, I. Experimental results, comparison, and application. *J. Phys. Chem.* **86**, 1808.

ae. Frank, P., Bhaskarun, K. A., and Just, Th. (1986). High temperature reaction of triplet methylene and ketene with radicals. *J. Phys. Chem.* **90**, 2226.

---

for the forward reaction paths (Karra and Senkan, 1988b). Reverse reaction rates can be calculated from the considerations of the detailed balancing between the forward and reverse rates through the use of the equilibrium constants. This mechanism was constructed by systematically considering *all* plausible elementary reactions of CH_3Cl and O_2, and their daughter species, and then eliminating those reactions that did not contribute to reaction rates and were determined to be unimportant by the sensitivity analysis. It must be recognized, however, that the development of any detailed chemical kinetic mechanism involves a considerable degree of chemical intuition, together with the use of principles of physical chemistry and chemical kinetics that has been the primary subject of this article.

The reactions describing CH_4 combustion were then appended to the CH_3Cl mechanism (Westbrook and Dryer, 1984). The CH_3Cl/CH_4 composite mechanism subsequently was combined with the chlorine-inhibited

CO oxidation submechanism developed and tested previously (Chang et al., 1987). At present, the mechanism extends only up to C_2 compounds because accurate quantitative experimental data exist only for those species (Karra and Senkan, 1987).

The rate parameters either correspond to those reported in the literature whenever available, or they were estimated using methods described in this article. It is particularly important to recognize that the recombination of CH_2Cl and CH_3 radicals is primarily responsible for the formation of C_2 hydrocarbons in the high-temperature reactions of CH_3Cl and CH_4, since C_1 species are the only sources of carbon. The chemically activated CH_2Cl/CH_2Cl and CH_2Cl/CH_3 recombination processes were analyzed by using the bimolecular quantum RRK (Rice–Rampsperger–Kassel), or QRRK, method discussed earlier. Fall-off corrections for simple unimolecular decomposition reactions were made either by using Troe's semiempirical formalism when both the high- and low-pressure rate parameters were available, or by the unimolecular quantum RRK method, again in accordance with the principles discussed earlier.

The reaction mechanism developed was then used to predict the experimental species profiles obtained in one-dimensional flames (Kee et al., 1985; Karra and Senkan, 1987). For one-dimensional, premixed, laminar flat flames, the energy and mass transport equations are given by the following:

$$mc_p dT/dx - d(kAdT/dx)/dx + A \sum_{k=1}^{N} \rho Y_k V_k c_{pk}(dT/dx)$$

$$+ A \sum_{k=1}^{N} w_k h_k M_k = 0.0, \qquad (159)$$

$$m dY_k/dx + d(\rho A Y_k V_k)/dx - A w_k M_k = 0.0 \qquad (k=1,\ldots,N), \qquad (160)$$

where $m = \rho u A$ is the mass flow rate, c_p is the mean specific heat of the mixture, T is the temperature, x is the distance along the flame (which is equivalent to time), A is the cross-sectional area of the stream tube enclosing the flame, u is the fluid velocity, k is the thermal conductivity of the mixture, V_k is the diffusion velocity of the kth species, $\rho = P(MW)/(RT)$ is the mass density, P is the total pressure, MW is the mean molecular weight, w_k is the net molar rate of generation of species k as a consequence of competition among various elementary reactions in the mechanism presented in Table XV, M_k is the molecular weight, h_k is the specific enthalpy, and Y_k is the mass fraction of species k.

Thermochemical information was acquired from standard sources, where these were available. However, such data for some of the species in the mechanism have not been documented; consequently, they were estimated using methods described in this paper. Physical molecular properties were either acquired from conventional sources (Hirschfelder et al., 1954; Svehla,

1962; Reid et al., 1977; Kee et al., 1983), whenever available, or were estimated using suitable methods discussed in these references.

The model described above was then used to simulate a flame with the following prereaction composition: $CH_3Cl = 13.4\%$, $CH_4 = 26.6\%$, $O_2 = 37.6\%$, and $Ar = 23.3\%$. For this, the Sandia flame code was utilized (Kee et al., 1983). In Fig. 23, calculated mole fraction profiles (indicated by lines) for CH_3Cl, CH_4, O_2, CO, and CO_2 are compared to those measured experimentally (indicated by symbols). Similar profiles for C_2H_2, C_2H_4, and C_2H_6 are presented in Fig. 24. The temperature profile along the flame, used in the calculations, is also presented in Figure 23. As evident from these figures, the agreement between the model and experimental data is reasonable, suggesting that the major features of the flame chemistry of CH_3Cl/CH_4 are described reasonably well by the mechanism, since uncertainties in the transport model are expected to be extremely small.

It is possible that even better agreement between model predictions and the experiments can be achieved if rate constants of influential reactions (whose rate parameters are uncertain) are adjusted within reasonable limits, i.e., the mechanism can be optimized for the particular experiment. This, however,

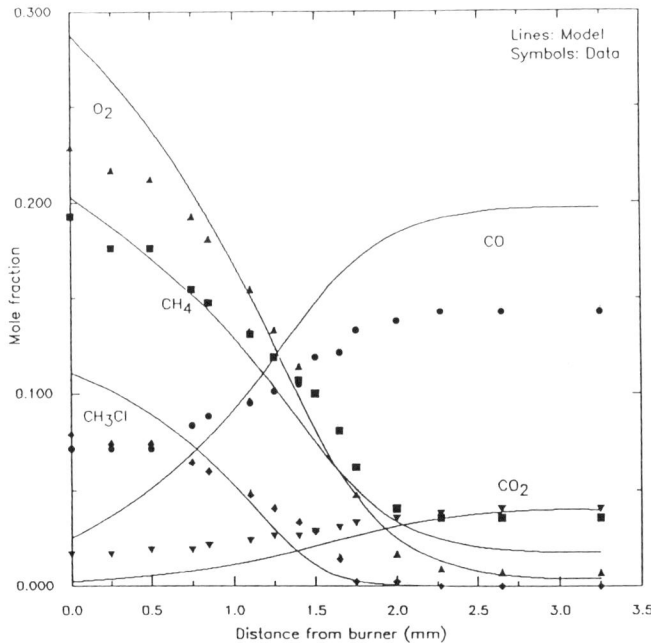

FIG. 23. Comparison of experimentally measured mole fraction profiles for CH_3Cl, CH_4, and O_2 with those determined using the DCKM presented in Table XIV.

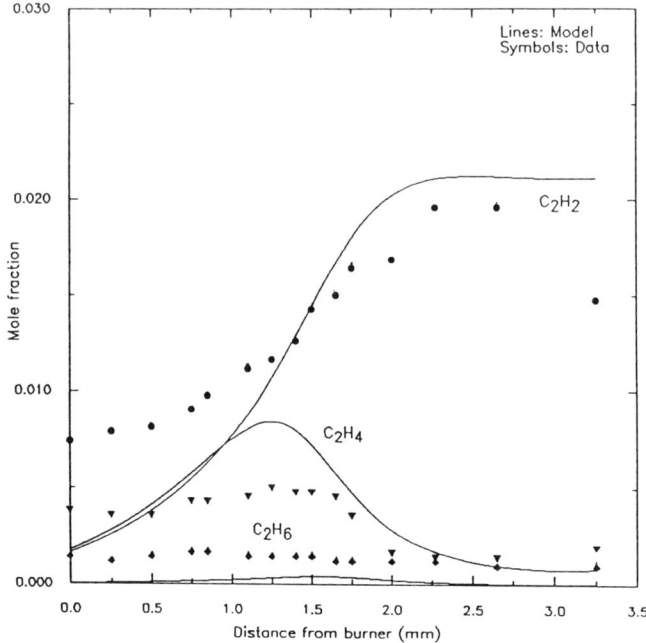

FIG. 24. Comparison of experimentally measured mole fraction profiles for C_2H_2, C_2H_4, and C_2H_6 with those determined using the DCKM presented in Table XIV.

was not done, because observed discrepancies serve an important purpose of highlighting the deficiencies of the mechanism. These deficiencies subsequently can be isolated, studied, and thus better understood for the continued improvement of DCKMs.

The reaction set presented in Table XV also was subjected to a full sensitivity analysis in order to determine the rank order of the reactions in the mechanism. For this, normalized first-order sensitivity gradients (S_{ij}) were calculated along the flame using the following definition:

$$S_{ij} = (k_j/C_i)[dC_i/dk_j] = d[\ln C_i]/d[\ln k_j], \quad (161)$$

where C_i is the molar concentration for species i, and k_j is the preexponential factor for the jth reaction at a particular location in the flame. It must be noted that because of the reversible nature of the elementary reactions, the signs of the S_{ij}s calculated according to Eq. (161) do not necessarily reflect the direction of the reaction. This is especially the case for reactions in which the forward and reverse rates are nearly balanced, i.e., when the elementary reaction is near equilibrium. Therefore, a detailed analysis based on the rates of individual reactions along the flame was also performed.

After the determination of species mole fraction profiles, the net rates of individual elementary reactions in the mechanism were calculated along the flame zone, and rate profiles for reactions that had the highest net rates are presented for CH_3Cl, O_2, and C_2H_4 in Figs. 25, 26, and 27, respectively, together with the first-order sensitivity gradients for the influential reactions. From the considerations of these rates and sensitivity gradients, useful conclusions were made concerning important reactions and species reaction pathways (Karra and Senkan, 1988b).

For example, as evident from the reaction rate profiles (Fig. 25a), the bulk of the CH_3Cl is destroyed primarily by H and Cl radical attack, i.e., $CH_3Cl + H \rightleftharpoons CH_3 + HCl$ and $CH_3Cl + Cl \rightleftharpoons CH_2Cl + HCl$. From Fig. 25b, the concentration of CH_3Cl was most sensitive to its unimolecular decomposition $CH_3Cl \rightleftharpoons CH_3 + Cl$ in the very early parts of the flame. This is not surprising, since free radical concentrations are not dominant in the early part of the flame; thus, rates of initiation reactions are expected to influence reactant concentrations. This result also illustrates the need to describe accurately the effects of density (pressure) on unimolecular reaction rates even in modeling the structure of flames.

As seen in Fig. 26a, oxygen is consumed in this flame primarily by the reactions $H + O_2 \rightleftharpoons OH + O$ and $C_2H_3 + O_2 \rightleftharpoons CH_2O + CHO$. The former reaction is the important chain-branching reaction in combustion, responsible for flame propagation. The latter reaction is important in this flame because of the fuel-rich nature of the mixture. These results are also supported by the sensitivity gradients presented in Fig. 26b.

Based on the calculation of reaction rate profiles presented in Fig. 27a, earlier formation of ethylene in the flame is due to the chemically activated recombination of CH_3 and CH_2Cl radicals, i.e., $CH_3 + CH_3 \rightleftharpoons C_2H_4 + H_2$ and $CH_3 + CH_2Cl \rightleftharpoons C_2H_4 + HCl$. However, a significant fraction of C_2H_4 also forms by the decomposition of C_2H_5 and C_2H_4Cl radicals, i.e., $C_2H_5 \rightleftharpoons C_2H_4 + H$ and $C_2H_4Cl \rightleftharpoons C_2H_4 + Cl$. The destruction of C_2H_4 occurs primarily via the H and Cl radical attack, i.e., $C_2H_4 + H \rightleftharpoons C_2H_3 + H_2$ and $C_2H_4 + Cl \rightleftharpoons C_2H_3 + HCl$. As evident from these reactions, the presence of chlorine provides additional paths for both the formation and the destruction of C_2H_4. Consequently, the C_2H_4 concentration profile remains relatively insensitive to the presence of chlorine in the system.

As is evident from the foregoing discussion, the detailed chemical kinetic mechanism presented in Table XV is of considerable value in interpreting experimental measurements. However, the real advantage of this mechanism lies in its capacity to predict the chemical reaction behavior of CH_3Cl/CH_4 mixtures well beyond the range of conditions investigated. For example, this mechanism was used for the better definition of operating conditions under which the formation of intermediates such as C_2H_4 and C_2H_2 is maximized.

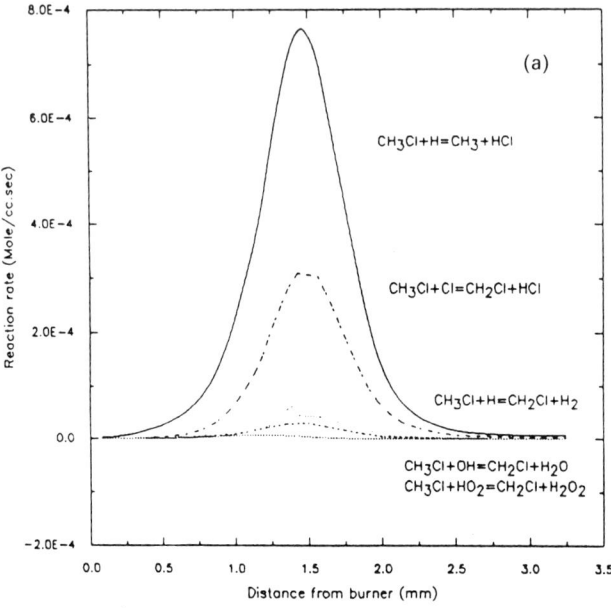

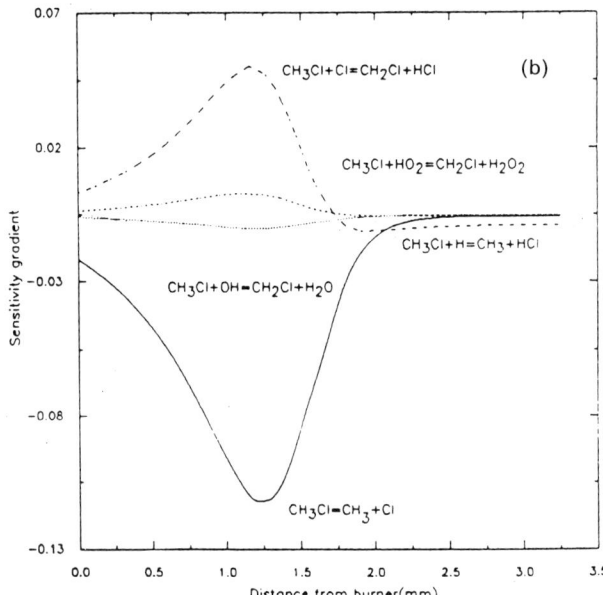

FIG. 25. (a) Reaction rate and (b) sensitivity gradient profiles for the most important elementary reactions involving CH_3Cl.

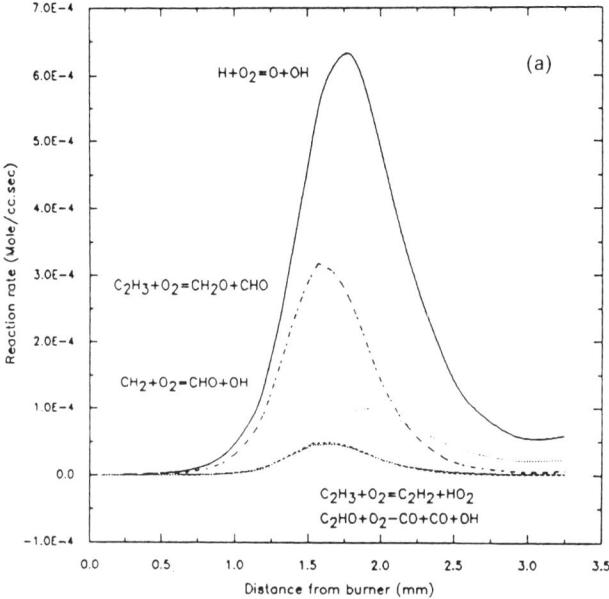

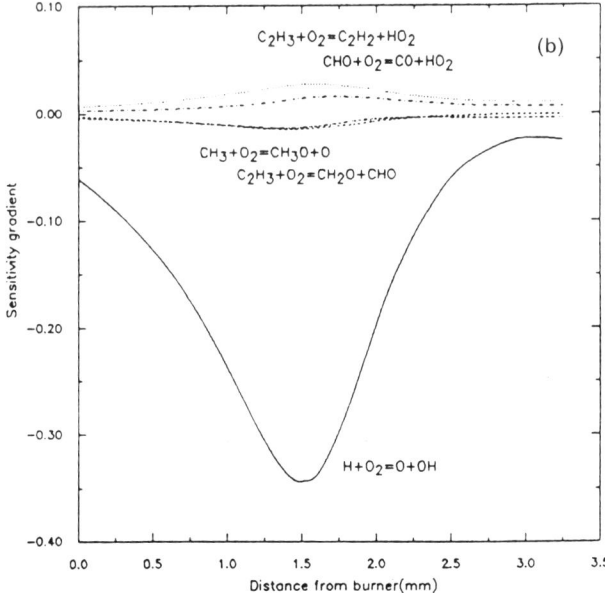

FIG. 26. (a) Reaction rate and (b) sensitivity gradient profiles for the most important elementary reactions involving O_2.

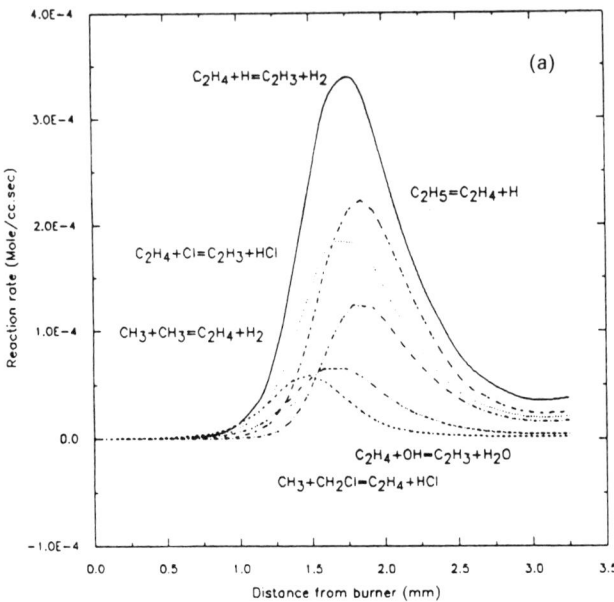
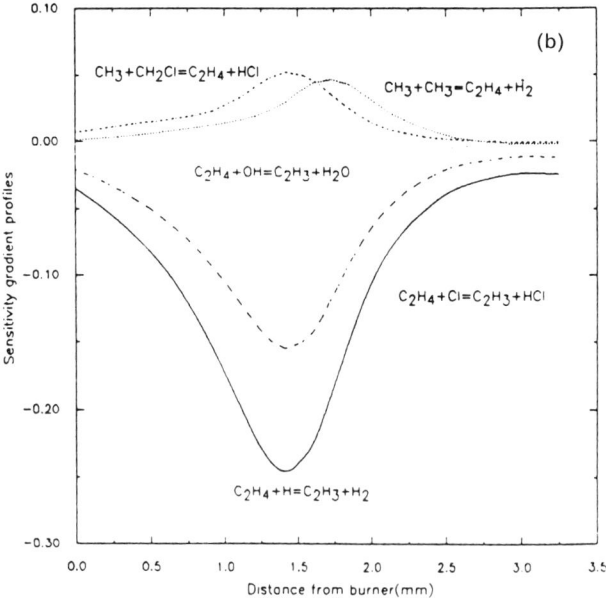

FIG. 27. (a) Reaction rate and (b) sensitivity gradient profiles for the most important elementary reactions involving C_2H_4.

In fact, such a study subsequently led to the development of the chlorine-catalyzed oxidative pyrolysis (CCOP) process to convert methane into more useful C_2 products (Granada et al., 1987; Senkan, 1987). Based on the sensitivity and reaction path analysis of the reaction mechanism discussed above, reactions that are important under the conditions of the CCOP process have been identified, and they are presented in Fig. 28 (Karra and Senkan, 1988b). As evident from this figure, the rank order of importance of reactions is clearly different in the CCOP process from those under flame conditions. For example, in the CCOP process, CH_3Cl is consumed primarily by the unimolecular decomposition route, i.e., $CH_3Cl \rightleftharpoons CH_3 + Cl$, as well as by Cl radical attack, i.e., $CH_3Cl + Cl \rightleftharpoons CH_2Cl + HCl$. Similarly, O_2 is consumed almost exclusively by the $C_2H_3 + O_2 \rightleftharpoons CH_2O + CHO$ reaction, as opposed to the chain-branching reaction, $H + O_2 \rightleftharpoons OH + O$.

As seen in Fig. 28, only a small fraction of the total number of reactions shown in Table XV were important under conditions of interest to the CCOP process. These reactions can be identified and used to construct a smaller reaction mechanism. Although smaller reaction mechanisms are

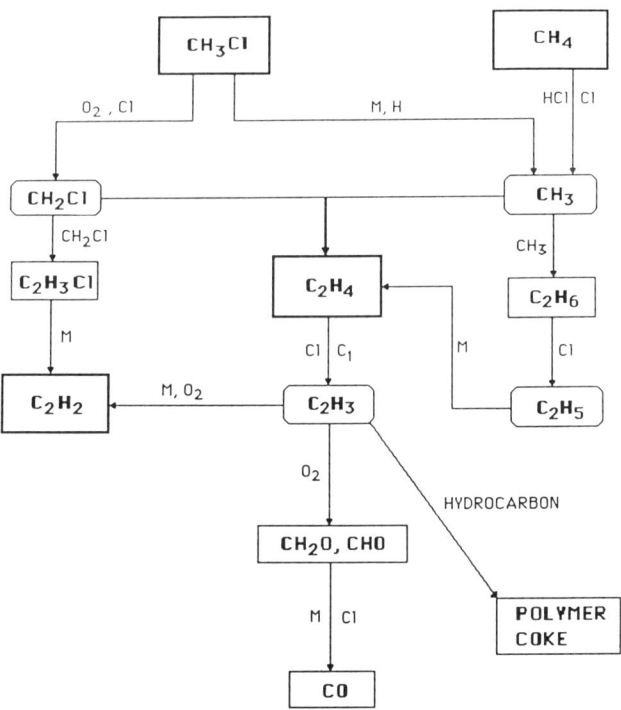

FIG. 28. Major reaction channels for the oxidative pyrolysis of CH_3Cl/CH_4 (from Karra and Senkan, 1988).

advantageous—for example, to illustrate major reaction pathways under a given set of operating conditions—the resulting model will have a far narrower range of applicability than the model that is based on the use of a DCKM. Consequently, for process research and development, the use of comprehensive DCKMs clearly would be necessary. However, for process design and control, smaller mechanisms may be more efficient to use.

It must also be recognized that the success of any detailed chemical kinetic mechanism in fitting available experimental data does not guarantee the accuracy of the mechanism. Our knowledge of the detailed chemical kinetic mechanism of complex reactions is always, in principle, *incomplete*. Consequently, mechanisms must continually be revised as new, more reliable information—both experimental and theoretical—becomes available. In fact, it is this aspect of detailed chemical kinetic modeling that renders the subject rich, full of surprises and opportunities for creative work.

Acknowledgments

This work was supported, in part, by the U.S. Environmental Protection Agency and the National Science Foundation. The author also would like to thank many of his current and past students for their input to the material presented in this article, to Professor John H. Seinfeld for his encouragement for the publication of this work, and to Professors Arthur Fontijn and Robert W. Carr, Jr., for their critical review of the manuscript.

Note

This article was originally written in 1988, and was intended to represent the state-of-the-art at that time. Naturally, new advances made in theoretical, computational and experimental chemical kinetics since then renders this piece somewhat out of date. However, although this article does not summarize and cite developments since 1988, it nonetheless should expose the reader to broader fundamental issues in modern chemical kinetics which continues to remain current.

References

Alfassi, Z. B., and Benson, S. W., "A simple empirical method for the estimation of activation energies in radical molecule metathesis reactions," *Int. J. Chem. Kinetics* **5**, 879 (1973).

Allara, D. L., and Edelson, D., "A computational analysis of a chemical switch mechanism. Catalysis-inhibition effects in a copper surface-catalyzed oxidation," *J. Phys. Chem.* **81**, 2443 (1977).

Almlof, J., and Luthi, P., "Theoretical methods and results for electronic structure calculations on very large systems," in "Supercomputer Research in Chemistry and Chemical Engineering," (Jensen, K. F., and Truhlar, D. G., eds.), ACS Symposium Series No. 353, 1987.

Arnaud, R., Barone, V., Olivella, S., and Sole, A., "*Ab initio* mechanistic studies of radical reactions, addition of methyl radical to acetylene and ethylene," *Chem. Phys. Lett.* **118**, 573 (1985).

Atkins, P. W., "Molecular Quantum Mechanics." Oxford University Press, 1986.

Baetzold, R. C., and Somorjai, G. A., "Preexponential factors in surface reactions," *J. Catalysis* **45**, 94 (1976).

Bahn, G. S., "Approximate thermochemical tables for some C–H and C–H–O species," NASA Report CR-2178 (1973).

Basevich, V. Ya., "Chemical kinetics in the combustion processes: A detailed kinetic mechanism and its implementation," *Prog. Energy. Combust. Sci.* **13**, 199 (1987).

Baulch, D. L., Duxbury, J., Grant, S. J., and Montague, D. C., "Evaluated kinetic data for high temperature reactions: Homogeneous gas phase reactions of halogen and cyanide containing species," *J. Phys. Chem. Ref. Data* **4**, 10, Supplement No. 1 (1981).

Baushlicher, C. W., Langhoff, S. R., Partridge, H., Halicioglu, T., and Taylor, P. R., "Theoretical approaches to metal chemistry," in "Supercomputer Research in Chemistry and Chemical Engineering" (Jensen, K. F., and Truhlar, D. G., eds.), ACS Symposium Series No. 353, 1987.

Benson, S. W., "Thermochemical Kinetics." John Wiley, New York, 1976.

Benson, S. W., and Hougen, G. R., "A simple, self-consistent electrostatic model for quantitative prediction of the activation energies of four-center reactions," *J. Amer. Chem. Soc.* **18**, 4036 (1965).

Benson, S. W., and O'Neal, H. E., "Kinetic data on gas-phase unimolecular reactions," *NSRDS-NBS* **21** (1970).

Berces, T., and Marta, F., "Activation energies for metathesis reactions of radicals," in "Chemical Kinetics of Small Organic Molecules: IV Reactions in Special Systems" (Alfassi, Z. B., ed.). CRC Press, Boca Raton, Florida, 1988.

Bird, R. B., Steward, W. E., and Lightfoot, E. N., "Transport Phenomena." John Wiley, New York, 1960.

Boudart, M., and Djega-Mariadassou, G., "Kinetics of Heterogeneous Catalytic Reactions." Princeton University Press, Princeton, New Jersey, 1984.

Brown, R. L., "Rate constants for H-atom transfer reactions by the BEBO method," *J. Res. Nat. Bur. Stds.* **86**, 605 (1981).

Butt, J. B., "Reaction Kinetics and Reactor Design." Prentice-Hall, New York, 1980.

Caracatsios, M., and Steward, W. E., "Sensitivity analysis of initial value problems including ODEs and algebraic equations," *Comp. Chem. Eng.* **9**, 359 (1985).

Carberry, J. J., "Chemical and Catalytic Reaction Engineering." McGraw-Hill, New York, 1976.

Carberry, J. J., "The contributions of heterogeneous catalysis to catalytic reaction engineering," *Chem. Eng. Prog.* **84**(2), 51 (1988).

Chang, W. D., Karra, S. B., and Senkan, S. M., "A computational study of chlorine inhibition of CO flames," *Combust. Flame* **69**, 113 (1987).

Chase, M. W., Davies, C. A., Downey, J. R., Frurip, D. J., McDonald, R. A., and Syverud, A. N., "JANAF thermochemical tables," *J. Phys. Chem. Ref. Data* **14**, Supplement No. 1 (1985).

Cohen, N., and Benson, S. W., "Transition-state-theory calculations for reactions of OH with haloalkanes," *J. Phys. Chem.* **91**, 162 (1987a).

Cohen, N., and Benson, S. W., "Empirical correlations for rate coefficients for reactions of OH with haloalkanes," *J. Phys. Chem.* **91**, 171 (1987b).

Coltrin, M. E., Kee, R. J., and Miller, J. A., "A mathematical model of silicon chemical vapor deposition." *J. Electrochem. Soc.* **133**, 1206 (1986).

Daudel, R., Leroy, G., Peeters, D., and Sana, M., "Quantum Chemistry." John Wiley, Chichester, Ireland, 1983.
Dean, A. M., "Predictions of pressure and temperature effects upon radical addition and recombination reactions," *J. Phys. Chem.* **89**, 4600 (1985).
Dean, A. M., and Westmoreland, P. R., "Bimolecular QRRK analysis of methyl radical reactions," *Int. J. Chem. Kinetics* **19**, 207 (1987).
DeMore, W. B., Molina, M. J., Watson, R. T., Golden, D. M., Hampson, R. F., Kurylo, M. J., Howard, C. J., and Ravishankara, A. R., "Chemical kinetics and photochemical data for use in stratospheric modeling," Evaluation No. 6, JPL Publication 85-37 (1985).
Dente, M. E., and Ranzi, E. M., "Mathematical modeling of hydrocarbon pyrolysis reactions," *in* "Pyrolysis: Theory and Industrial Practice" Albright, L. F., Crynes, B. L., and Corcoran, W. H., eds.). Academic Press, New York, 1983.
Dewar, M. J. S., "Quantum organic chemistry," *Science* **187**, 1037 (1975).
Dewar, M. J. S., and Olivella, S., "Ground states of molecules. 48. MINDO/3 study of some radical addition reactions," *J.A.C.S.* **100**, 5290 (1978).
Dewar, M. J. S., and Storch, D. M., "Comparative tests of theoretical procedures for studying chemical reactions," *J.A.C.S.* **107**, 3898 (1985).
Dewar, M. J. S., and Thiel, W., "Ground states of molecules. 38. The MNDO method. Approximations and parameters," *J.A.C.S.* **99**, 4899 (1977).
Dewar, M. J. S., Healy, E. F., and Stewart, J. J. P., "Location of transition states in reaction mechanisms," *J. Chem. Soc. Faraday Trans.* **80**, 227 (1984).
Dewar, M. J. S., Zoebisch, E. G., Healy, E. F., and Stewart, J. J. P., "AMl: A new general purpose quantum mechanical molecular model," *J.A.C.S.* **107**, 3902 (1985).
Dixon-Lewis, G., "Computer modeling of combustion reactions in flowing systems with transport," *in* "Combustion Chemistry" (W. C. Gardiner, Jr., ed.). Springer-Verlag, New York, 1984.
Dudukovic, M. P., "Chemical engineering: Current status and future directions," *Chem. Eng. Ed.* **21**(4), 210 (1987).
Dumesic, J. A., Milligan, B. A., Greppi, L. A., Balse, V. R., Sarnowski, K. T., Beall, C. E., Kataoka, T., and Rudd, D. F., "A kinetic modeling approach to the design of catalysts: Formulation of a catalyst design advisory program," *Ind. Eng. Chem. Res.* **26**, 1399 (1987).
Dunning, T. H., Harding, L. B., Wagner, A. F., Schatz, G. C., and Bowman, J. M., "Theoretical studies of the energetics and dynamics of chemical reactions," *Science* **240**, 453 (1988).
Eyring, H., "The activated complex and the absolute rate of chemical reactions," *Chem. Rev.* **17**, 65 (1935).
Filby, W. G., "The participation of free radicals in atmospheric chemistry," *in* "Chemical Kinetics of Small Organic Molecules: IV Reactions in Special Systems" (Alfassi, Z. B., ed.). CRC Press, Boca Raton, Florida, 1988.
Firsch, M., "Gaussian 88 User's Guide and Programmer's Reference." Gaussian Inc., Pittsburgh, Pennsylvania, 1988.
Fontijn, A., and Zellner, R., "Influence of temperature on rate coefficients of bimolecular reactions," *in* "Reactions in Small Transient Species (A. Fontijn and M. A. A. Clyne, eds.). Academic Press, New York, 1983.
Frenklach, M., Clary, D. W., Yuan, T., Gardiner, W. C., Jr., and Stein, S. E., "Mechanism of soot formation in acetylene–oxygen mixtures," *Combust. Sci. Tech.* **50**, 79 (1986).
Gaffney, J. S., and Bull, K., "Prediction of the rate constants for radical reactions using correlational tools," *in* "Chemical Kinetics of Small Organic Molecules: IV Reactions in Special Systems" (Alfassi, Z. B., ed.). CRC Press, Boca Raton, Florida, 1988.
Gardiner, W. C., and Troe, J., "Rate coefficients of thermal dissociation, isomerization, and recombination reactions," *in* "Combustion Chemistry" (W. C. Gardiner, Jr., ed.). Springer-Verlag, New York, 1984.

Glasstone, S., Laidler, K. J., and Eyring, H., "The Theory of Rate Processes." McGraw-Hill, New York, 1941.
Golden, D. M., and Larson, C. W., "Rate Constants for use in modeling," in "20th Symposium (Int'l) on Combustion." The Combustion Institute, Pittsburgh, Pennsylvania, 1984, p. 595.
Gonzales, C., Sosa, C., and Schlegel, H. B., "Ab initio study of addition reaction of the methyl radical to ethylene and formaldehyde," J. Phys. Chem. 93, 2435 (1989).
Gordon, S., and McBride, B. J., computer program for calculation of complex chemical equilibrium compositions, rocket performance, incident and reflected shocks and Chapman-Jougonet detonations, NASA SP-273 (1976).
Granada, A., Karra, S. B., and Senkan, S. M., "Conversion of CH_4 into C_2H_2 and C_2H_4 by the chlorine-catalyzed oxidative-pyrolysis (CCOP) process: Oxidative pyrolysis of CH_3Cl," Ind. Chem. Eng. Res. 26, 1901 (1987).
Hanson, R. K., and Salimian, S., "Survey of rate constants in the N/H/O system," in "Combustion Chemistry" (W. C. Gardiner, Jr., ed.). Springer-Verlag, New York, 1984.
Heneghan, S. P., Knoot, P. A., and Benson, S. W., "The temperature coefficient of the rates in the system $Cl + CH_4 \rightleftharpoons CH_3 + HCl$, thermochemistry of the methyl radical," Int. J. Chem. Kinetics 13, 677 (1981).
Hindmarsh, A. C., "LSODE: Livermore solver for ordinary differential equations." Lawrence Livermore Laboratory, Tech. Report No. 3342 (1980).
Hirschfelder, J. O., Curtiss, C. F., and Bird, R. B., "Molecular Theory of Gases and Liquids." John Wiley, New York, 1954.
Hoyland, J. R., "MINDO/2 calculations of the reaction of methyl radicals with ethylene and butadiene," Theor. Chim. Acta (Berl.) 22, 229 (1971).
Hwang, J.-T., "Sensitivity analysis in chemical kinetics by the method of polynomial approximations," Int. J. Chem. Kinetics 15, 959 (1983).
Ibrahim, M. R., and Schleyer, P. V. R, "Atom equivalents for relating ab initio energies to enthalpies of formation," J. Comp. Chem. 6, 157 (1985).
Ibrahim, M. R., Fataftah, Z. A., Schleyer, P. V. R., and Stout, P. D., "Calculation of enthalpies of hydrogenation of hydrocarbons," J. Comp. Chem. 8, 1131 (1987).
Johnston, H. S., "Gas Phase Reaction Rate Theory." Ronald Press, New York, 1966.
Johnston, H. S., and Parr, C., J. Amer. Chem. Soc. 85, 2544 (1963).
Jones, W. P., and Lindstedt, R. P., "Global reaction schemes for hydrocarbon combustion," Combust. Flame 73, 233 (1988).
Karra, S. B., and Senkan, S. M., "Chemical structures of sooting $CH_3Cl/CH_4/O_2/Ar$ and $CH_4/O_2/Ar$ flames," Combust. Sci. Tech. 54, 333 (1987).
Karra, S. B., and Senkan, S. M., "Analysis of the chemically activated CH_2Cl/CH_2Cl and CH_3/CH_2Cl recombination reactions at elevated temperatures using the QRRK method," Ind. Eng. Chem. Research 27, 447 (1988b).
Karra, S. B., and Senkan, S. M., "A detailed chemical kinetic mechanism for the oxidative pyrolysis of CH_3Cl," Ind. Eng. Chem. Res. 27, 1163 (1988b).
Kee, R. J., Warnatz, J., and Miller, J. A., "A Fortran computer code package for the evaluation of gas-phase viscosities, conductivities, and diffusion coefficients," Sandia Report, SAND83-8209 (1983).
Kee, R. J., Grear, J. F., Smooke, M. D., and Miller, J. A., "A Fortran program for modeling steady state laminar one-dimensional premixed flames," Sandia Report (1985).
Kerr, A. J., and Drew, R. M., "Handbook of Bimolecular and Termolecular Gas Reactions," Vol. III and IV. CRC Press, Boca Raton, Florida, 1987.
Kerr, A. J., and Moss, S. J., "Handbook of Bimolecular and Termolecular Gas Reactions," Vols. I and II. CRC Press, Boca Raton, Vlorida, 1981.
Kondratiev, V. N. "Rate constants of gas phase reactions," COM-72-10014. National Bureau of Standards, Washington, D.C., 1972.

Kramer, M. A., Kee, R. J., and Rabitz, H., "CHEMSEN: A computer Code for Sensitivity Analysis of Elementary Chemical Reaction Models." SAND82-8230, Sandia National Laboratories, Livermore, August, 1982a.

Kramer, M. A., Calo, J. M., Rabitz, H., and Kee, R. J., "AIM: The Analytically Integrated Magnus Method for Linear and Second Order Sensitivity Coefficients." SAND82-8231, Sandia National Laboratories, Livermore, August, 1982b.

Laidler, K. J., "Chemical Kinetics," 3rd Ed. Harper and Row, New York, 1987.

Lowe, J. P., "Quantum Chemistry." Academic Press, New York, 1978.

Lutz, A. E., Kee, R. J., and Miller, J. A., "SENKIN: A Fortran Program for Predicting Homogeneous Gas Phase Kinetics with Sensitivity Analysis." SAND87-8248, Sandia National Laboratories, Livermore, February, 1988.

Martinez, M. R., "Estimation of Gas-Phase Thermokinetic Parameters: Thermochemical Group Additivity Computer Program (TGAP)." NTIS AD-762-614 and AD-769-631, 1973.

Mayer, S. W., Schieler, L., and Johnston, H. S., "Computation of high-temperature rate constants for bimolecular reactions of combustion products," in 11th Symposium (Int'l) on Combustion." The Combustion Institute, Pittsburgh, 1967, p. 837.

McNesby, J. R., Drew, C.M., and Gordon, A. S., "Mechanism of the decomposition of primary and secondary n-butyl free radicals," $J.\ Chem.\ Phys.$ **24**, 1260 (1956).

McQuarrie, D. A., "Statistical Mechanics." Harper and Row, New York, 1976.

Melius, C. F., "Thermochemistry of hydrocarbon intermediates in combustion: Application of the BAC-MP4 method," private communication (1990).

Melius, C. F., and Binkley, J. S., "Thermochemistry of the decomposition of nitramines in the gas phase.. in 21st Symposium (Int'l) Combustion." The Combustion Institute, Pittsburgh, 1986, p. 1953.

Miller, J. A., and Fisk, G. A., "Combustion chemistry," $Chem.\ Eng.\ News-Special\ Report$, August 31, p. 22 (1987).

Moore, J. M., and Pearson, R. G., "Kinetics and Mechanism," 3rd Ed. Wiley-Interscience, New York, 1981.

Mozurkewich M., and Benson, S. W., "Negative activation energies and curved Arrhenius plots. 1. Theory of reactions over potential wells," $J.\ Phys.\ Chem.$ **88**, 6429 (1984).

Neta, P., "Reactions of radicals with biologically important molecules," in "Chemical Kinetics of Small Organic Molecules: IV Reactions in Special Systems" (Alfassi, Z. B., ed.). CRC Press, Boca Raton, Florida, 1988.

Nicholas, J., "Chemical Kinetics: A Modern Survey of Gas Reactions." John Wiley, New York, 1976.

O'Neil, H. E., and Benson, S. W., "A method for estimating Arrhenius A factors for four- and six-center unimolecular reactions," $J.\ Phys.\ Chem.$ **71**, 2903 (1967).

Oppelt, T. E., "Incineration of Hazardous Waste," $J.\ Air\ Poll.\ Cont.\ Assoc.$, **37**, 558 (1987).

Paczko, G., Lefdal, P. M., and Peters, N., "Reduced reaction schemes for methane, methanol, and propane flames" in "21st Symposium (Int'l) Combustion." The Combustion Institute, 1986, p. 739.

Pauling, L., "The Nature of the Chemical Bond," 3rd Ed. Cornell University Press, Ithaca, New York, 1960.

Pedley, J. B., Naylor, R. D., and Kirby, S. P., "Thermochemical Data of Organic Compounds." Chapman and Hall, London, 1986.

Petzold, L. R., "A Description of DASSL: A Differential/Algebraic System Solver." Sandia National Laboratories Report, SAND82-8637 (1982).

Pitchai, P., and Klier, K., "Partial oxidation of methane," $Catal.\ Rev.\ Sci.\ Eng.$ **28**, 13 (1986).

Pople, J. A., and Beveridge, D., "Approximate Molecular Orbital Theory." McGraw-Hill, New York, 1970.

Rabitz, H., Kramer, M. A., and Dacol, D., "Sensitivity analysis in chemical kinetics," *Ann. Rev. Phys. Chem.* **34**, 419 (1983).

Reid, R. C., Prausnitz, J. M., and Sherwood, T. K., "Properties of Gases and Liquids," 3rd Ed. McGraw-Hill, New York, 1977.

Robinson, P. J., and Holbrook, K. A., "Unimolecular Reactions." John Wiley, New York, 1972.

Roenigk, K. F., Jensen, K. F., and Carr, R. W., "Rice–Rampsperger–Kassel–Marcus theoretical prediction of high-pressure Arrhenius parameters by nonlinear regression: Application to silane and disilane decomposition," *J. Phys. Chem.* **91**, 5732 (1987).

Rogowski, D. F., Marshall, P., and Fontijn, A., "High-temperature fast-flow reactor kinetics studies of the reactions of Al with Cl_2, Al with HCl, and AlCl with Cl_2 over wide temperature ranges," *J. Phys. Chem.* **93**, 1118 (1989).

Rosner, D. E., "Transport Processes in Chemically Reacting Flow Systems." Butterworths, Boston, 1986.

Russel, J. J., Seetula, J., Senkan, S. M., and Gutman, G., "Kinetics and thermochemistry of the methyl radical: Study of the CH_3 + HCl reaction," *Int. J. Chem. Kinet.*, in press (1988a).

Russel, J. J., Seetula, J., Senkan, S. M., and Gutman, G., "Kinetics of reactions of chlorinated vinyl radicals (CH_2CCl and C_2Cl_3) with molecular oxygen," *J. Phys. Chem.*, in press (1988b).

Schmidt, M. W., Boatz, J. A., Baldridge, K. K., Koseki, S., Gordon, M. S., Elbert, S. T., and Lam, B., "GAMESS: General atomic and molecular structure system," *QCPE Bull.* **7**, 115 (1987).

Seaton, W. H., Freedman, E., and Trevek, B., "CHETAH—The ASTM Chemical Thermodynamic and Energy Release Evaluation Program." ASTM D51. Philadelphia, Pennsylvania, 1974.

Seinfeld, J. H., "Atmospheric Chemistry and Physics of Air Pollution." John Wiley, New York, 1986.

Semenov, N. N., "Some Problems in Chemical Kinetics and Reactivity." Princeton University Press, 1958.

Senkan, S., "Conversion of Methane by Chlorine Catalyzed Oxidative Pyrolysis," *Chem. Eng. Prog.*, **12**, 58 (1987).

Shimanouchi, T., "Tables of molecular vibration frequencies: Consolidated volume I," *NSRDS-NBS*, 39 (1972).

Siegneur, C., Stephanopoulos, G., and Carr, R. W., Jr., "Dynamic sensitivity analysis of chemical reaction systems: A variational method," *Chem. Eng. Sci.* **37**, 845 (1982).

Slagle, I., Park, P. Y., Heaven, M., and Gutman, D., "Kinetics of Polyatomic Free Radicals. Reaction of Vinyl Radicals with Molecular Oxygen," *J. Am. Chem. Soc.*, **106**, 4356 (1984).

Stein, S. E., and Fahr, A., "High-temperature stabilities of hydrocarbons," *J. Phys. Chem.* **89**, 3714 (1985).

Stein, S. E., Golden, D. M., and Benson, S. W., "Predictive scheme for thermochemical properties of polycyclic aromatic hydrocarbons," *J. Phys. Chem.* **81**, 314 (1977).

Stewart, J. J. P., "MOPAC: A general molecular orbital package," *QCPE*, 455 (1987).

Stewart, J. J. P., "Optimization of Parameters for Semiempirical Methods I. Method", *J. Comput. Chemistry*, **10**, 209 (1989a).

Stewart, J. J. P., "Optimization of Parameters for Semiempirical Methods II. Applications", *J. Comput. Chemistry*, **10**, 221 (1989b).

Svehla, R. A., "Estimated viscosities and thermal conductivities of gases at high temperatures," NASA Tech. Report R-132 (1962).

Tilden, J. W., Costanza, V., McRae, G. J., and Seinfeld, J. H., "Sensitivity analysis of chemically reacting systems," *in* "Modeling of Chemical Reaction Systems" (K. H. Ebert, P. Deuflhard, and W. Jaeger, eds.). Springer-Verlag, New York, 1981.

Till, W. C., and Luxon, J. T., "Integrated Circuits: Materials, Devices, and Fabrication." Prentice-Hall, New York, 1982.

Toong, T-Y., "Combustion Dynamics: The Dynamics of Chemically Reacting Fluids." McGraw-Hill, New York, 1983.

Toyoshima, I., and Somorjai, G. A., "Heats of chemisorption of O_2, H_2, CO, CO_2 and N_2 on polycrystalline and single crystal transition metal surfaces," *Catal. Rev. Sci. Eng.* **19**, 105 (1979).

Tsang, W., and Hampson, R. F., "Chemical kinetic data base for combustion chemistry. Part I. Methane and related compounds," *J. Phys. Chem. Ref. Data* **15**, 1087 (1986).

Tully, F. P., and Ravishankara, A. R., *J. Phys. Chem.* **84**, 3126 (1980).

Veillard, A., ed., "Quantum Chemistry: The Challenge of Transition Metals and Coordination Chemistry." Reidel Publishing Co., Dordrecht, The Netherlands, 1986.

Warnatz, J., "Rate coefficients in the C/H/O system," in "Combustion Chemistry" (W. C. Gardiner, Jr., ed.). Springer-Verlag, New York, 1984.

Wentrup, C., "Reactive Molecules." John Wiley, New York, 1984.

Westbrook, C. K., "Numerical simulation of chemical kinetics of combustion," in "Chemical Kinetics of Small Organic Molecules: IV Reactions in Special Systems" (Alfassi, Z. B., ed.). CRC Press, Boca Raton, Florida, 1988.

Westbrook, C. K., and Dryer, F. A., "Chemical kinetic modeling of hydrocarbon combustion," *Prog. Energy Combust. Sci.* **10**, 1 (1984).

Westley, F., "Chemical Kinetics of the Gas Phase Combustion of Fuels." NBS Special Publication No. 449, 1976.

Westley, F., "Table of Recommended Rate Constants for Chemical Reactions Occurring in Combustion." NBS Report, NSRDS-NBS 67, 1980.

Westmoreland, P. R., Howard, J. B., Longwell, J. P., and Dean, A. M., "Prediction of rate constants for combustion and pyrolysis reactions by bimolecular QRRK," *AIChE J.* **32**, 1971 (1986).

Wiberg, K. B., "Group equivalents for converting *ab initio* energies to enthalpies of formation," *J. Comp. Chem.* **5**, 197 (1984).

Willems, P. A., and Froment, G. F., "Kinetic modeling of the thermal cracking of hydrocarbons. 1. Calculation of frequency factors," *Ind. Eng. Chem. Res.* **27**, 1959 (1988a).

Willems, P. A., and Froment, G. F., "Kinetic modeling of the thermal cracking of hydrocarbons. 1. Calculation of activation energies," *Ind. Eng. Chem. Res.* **27**, 1966 (1988b).

Zhdanov, V. P., Pavlicek, J., and Knor, Z., "Preexponential factors for elementary surface processes," *Catal. Rev. Sci. Eng.* **30**, 501 (1988).

OPTIMIZATION STRATEGIES FOR COMPLEX PROCESS MODELS

Lorenz T. Biegler

Department of Chemical Engineering
Carnegie Mellon University
Pittsburgh, Pennsylvania

I. Introduction 197
II. Development of Newton-Type Optimization Algorithms 199
   A. Large-Scale Successive Quadratic Programming 202
   B. SQP for Parameter Estimation 206
III. Flowsheet Optimization 207
   A. Case Study for Process Optimization 210
   B. Advanced Applications of Process Optimization 211
IV. Differential-Algebraic Optimization 216
   A. Variational Methods 217
   B. NLP Methods with an Embedded ODE Model 218
V. Simultaneous Approaches for Differential/Algebraic Optimization 220
   A. Accuracy of Representation of ODEs 222
   B. Simultaneous Formulations for Profile Optimization 238
   C. Generalization to Large-Scale Problems 244
VI. Summary and Conclusions 249
   References 251

I. Introduction

Process modeling is usually performed for two reasons. For fundamental and scientific studies a process model serves to explain and predict the quantitative behavior of physical or chemical phenomena in the process. The predictive capability of the model, however, is usually exploited by the engineer in order to improve the process. Once the model, the process and problem limitations, and the criterion for improvement are clearly and quantitatively defined, this task can be viewed as an optimization problem, i.e., *find the best value of the improvement criterion, or objective function, subject to problem limitations and consistent with the predictions of the process model.*

Unfortunately, once a complex model is constructed, the optimization task is often handled by tedious and time-consuming case studies. Consequently, such tasks are performed infrequently and are often not considered fruitful activities for research. Instead, much research in process design concentrates on developing simpler, low-dimensional representations of the process model (often through geometrical or intuitive arguments) that render the specific optimization problem easier to understand. These approaches include pinch technology for energy integration (Linnhoff et al., 1982), synthesis of separation systems (Westerberg, 1985), chemical reactor network synthesis (Glasser et al., 1987) and the overall design of process flowsheets (Douglas, 1988). The value of the insights gained by these methods cannot be questioned. However, because such methods are limited by the problem, they frequently cannot be extended to problems with additional limitations (constraints) or interactions among subsystems.

It is therefore the purpose of the paper to show that, through application of powerful optimization strategies and careful tailoring to the process model, direct optimization of complex process models does not need to be time-consuming or difficult. Indeed the simultaneous approach discussed in the next section often requires little more effort for optimization than solution of the model itself. Because the optimization and modeling task are closely coupled, successful application of these strategies usually requires knowledge of the advantages and limitations of the optimization strategies, as well as knowledge of the process model and its solution strategy. Frequently, different mathematical representations of an equivalent model can make a difficult optimization problem easy or even trivial to solve.

In the next section we state the process optimization problem and develop a Newton-type approach to solving nonlinear programs (NLPs). We concentrate on these approaches because they are responsible for the recent widespread application of optimization to process models. Earlier approaches based on direct search procedures or direct application of penalty functions are generally undesirable for large process applications (see, for example, Edgar and Himmelblau, 1988). Here special attention is also paid to extending Newton-type approaches to large-scale problems. Section III deals with the application of this approach to flowsheet optimization. Here efficient strategies have been developed over the past decade that have now been integrated into commercially available modeling tools. Moreover, as will be shown, this optimization capability now allows the consideration of many interesting and difficult process design problems. Section IV extends this approach to differential/algebraic equation (DAE) models and compares this approach to a more conventional approach where the model is treated as a black box called by an optimization algorithm. Section V then discusses theoretical properties and presents applications of the simultaneous approach for reactor optimization and determination of optimal operating profiles. Finally, section VI summarizes the paper and outlines directions for future research.

II. Development of Newton-Type Optimization Algorithms

Newton-type algorithms have been applied to process optimization over the past decade and have been well studied. In this section we provide a concise development of these methods, as well as their extension to handle large-scale problems. First, however, we consider the following, rather general, optimization problem:

$$\begin{align} \text{Min } & F(x, y) \\ \text{subject to (s.t.)} & \\ & g(x, y) \leqslant 0 \\ & h(x, y) = 0 \end{align} \quad (1)$$

where F is a scalar objective function and g and h are vector functions that represent inequality and equality constraints, respectively. All of these are, generally, nonlinear algebraic functions of the vectors x and y. Here the x variables vary continuously in the optimization problem, while the y variables are defined as taking only discrete (e.g., integer) values for the optimization problem. The problem could also be extended to deal with differential equations and (infinite-dimensional) variables that are functions of time. However, we defer consideration of these systems until Section IV. In addition, the more general multicriterion problem where $F(x, y)$ is a vector function will not be covered here although it is based on consideration of (1). An interesting survey of multicriterion optimization with process applications can be found in Clark and Westerberg (1983).

When the y variables cannot be relaxed to continuous variables, e.g., where they represent yes/no (0–1) decisions, then (1) is termed a mixed integer nonlinear program (MINLP). MINLP formulations have been used quite successfully to represent a variety of structural optimization problems that occur in process design and synthesis (see Kocis and Grossmann, 1988, and Floudas et al., 1989). Solution strategies for MINLPs usually proceed on two levels. First, an inner or primal problem is solved where the y variables are fixed and optimal values of x are determined (through nonlinear programming). This determines an upper bound on the optimal value of $F(x, y)$ in (1). Next, a master problem is formulated that updates the values of y for the primal problem (this is frequently a mixed integer *linear* programming problem). Solution of the master problem provides, under conditions of convexity, a lower bound to the optimum of (1), and the solution to the MINLP is found when the upper bound (from the NLP) and lower bound (from the master problem) meet. While the master problem can be formulated in several ways (e.g., generalized Benders decomposition, outer approximation/equality relaxation), an efficient nonlinear programming strategy is always required for the inner problem. Thus, for MINLPs, efficient nonlinear programming strategies are essential elements for process optimization.

Moreover, the nonlinear program itself,

$$\text{Min } F(x)$$
$$\text{s.t. } g(x) \leqslant 0 \tag{2}$$
$$h(x) = 0$$

represents a wide variety of problem formulations for the process optimization task. Consequently, this study considers process applications of nonlinear programming and the development of efficient methods for solution of (1). In particular, we concentrate on Newton-type algorithms, as these approaches are currently the most efficient nonlinear programming strategies and are responsible for the recent widespread application of optimization to process models.

Necessary conditions for a local solution to (2) are given by the following Kuhn–Tucker conditions:

$$\begin{aligned} \nabla L(x, \mu, \lambda) &= \nabla F + \nabla g \mu + \nabla h \lambda = 0, \\ g^T \mu &= 0, \\ g(x) &\leqslant 0, \quad \mu \geqslant 0, \\ h(x) &= 0, \end{aligned} \tag{3}$$

where μ and λ are multipliers on g and h, respectively, and the Lagrange function is $L(x, \mu, \lambda) = F(x) + g(x)\mu + h(x)\lambda$. (In addition to these, sufficient conditions for optimality require that the Hessian [second derivative] matrix of the Lagrange function, projected into the tangent space of the active constraints, must also be positive definite. However, these conditions are much harder to check.) Now assume for the moment that we know which inequalities are active at the optimum and denote these as $g_A = 0$. Correspondingly, multipliers μ for g_A are positive, and Kuhn–Tucker conditions thus simplify to

$$\begin{aligned} \nabla L(x, \mu, \lambda) &= \nabla F(x) + \nabla g_A(x)\mu + \nabla h(x)\lambda = 0, \\ g_A(x) &= 0, \\ h(x) &= 0, \\ (\mu &> 0). \end{aligned} \tag{4}$$

OPTIMIZATION STRATEGIES FOR COMPLEX PROCESS MODELS   201

Applying Newton's method at iteration k to this set of equations leads to

$$\begin{bmatrix} \nabla_{xx}L & \nabla g_A & \nabla h \\ \nabla g_A^T & 0 & 0 \\ \nabla h^T & 0 & 0 \end{bmatrix} \begin{bmatrix} d \\ \Delta\mu \\ \Delta\lambda \end{bmatrix} = -\begin{bmatrix} \nabla L(x^k, \mu^k, \lambda^i) \\ g_A(x^k) \\ h(x^k) \end{bmatrix}, \quad (5)$$

where the search direction d provides a change in x, and $\Delta\mu$ and $\Delta\lambda$ provide changes in the multipliers. The preceding Newton step can also be obtained by solving the following equivalent, equality constrained quadratic program (QP):

$$\operatorname*{Min}_{d} \nabla F(x^k) + \frac{1}{2} d^T \nabla_{xx} L(x^k, \mu^k, \lambda^k) d$$

$$\text{s.t.} \quad h(x^k) + \nabla h(x^k)^T d = 0 \quad (6)$$

$$g_A(x^k) + \nabla g_A(x^k)^T d = 0$$

This form is convenient in that the active inequality constraints can now be replaced in the QP by all of the inequalities, with the result that g_A is determined directly from the QP solution. Finally, since second derivatives may often be hard to calculate and a unique solution is desired for the QP problem, the Hessian matrix, $\nabla_{xx}L$, is approximated by a positive definite matrix, B, which is constructed by a quasi-Newton formula and requires only first-derivative information. Thus, the Newton-type derivation for (2) leads to a nonlinear programming algorithm based on the successive solution of the following QP subproblem:

$$\operatorname{Min} \nabla F(x^k) + \frac{1}{2} d^T B^k d$$

$$\text{s.t.} \quad h(x^k) + \nabla h(x^k)^T d = 0 \quad (7)$$

$$g(x^k) + \nabla g(x^k)^T d \leqslant 0$$

The successive quadratic programming (SQP) algorithm (Han, 1977; Powell, 1977) has been the algorithm of choice for model-based process optimization. In several case studies (Hock and Schittkowski, 1981), it has been shown to require the fewest function (model) evaluations. The SQP algorithm also requires a line search or trust region strategy (to restrict the steplength) in order to ensure convergence from poor starting points. A number of commercially available SQP codes have been developed for solving nonlinear programs (2) by repeated application of (7). These include the VF02AD and VF13AD subroutines in the Harwell library, NCONF and NCONG in the IMSL library, and NPSOL from Stanford University. More detailed information on these programs has been compiled by Edgar and Himmelblau (1988).

It should be mentioned, however, that the performance of SQP, or any other gradient-based algorithm, is strongly dependent on the accuracy of the derivative calculations for the objective and constraint functions. Here, expected rates of convergence deteriorate, the method may fail to find an improved point (e.g., a line search failure), and, in fact, the method may not even recognize a local optimum. This last case occurs because the optimality conditions (3) are based on gradient information. In the applications described in this paper we assume that all gradients can be calculated accurately. Of course, when analytical gradients are available there should be few problems (e.g., for equation-based models). On the other hand, if "black-box" procedures are evaluated (e.g., in modular flowsheeting), then gradients frequently need to be approximated by finite difference; a suitable stepsize needs to be carefully determined to balance approximation errors. Here, truncation errors (due to truncated Taylor series terms) increase directly with increasing stepsize, while roundoff errors (due to machine precision and internal convergence loops) increase inversely with the stepsize. Thus, while approximation error cannot be eliminated with finite differences, it can be minimized with additional function evaluations and analysis for each particular application. Workable gradient calculation strategies for flowsheet optimization have been described by Kaijaluoto (1985) and Vasantharajan and Biegler (1988b).

Finally, a great advantage to SQP is that it does not require convergence of the equality constraints, $h(x) = 0$, at intermediate points. Consequently, the process model (or at least the part directly incorporated into the optimization problem) can be solved simultaneously with the optimization problem. In the next section we discuss the application of the SQP algorithm to flowsheet optimization. Here, if the number of variables in the optimization problem is small, application is straightforward. On the other hand, when the number of variables, n, becomes large ($n > 100$, say), special-purpose extensions to SQP are required. These are discussed in the remainder of this section.

A. LARGE-SCALE SUCCESSIVE QUADRATIC PROGRAMMING

Derived from the perspective of a Newton method, SQP has superlinear convergence properties (under mild conditions) and therefore generally requires only a few function evaluations. However, when the size of the nonlinear program becomes large, the cost of the optimization is dominated by the Newton (or QP) step. Here, the basic SQP method exhibits the following inefficiencies. First, the quasi-Newton Hessian approximation for the B matrix is dense; for a large number of variables, considerable storage and computational overhead are required. Moreover, many current QP solvers are also based on dense matrix factorizations. These can be inefficient for large problems.

OPTIMIZATION STRATEGIES FOR COMPLEX PROCESS MODELS 203

To extend SQP to large systems, two approaches have emerged recently. First, advantage has been taken of quasi-Newton updates so that they are stored and evaluated in an efficient manner. Also, quadratic programming methods have recently been tailored to take advantage of system sparsity. Both Nickel and Tolle (1989) and Mahidhara and Lasdon (1989) describe the development and testing of such methods by using positive definite BFGS (Broyden–Fletcher–Goldfarb–Shanno) updates (see Gill *et al.*, 1981). These updates can also be exploited by limited memory algorithms (Liu and Nocedal, 1988) where update vectors are stored only for the last few interations. The dense Hessian matrix is therefore never constructed directly; instead, vector-matrix multiplications are reconstructed from the update vectors whenever they are required in the QP step. Mahidhara and Lasdon (1989) applied this approach to a number of examples drawn from economic models, as well as problems in optimal control with many degrees of freedom.

For process optimization problems, the sparse approach has been further developed in studies by Kumar and Lucia (1987), Lucia and Kumar (1988), and Lucia and Xu (1990). Here they formulated a large-scale approach that incorporates indefinite quasi-Newton updates and can be tailored to specific process optimization problems. In the last study they also develop a sparse quadratic programming approach based on indefinite matrix factorizations due to Bunch and Parlett (1971). Also, a trust region strategy is substituted for the line search step mentioned above. This approach was successfully applied to the optimization of several complex distillation column models with up to 200 variables.

The second approach exploits the fact that while the optimization problem with m active constraints and n variables is generally large, few degrees of freedom are generally present. Thus, we can apply decomposition strategies to reduce the size of the quadratic programming subproblem (7). Decomposed or reduced SQP algorithms based on range and null space factorizations have been considered in the optimization literature over the past decade (Murray and Wright, 1978; Gill *et al.*, 1981; Coleman and Conn, 1982; Gabay, 1982). This approach considers a much smaller quadratic program in the reduced space, where linearizations of the active constraints are always satisfied. Here the number of variables in this QP is equal to the degrees of freedom of the problem, $n - m$. Linear equations are then solved to compute the range space move to a feasible point in the linear subspace of the m active constraints. Several studies have shown (Powell, 1978; Gabay, 1982; Nocedal and Overton, 1985) that these decomposed methods can have the same theoretical convergence properties as full space methods that use (7). Finally, Gurwitz and Overton (1989) give a numerical comparison of several variations of this approach on small optimization problems with inequality constraints.

Decomposition algorithms can be derived from the linear system that results from the optimality conditions for (7). Considering only m equality constraints and n variables for the moment, we have the following linear equations:

$$B^k d + \nabla h(x^k)\lambda = -\nabla F(x^k),$$
$$\nabla h(x^k)^T d = -h(x^k). \tag{8}$$

Now consider an $n \times (n - m)$ projection matrix Z that has the property $Z^T \nabla h = 0$. This matrix can be created either by an orthonormal factorization of ∇h or simply by partitioning the x variables into u and v (decision and dependent variables, respectively) and ∇h^T into $[\nabla_u h^T \nabla_v h^T]$. Z^T is then given by $[I \ - \nabla_u h(\nabla_v h)^{-1}]$. We also specify an $n \times m$ matrix Y so that $Q = [Z \ Y]$ is a nonsingular $n \times n$ matrix. We can now substitute the search direction by

$$d = Z d_z + Y d_y$$

and premultiply the first row in Eq. (8) by Q^T. The resulting system of linear equations at iteration k is (superscript k suppressed for brevity):

$$\begin{bmatrix} Y^T B Y & Y^T B Z & Y^T \nabla h \\ Z^T B Y & Z^T B Z & 0 \\ \nabla h^T Y & 0 & 0 \end{bmatrix} \begin{bmatrix} d_y \\ d_z \\ \lambda \end{bmatrix} = - \begin{bmatrix} Y^T \nabla F \\ Z^T \nabla F \\ h \end{bmatrix}. \tag{9}$$

For ∇h with linear independent columns the $m \times m$ matrix $Y^T \nabla h$ is nonsingular. Note that each vector can be solved sequentially starting from the bottom row. Only $Y^T \nabla h$ and h are required for d_y, and only $Z^T B Y, Z^T B Z$, and $Z^T \nabla F$ are required for d_z once d_y is known. Finally, the exact value of the λ vector is only meaningful at the optimum, i.e., when both d_z and d_y are zero and therefore $Y^T B Y$ and $Y^T B Z$ become unimportant. Several studies (Wright, 1976; Nocedal and Overton, 1985) therefore show that first-order estimates of λ (with $Y^T B Y$ and $Y^T B Z$ set to zero) do not affect the rate of convergence, and thus the large matrices $Y^T B Y$ ($m \times m$) and $Y^T B Z$ ($m \times (n - m)$) need not be supplied.

Also, when the constraints are linear, the method always remains in the subspace of the constraints (for a feasible starting point) and d_y remains zero. Consequently, for this case the $Z^T B Y$ matrix is unimportant and only the $(n - m) \times (n - m)$ $Z^T B Z$ matrix needs to be supplied or calculated. While this is not true for nonlinear constraints, it is still convenient to deal only with $Z^T B Z$ and set $Z^T B Y$ to zero. Moreover, as long as d_y remains relatively small, a reasonable superlinear rate of convergence can still be maintained for this decomposition (Nocedal and Overton, 1985).

Dealing with $Z^T B Z$ directly has several advantages if $n - m$ is small. Here the matrix is dense and the sufficient conditions for local optimality require that $Z^T B Z$ be positive definite. Hence, the quasi-Newton update formula can be applied directly to this matrix. Several variations of this basic algorithm

OPTIMIZATION STRATEGIES FOR COMPLEX PROCESS MODELS   205

have been proposed over the past decade. In the optimization literature, orthonormal factorizations of Z and Y have been applied (where $Z^T Y = 0$, $Z^T Z = I$ and $Y^T Y = I$) to simplify the analysis. Also, most of these reduced space methods choose the active constraints prior to decomposition (the EQP approach) and simply solve them as linear equations that result from (9).

In process engineering, Berna et al. (1980) constructed a reduced SQP method but maintained factorized updates of the full Hessian matrix B. Here Z^T was given implicitly by $[I \ -\nabla_u h (\nabla_v h)^{-1}]$, which is easily constructed using a sparse matrix algorithm. Locke et al. (1983) used the same Z matrix and applied the decomposition outlined above by discarding $Y^T BY$, $Y^T BZ$, and $Z^T BY$. For simplicity they also set $Y = [0 \ I]^T$. Both studies demonstrated this approach on small example problems. Vasantharajan and Biegler (1988a) also tailored this algorithm to deal with L/U factorizations, but required that the columns of Y and Z be orthogonal to each other, i.e., $Y^T Z = 0$. This modification ensures that d_y is an orthogonal projection into the space of the linearized active constraints. Unlike the previous choice for Y, this "least squares" projection ensures that Yd_y is as small as it can be. Hence, dropping the $Z^T BY$ matrix in (9) is not as serious with this approach. Also, the least squares projection is not as sensitive to the partitioning of independent and dependent variables as the approach by Locke et al. (1983).

These studies apply decompositions that give SQP the potential to solve large, sparse problems with few degrees of freedom. It should also be mentioned that an alternative to SQP that has been applied to sparse, mostly linear problems is the MINOS/Augmented package (Murtagh and Saunders, 1978). In this approach, the nonlinearly constrained problem is solved as a sequence (major iterations) of linearly constrained subproblems, which have as their objective function a projected augmented Lagrangian function:

$$L^*(x, \mu, \lambda) = F(x) + g_A(x)^T \mu + h(x)^T \lambda + \frac{1}{2}\rho(h(x)^T h(x) + g_A(x)^T g_A(x)). \quad (10)$$

The solution of each linearly constrained problem can be found by applying unconstrained algorithms (minor iterations), such as quasi-Newton methods, projected in the space of the active linear constraints. Because a reduced space nonlinear program is solved for each major interation, instead of its quadratic approximation, this strategy generally requires more function and gradient evaluations than SQP for nonlinear problems. However, it incorporates efficient sparse implementations, especially with respect to matrix factorizations for updating Z from one active set to another. Consequently, the MINOS algorithm often performs well on large-scale process optimization problems, especially if the constraints are "nearly" linear.

In addition to the studies by Nocedal and Overton (1985) and Gurwitz and Overton (1989), Vasantharajan and Biegler (1988a) and Vasantharajan et al.

(1990) provide a demonstration of the effectiveness of reduced SQP strategies for nonlinear programming. In the last study this approach is very effective on nonlinear process optimization problems of up to 1,000 variables. This approach has also been applied to large dynamic optimization problems described in Section V. Before leaving this section, we note that there is considerable research activity both in the development of sparse SQP implementations and in decomposition strategies. The former methods are especially well suited for problems with many degrees of freedom and structured sparse Hessian matrices. Application of decomposition strategies here would destroy this structure.

For many process optimization problems with few degrees of freedom, on the other hand, decomposition strategies seem to be much easier to apply and have the same theoretical convergence rate (two-step Q-superlinear) as the full space method given by (7) (see Nocedal and Overton, 1985). Vasantharajan and Biegler (1988a) support this property with a numerical comparison that shows that reduced space methods generally require fewer iterations than conventional full space methods. Moreover, the decomposition approach can be tailored to the specific problem through the appropriate choices of Y and Z projection matrices. Note from (9) that many choices of Y and Z will yield equivalent search directions as long as both the vector $Z^T B Y d_y$ and the matrix $Z^T B Z$ are provided. Since d_y is available before calculation of d_z and $Z^T B Y d_y$ is the directional derivative of $Z^T \nabla L$ along $Y d_y$, this term can be estimated even if the $Z^T B Y$ matrix is not directly available.

B. SQP FOR PARAMETER ESTIMATION

A special case of these decomposition properties occurs for parameter estimation problems with small residuals (good model/measurement agreement). Using a least squares objective function of the form

$$F(x) = \sum_j b_j e_j^T W e_j, \qquad (11)$$

where

$e_j = y_j - \bar{y}_j$,
$\bar{y}_j = j$th data point,
$y_j = f(y, \theta)$, calculated value,
$x^T = [y^T, \theta^T]$,
θ = parameters to be estimated,
W = weighting matrix for residuals,

one can now exploit the property that $\partial F/\partial y \approx 0$ at the optimum (from

$e_j \approx 0$) and the estimate of λ at the optimum (see (9)) is also zero. Consequently, the Hessian of the Lagrange function can be given by the Hessian of the objective function:

$$\nabla_{xx}L = \nabla_{xx}F + \sum_{i=1}^{m}\nabla_{xx}h_i\lambda_i \approx \nabla_{xx}F, \qquad (12)$$

which is readily available. Consequently, Z^TBZ and Z^TBYd_y are easily calculated and yield substantial improvements even over the general-purpose reduced SQP method. Here Tjoa and Biegler (1991) have also modified this formulation for examples where the residuals are not necessarily small. An adaptive algorithm thus results for both cases that is based on the unconstrained algorithm of Fletcher and Xu (1987).

As a result of this approach, a fast rate of convergence (quadratic, in some cases) can be achieved. Tjoa and Biegler (1991) applied this approach to parameter estimation of ODE models. This task occurs most often in the determination of rate constants from batch reactor data. Here the orthogonal collocation approach of Section V is applied to discretize the model, and a constrained nonlinear program results. As a result of the decomposition and tailored B matrix, Tjoa and Biegler (1991) developed an algorithm that leads to drastic reductions (60 to 80%) in the computational effort when compared to the method in Vasantharajan *et al.* (1990). Moreover, since a Hessian matrix is used that is structured to the optimization problem, the method is also more reliable and less sensitive to poor problem scaling.

Nevertheless, in process design problems where model evaluations are frequently expensive, application of the SQP algorithm alone has provided great impetus toward solving optimization problems efficiently. Flowsheet optimization, one class of problems in this category, is discussed in the next section.

III. Flowsheet Optimization

Steady-state process simulation or process flowsheeting has become a routine activity for process analysis and design. Such systems allow the development of comprehensive, detailed, and complex process models with relatively little effort. Embedded within these simulators are rigorous unit operations models often derived from first principles, extensive physical property models for the accurate description of a wide variety of chemical systems, and powerful algorithms for the solution of large, nonlinear systems of equations.

Process flowsheeting tools can usually be classified within two categories. In the *equation-oriented* framework, the simulation system creates and collects a single large set of equations that is then passed to a large, sparse equation solver. Because the equations are solved with a general-purpose algorithm, this approach is very flexible for problem specifications. Provided that the desired process variables can be accessed, virtually any relations can be incorporated among them.

The older *modular simulation mode*, on the other hand, is more common in commerical applications. Here process equations are organized within their particular unit operation. Solution methods that apply to a particular unit operation solve the unit model and pass the resulting stream information to the next unit. Thus, the unit operation represents a procedure or module in the overall flowsheet calculation. These calculations continue from unit to unit, with recycle streams in the process updated and converged with new unit information. Consequently, the flow of information in the simulation systems is often analogous to the flow of material in the actual process. Unlike equation-oriented simulators, modular simulators solve smaller sets of equations, and the solution procedure can be tailored for the particular unit operation. However, because the equations are embedded within procedures, it becomes difficult to provide problem specifications where the information flow does not parallel that of the flowsheet. The earliest modular simulators (the sequential modular type) accommodated these specifications, as well as complex recycle loops, through inefficient iterative procedures. The more recent simultaneous modular simulators now have efficient convergence capabilities for handling multiple recycles and nonconventional problem specifications in a coordinated manner.

Currently, many commercial simulators are neither entirely equation-oriented nor sequential modular. All of them contain some procedures, whether for physical property or unit operations calculations. For process optimization, however, the structure of the process simulator needs to be considered carefully. For simulators that apply simultaneous convergence methods in an equation-oriented mode, the large-scale SQP algorithm can be applied directly, since all of the information required for optimization is readily available. Since the resulting optimization problem is rather large with this mode, some large-scale form of SQP is required, or the large-scale MINOS approach is applied. In fact, both the ASCEND (Piela *et al.*, 1991) and SPEEDUP (Pantelides *et al.*, 1988) equation-oriented simulators are interfaced to MINOS; the related QUASILIN (Hutchison *et al.*, 1988) simulator incorporates an SQP decomposition strategy for optimization.

Modular simulators are frequently constructed on three levels. The lowest level consists of thermodynamics and other physical property relations that are accessed frequently for a large number of flowsheeting utilities (flash calculations, enthalpy balances, etc.). The next level consists of unit operations models as described above. The highest level then deals with the sequencing and convergence of the flowsheet models. Here, simultaneous optimization and simulation can be applied to deal with the flowsheet topology. Consider the block diagram given in Fig. 1. The simulation problem written at the level of flowsheet topology is given by

$$h(u, v) = v - w(u, v) = 0,$$
$$c(u, v) = 0, \qquad (13)$$

where the first equation refers to convergence of the recycle streams and the second refers to a user-specified design specification. The variables x are partitioned into u and v such that u refers to flowsheet decision variables, v is made up of the guessed recycle stream, and $w(u, v)$ is the calculated recycle stream given inputs u and v. The optimization problem can be written as a direct extension of this problem:

$$\begin{align} \text{Min} \quad & F(u, v) \\ \text{s.t.} \quad & h(u, v) = v - w(u, v) = 0 \\ & c(u, v) = 0 \\ & g(u, v) \leq 0 \end{align} \qquad (14)$$

Since the number of variables is generally not large and function evaluations from the flowsheeting program are frequently the dominant computational cost, the SQP algorithm can be applied directly to this problem. This approach to flowsheet optimization has been reported in numerous studies. Biegler and Hughes (1982), Chen and Stadtherr (1985), Biegler and Cuthrell (1985), and Kaijaluoto (1984) describe similar SQP approaches for

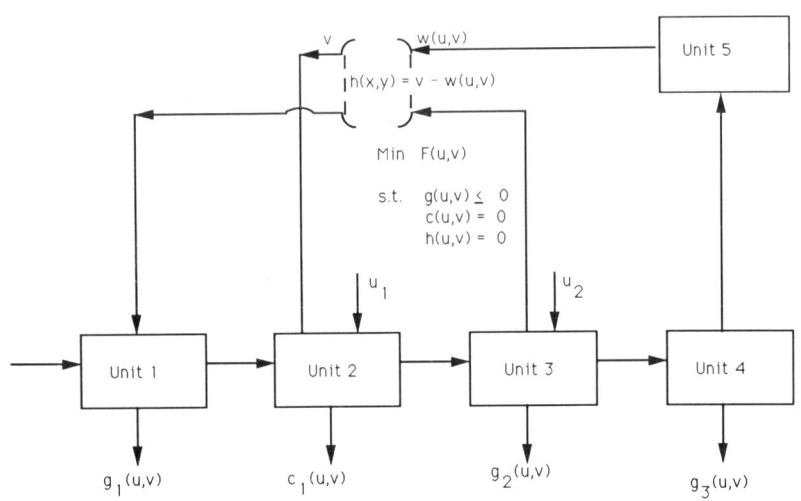

FIG. 1. Structure and variables for fixed flowsheet optimization (NLP).

(14). Further extensions to improve reliability of the algorithm were discussed by Lang and Biegler (1987) and Kisala *et al.* (1987). In both of these studies, the importance of preventing large violations of the recycle constraints was considered. This led to a hybrid algorithm where partial recycle convergence was applied (when necessary) over the course of the optimization. The algorithm of Lang and Biegler has recently been implemented in the FLOWTRAN simulator (Seader *et al.*, 1987) for educational use; numerous commercial simulators have similar optimization capabilities. See Biegler (1989) for a review.

A. Case Study for Process Optimization

To demonstrate the effectiveness of this approach, consider the optimization of the ammonia process given in Fig. 2. This process is a single-loop design with a three-stage adiabatic flash separation. Further details of the process can be found in Lang and Biegler (1987).

Nitrogen and hydrogen-rich feeds are mixed, compressed and further combined with a high-pressure recycle stream. The mixture is then reacted at high pressure and temperature. Here the reactor model is assumed to be

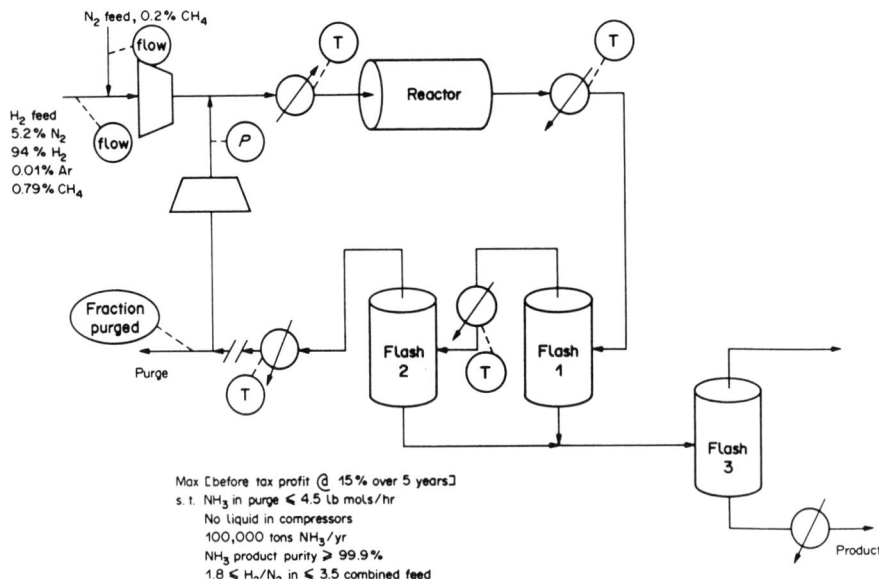

FIG. 2. Ammonia process case study. Reprinted with permission from *Comp. & Chem. Eng.*, **11**, 143–152, Y.-D. Lang and L. T. Biegler, "A Unified Algorithm for flowsheet optimization," Copyright 1987, Pergamon Press PLC.

equilibrium-limited. A description of the modeling equations and the corresponding FLOWTRAN subroutine can be found in Seader et al. (1987). The reactor effluent is cooled in several stages (not shown) and flashed in two stages at high pressure; overhead vapor is recycled back to the reactor. Liquid from both flash units is then sent to a low-pressure flash separation to obtain pure final product. This problem was simulated with the FLOWTRAN simulator using the SCOPT optimization block.

Statement of the optimization problem is also given in Fig. 2. Here the objective function is the before-tax profit over a five-year life and a 15% rate of return. Capital, raw material, and utility costs were estimated using FLOWTRAN cost blocks. In addition a number of design constraints are placed on the optimization problem. The most interesting one is the specification of the production rate (100,000 tons/year) rather than a fixed feed. In the modular framework, this would normally require an iteration loop around the entire flowsheet, a very expensive calculation. However, in the optimization problem, the feed flowrates are given as decision variables, and imposition of this constraint is straightforward. Finally, a list of the eight decision variables and the tear variables is given in Table I.

The optimization problem was solved in only five SQP iterations and only 139.3 CPU seconds on a VAX 8650, by using the simultaneous approach. Lang and Biegler also present a comparison of related methods and indicate that slight improvements can be achieved by using partial flowsheet convergence strategies. The results of this optimization problem are also presented in Table I. Here the objective function increases from an initial value of \$20.66 × 10^6 to \$24.93 × 10^6. Note that the inlet temperatures of the reactor and the flash units remain at their lower bounds. This example is typical of the performance that can be achieved. As shown in numerous studies (Lang and Biegler, 1987; Kisala et al., 1987; Chen and Stadtherr, 1985), optimization with process simulators generally requires about an order of magnitude more computational effort than the simulation problem. In fact, for many problems, such as for this example, the effort is much less than this estimate.

B. ADVANCED APPLICATIONS OF PROCESS OPTIMIZATION

As shown in this example, flowsheet optimization, which previously required several hundred simulation time equivalents (Gaines and Gaddy, 1971), now requires about 10. Thus, many complex design problems, which can be treated through formulation and solution of an optimization problem, can now be handled routinely. In the remainder of this section, we briefly summarize three important problems of this type:

- flowsheet optimization and heat integration,
- structural design and optimization,
- optimization of reactor-based flowsheets.

TABLE I
Results of Ammonia Flowsheet Optimization Problem*

No. Item	Optimum	Starting Point	Lower Bound	Upper Bound
A. Objective Function (10^6/yr)	26.9286	20.659		
B. Design Variables				
1. Inlet temperature of reactor (°F)	400	400	400	600
2. Inlet temperature of first flash (°F)	65	65	65	100
3. Inlet temperature of second flash (°F)	35	35	35	60
4. Inlet temperature of recycle compressor (°F)	80.52	107	60	400
5. Purge fraction (%)	0.0085	0.01	0.005	0.1
6. Inlet pressure of reactor (psia)	2,163.5	2,000	1,500	4,000
7. Flowrate of feed 1 (lb mol/h)	2,629.7	2,632.0	2,461.4	3,000
8. Flowrate of feed 2 (lb mol/h)	691.78	691.4	643	1,000
C. Tear Variables				
1. Flow rate (lb mol/h)				
N_2	1,494.8	1,648		
H_2	3,618.4	3,676		
NH_3	524.2	424.9		
Ar	175.3	143.7		
CH_4	1,989.1	1,657		
2. Temperature (°F)	80.52	60		
3. Pressure (psia)	2,080.4	1,930		

*Reprinted with permission from *Comp. & Chem. Eng.*, 11, 143–152, Y.-D. Lang and L. T. Biegler, "A Unified Algorithm for Flowsheet Optimization," Copyright 1987, Pergamon Press PLC.

Heat exchanger network synthesis (HENS) has been an important research problem for the last two decades and has a rich literature devoted to it (Linnhoff et al., 1982; Britt et al., 1990). In particular, a useful concept for HENS is the derivation of the minimum utility cost before deriving the actual network. However, the task of designing or retrofitting a heat exchanger network is often considered after the flowsheet has been designed and optimized. This is clearly suboptimal, since utility requirements and heat duties are considered in the optimization problem and the HENS problem is strongly influenced by the overall design problem.

Duran and Grossmann (1986) first considered this problem through an

NLP formulation where constraints were added to allow for the targeting of the heat exchange network. Lang et al. (1988) applied this approach to complex flowsheeting optimization problems, including the ammonia process described above. Here it was shown that consideration of heat integration at the optimization stage leads to greater conversion of raw material to product and smaller raw materials cost. Conceptually, the simultaneous problem can be formulated by considering the targeting procedure for HENS as a module embedded within the flowsheeting program; the resulting optimization problem is thus straightforward. However, because this procedure is not a differentiable function, it needs to be recast as a set of inequalities where the kinks are identified explicitly and approximated by smooth functions. Most recently, Kravanja and Grossmann (1990) developed an integrated targeting model for HENS that also accounts for capital cost of the heat exchange network as well as the utility costs.

Flowsheet optimization is also regarded as a key task in the structural optimization of a flowsheet. As a described in the introduction, structural optimization for process design can be formulated as a mixed integer nonlinear program (MINLP). This then allows for addition or replacement of existing units, and consideration of a number of design options simultaneously. In these formulations individual units are turned on and off over the course of the optimization, as suggested by the MINLP master problem.

Harsh et sl. (1989) applied the MINLP approach to debottlenecking a flowsheet for retrofit designs. Here integer decisions were used to model the expansion or replacement of existing equipment. Using the FLOWTRAN simulator for an ammonia process retrofit, an improvement in profit of over seven million dollars was realized with an investment of only about $100,000. Similarly, Caracotsios and Petrellis (1989) developed the ACCOPT system, where an MINLP approach was applied to the ASPEN simulator for general-purpose flowsheeting problems. This study includes a number of flowsheet optimization problems, including determination of an optimal feed tray location for a distillation column. Despite these studies, the application of MINLP to flowsheeting systems is not entirely straightforward. First, convexity assumptions that are required for solution of the master problem (for discrete decisions) cannot always be satisfied for flowsheet optimization. Thus, some relaxation of the master problem is often required. Second, the elimination of existing units and addition of new units is difficult to incorporate into the master problem when no prior information is available for these units. Consequently, an additional set of optimization calculations may be required to provide reliable lower-bound information for the master problem. Here Benders decomposition or the modeling decomposition strategy of Kocis and Grossmann (1988) seems especially useful. Finally, since the MINLP is solved as a series of NLPs, the overall solution can still be time-consuming. Consequently, the further improvement of MINLP (master problem) and NLP algorithms is still an important research activity.

Before leaving this section we consider a slightly different optimization problem that may also be expensive to solve. In flowsheet optimization, the process simulator is based almost entirely on equilibrium concepts. Separation units are described by equilibrium stage models, and reactors are frequently represented by fixed conversion or equilibrium models. More complex reactor models usually need to be developed and added to the simulator by the engineer. Here the modular nature of the simulator requires the reactor model to be solved for every flowsheet pass, a potentially expensive calculation. For simulation, if the reactor is relatively insensitive to the flowsheet, a simpler model can often be substituted. For process optimization, a simpler, insensitive model will necessarily lead to suboptimal (or even infeasible) results. The reactor and flowsheet models must therefore be considered simultaneously in the optimization.

The large-scale optimization tools discussed above become important for this problem because these simultaneous approaches necessarily require that the models be "opened up" to the optimization problem. Moreover, for reactor optimization, the ordinary differential equation (ODE) model needs to be represented by a set of algebraic equations; hence, some level of discretization needs to be performed. To demonstrate this concept Cuthrell and Biegler (1986) and Vasantharajan and Biegler (1988b) combined a plug flow reactor model with a process flowsheeting system by applying orthogonal collocation to the reactor and adding the resulting equations to the optimization problem. This is illustrated in Fig. 3. Here the original optimization problem,

$$\begin{aligned}
\text{Min} \quad & F(u, v, Z(t)) \\
\text{s.t.} \quad & g(u, v, Z(t)) \leqslant 0 \\
& h_t(u, v) = v_t - w_t(u, v) = 0 \\
& h_f(u, v) = v_f - w_f(u, v) = 0 \\
& h_e(u_r, v, Z(t)) = v_e - w_e(u_r, v_f, Z(t)) = 0 \\
& c(u, v, Z(t)) = 0 \\
& \dot{Z}(t) = f(u_r, v_f, Z(t)) \\
& Z(t_0) = Z_0(v_f) \\
& u^L \leqslant u \leqslant u^U \\
& v^L \leqslant v \leqslant v^U
\end{aligned} \quad (15)$$

is rewritten with $Z(t)$ parameterized as a polynomial function; v and u are the stream vector and decision variables, respectively. The differential equations are rewritten as algebraic constraints at a discrete set of collocation points, after substituting the expression for $Z(t)$. Thus, variables u, v and the parameterization for $Z(t)$ are incorporated into the NLP along with algebraic

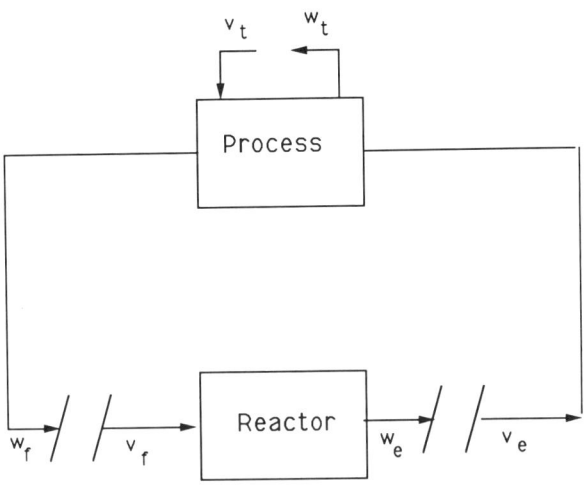

FIG. 3. Reactor-based flowsheet torn for simultaneous optimization. Reprinted with permission from *Chem. Eng. Res. Des.* **66**, 396, S. Vasantharajan and L. T. Biegler, "Simultaneous Solution of Reactor Models within Flowsheet Optimization," 1988.

constraints from the ODEs. A detailed presentation of this approach is deferred to Section V.

Note that this formulation illustrates an interesting trade-off for the optimization problem. In the modular mode the optimization problem remains fairly small and function evaluations (e.g., the reactor model) are expensive. With the simultaneous formulation, the model becomes a set of equations whose right-hand sides are much cheaper to evaluate, but the size of the optimization problem increases. Nevertheless, Vasantharajan and Biegler (1988b) showed that, even without SQP decomposition, the simultaneous approach for the reactor was 38% cheaper for the *entire flowsheet optimization* than the modular approach. Moreover, the number of function evaluations for the reactor model decreased by over an order of magnitude.

Current progress with flowsheet optimization not only demonstrates the advantage of successive quadratic programming and simultaneous formulations, but also raises many interesting problem formulations for future research. The three applications discussed above are now starting to be considered by industrial practitioners (Britt *et al.*, 1990, and Caracotsios and Petrellis, 1989). Future work will also deal with the consideration of optimization under uncertainty, as well as optimal process designs to handle a variety of production rates or feedstocks. These can be treated as multiperiod problems, which are currently quite expensive to solve, but can be made tractable through reformulation and tailored decomposition procedures.

The application of simultaneous optimization to reactor-based flowsheets leads us to consider the more general problem of differentiable/algebraic optimization problems. Again, the optimization problem needs to be reconsidered and reformulated to allow the application of efficient nonlinear programming algorithms. As with flowsheet optimization, older conventional approaches require the repeated execution of the differential/algebraic equation (DAE) model. Instead, we briefly describe these conventional methods and then consider the application and advantages of a simultaneous approach. Here, similar benefits are realized with these problems as with flowsheet optimization.

IV. Differential-Algebraic Optimization

Differential-algebraic models appear in all aspects of chemical and process engineering. Here the optimization problem can be as fundamental as the design of a single catalyst pellet or as complex as the optimal operation and control of a chemical plant (Renfro et al., 1987). However, like flowsheeting problems, without proper formulation, even the simplest and smallest of these optimization problems can be difficult and time-consuming to solve.

The general optimization problem can be represented as follows:

$$\begin{aligned}
\min_{x, U(t), Z(t)} \quad & F(x, U(t), Z(t)) \\
\text{s.t.} \quad & c(x, U(t), Z(t)) = 0 \\
& g(x, U(t), Z(t)) \leq 0 \\
& \dot{Z}(t) = f(x, U(t), Z(t), t), \quad t \in [0, t_f] \\
& Z(0) = Z_0 \\
& x^L \leq x \leq x^U \\
& U^L \leq U(t) \leq U^U \\
& Z^L \leq Z(t) \leq Z^U
\end{aligned} \quad (16)$$

where

F = objective function,
g = algebraic inequality constraint vector,
c = algebraic equality constraint vector,
x = parameter, decision variable vector,
$Z(t)$ = state profile vector,
$U(t)$ = control profile vector,
x^L, x^U = variable bounds,
U^L, U^U = control profile bounds,
Z^L, Z^U = state profile bounds.

Here the problem is given as an initial value problem, although the concepts can easily be generalized to boundary value problems and even partial differential equations. Note also that both continuous variables, x (parameters), and functions of time, $U(t)$ (control profiles), are included as decision variables. Constraints can also be enforced over the entire time domain and at final time.

Problems in reactor or separator design generally involve only parameters, x, that determine equipment sizes. This usually generalizes to many process design problems as well. Similarly, other problem classes, such as reactor network optimization (Achenie and Biegler, 1990; Kokossis and Floudas, 1990) and parameter estimation (Biegler et al., 1986; Tjoa and Biegler, 1991) are described by parameter optimization problems. On the other hand, optimal control profiles are encountered most often in process operation and control where the independent variable in the ODE model is time. These may also occur in steady-state applications to reactor design or specialized unit operations where length is the independent variable. Also, unless the variational conditions yield an explicit condition for $U(t)$, the control profile is usually redefined as a function of a finite set of parameters, such as a piecewise constant function. Although it is suboptimal, the parameterization of $U(t)$ allows for the straightforward application of nonlinear programming methods. Finally, problem (16) becomes especially difficult to solve in the presence of state variable inequality and equality constraints. Often, these can only be handled by simultaneous methods.

Methods for solving at least subclasses of (16) can be divided into three basic types:

- iterative methods based on *variational conditions*,
- *feasible path* nonlinear programming methods,
- *simultaneous* nonlinear programming methods.

Here we briefly describe the first two approaches with the aim of contrasting them to simultaneous methods presented in Section V. In that section, the advantages of simultaneous methods also lead us to consider some open questions and concentrate on current work in this area.

A. Variational Methods

For engineering systems, variational methods for (16) have usually been described in the context of optimal control. Considerable research activity was devoted to this topic over the past three decades. Perhaps the most familiar engineering text on optimal control is Bryson and Ho (1975). While written from the perspective of mechanical and aeronautical problems, its theory mirrors the large amount of analytical research in chemical engineering that took place in the 1960s and early 1970s, with many applications in reactor design and nonsteady-state processing.

This analytical approach to solving small design problems naturally led to iterative algorithms based on variational conditions. Using Pontryagin's maximum principle, various *control vector iteration* algorithms were proposed that involved the solution of model equations forward in time, adjoint equations backward in time, and intermittent updating of the control profile to maximize the Hamiltonian function. Using the gradient of the Hamiltonian with respect to $U(t)$, familiar unconstrained algorithms for nonlinear programming were extended to the infinite dimensional case, e.g.,

- steepest descent (Bryson and Denham, 1962), *et al.*,
- conjugate gradients (Lasdon, *et al.*, 1967),
- quasi-Newton method (Ladson, 1970; Kelley and Sachs, 1987),
- Newton method (Luus and Lapidus, 1967).

Jones and Finch (1984) provide an extensive comparison of both first- and second-order methods on several process examples.

These methods are efficient for problems with initial-value ODE models without state variable and final time constraints. Here solutions have been reported that require from several dozen to several hundred model (and adjoint equation) evaluations (Jones and Finch, 1984). Moreover, any additional constraints in this problem require a search for their appropriate multiplier values (Bryson and Ho, 1975). Usually, this imposes an additional outer loop in the solution algorithm, which can easily require a prohibitive number of model evaluations, even for small systems. Consequently, control vector iteration methods are effective only when limited to the simplest optimal control problems.

B. NLP Methods with an Embedded ODE Model

On the other hand, the optimal control problem with a discretized control profile can be treated as a nonlinear program. The earliest studies come under the heading of control vector parameterization (Rosenbrock and Storey, 1966), with a representation of $U(t)$ as a polynomial or piecewise constant function. Here the model is solved repeatedly in an inner loop while parameters representing $U(t)$ are updated on the outside. While "hill climbing" algorithms were used initially, recent efficient and sophisticated optimization methods require techniques for accurate gradient calculation from the DAE model.

Using this nonlinear programming approach (also termed the embedded model or feasible path approach), we denote as x the vector of parameters representing $U(t)$ as well as the parameters x. For example, if $U(t)$ is assumed piecewise constant over a variable distance, we include u_i and t_i in x. Problem

(16) then becomes the following nonlinear programming problem:

$$\text{Min}_{x} \ F(x)$$
$$\text{s.t.} \ g(x) \leq 0$$
$$x^L \leq x \leq x^U$$

and the ODE model:

$$\dot{Z}(t) = f(x, U(t), Z(t), t), \qquad t \in [0, t_f], \tag{17}$$
$$Z(0) = Z_0,$$

is solved implicitly in order to evaluate the $F, c,$ and g at final time.

Note that independent variable, time, disappears from this problem. While final time constraints in (16) appear naturally in (17), other constraints that need to be enforced over the time domain are difficult to handle. For example, Sargent and Sullivan (1977) converted these to final time constraints by integrating the square of the constraint violations and forcing these to be less than a tolerance at final time; this however, leads to degeneracies in solving the nonlinear program.

Gradient calculations for the x variables are obtained from implicit reformulations of the DAE model. Clearly the easiest, but least accurate, way is simply to re-solve the model for each perturbation of the parameters. Sargent and Sullivan (1977, 1979) derived these gradients using an adjoint formulation. In addition, they were able to accelerate the adjoint computations by retaining the information from the model solution (the forward step) for the adjoint solution in the backward step. This approach was later refined for variable stepsize methods by Morison (1984). The adjoint approach to parameterized optimal control was also used by Jang et al. (1987) and Goh and Teo (1988).

Using similar parameterizations for the control profile, a number of researchers evaluated gradients with respect to x through sensitivity equations from the ODE model. This was considered by Caracotsios and Stewart (1985) in solving parameter estimation problems. In describing universal dynamic matrix control (UDMC), Morshedi (1986) applied a simplified sensitivity equation approach, which relies on local model linearizations to develop state variable sensitivities analytically (see Morshedi et al., 1986). Also, for the optimization problem, Morshedi enforced state variable constraints only at the end of each time step, and did not require the constraint transformation of Sargent and Sullivan. This latter approach was implemented in a moving time horizon approach for on-line optimization and control.

Either approach results in gradient calculations with costs proportional to problem size; effort for evaluating gradients with adjoint approaches is proportional to the number of (objective and constraint) functions evaluated at final time, while effort for sensitivity equations is proportional to the dimension of x. Consequently, the choice of gradient calculation approach depends on the structure of the particular problem. Nevertheless, both Caracotsios and Stewart (1985) and Sargent and Sullivan (1979) demonstrated that gradient calculation approaches can be accelerated considerably by tailoring the ODE solver to include sensitivity or adjoint equations.

The embedded model approach represented by problem (17) has been very successful in solving large process problems. Sargent and Sullivan (1979) optimized feed changeover policies for a sequence of distillation columns that included seven control profiles and 50 differential equations. More recently, Mujtaba and Macchietto (1988) used the SPEEDUP implementation of this method for optimal control of plate-to-plate batch distillation columns.

Despite its success, the embedded model approach still requires repeated solution of the process model (and sensitivities). For large processes or for processes that require the solution of rigorous underlying procedures, this approach can become expensive. Moreover, for stiff or otherwise difficult systems, this approach is only as reliable as the ODE solver. The embedded model approach also offers only indirect ways of handling time-dependent constraints. Finally, the optimal solution of this approach is only as good as its control variable parameterization, which often can only be improved by *a priori* information about the specific problem. Consequently, we now consider the simultaneous approach to (16) as an alternative to solution methods for (17).

V. Simultaneous Approaches for Differential/Algebraic Optimization

Instead of parameterizing the control profile and solving (16) as a nonlinear program, the simultaneous approach begins with a parameterization of both the control *and* state variable profiles in order to solve a mathematical programming problem consisting of algebraic equations. Canon *et al.* (1970) applied this approach to simple difference equations, while Lynn *et al.* (1971) used the method of weighted residuals (MWR) to parameterize their optimal control problems. Specifically, they employed Galerkin's method to reduce a simple optimal control problem to a set of algebraic equations, which could then be solved with Newton's method. Collocation, another MWR method, was used by Neuman and Sen (1973) to solve a state variable constrained linear-quadratic (LQ) control problem. This parameterization led to the straightforward solution of a quadratic program. Tsang et al. (1977) and Oh and Luus (1977) applied this approach to more general nonlinear problems, with the former study using the GRG algorithm to solve the resulting nonlinear program.

These early approaches suffered from two drawbacks. First, simultaneous approaches lead to much larger nonlinear programs than embedded model approaches. Consequently, nonlinear programming methods available at that time were too slow to compete with smaller feasible path formulations. Second, care must be taken in the formulation in order to yield an accurate algebraic representation of the differential equations.

Nevertheless, Neuman and Sen (1973) were able to solve general, state-constrained LQ problems as quadratic programs. Thus, an extension to nonlinear problems can be made by considering successive quadratic programming. Biegler (1984) applied SQP and orthogonal collocation to a small optimal control problem; this approach showed considerably better performance than control vector iteration methods. As mentioned in Section III, Cuthrell and Biegler (1986) and Vasantharajan and Biegler (1988a) applied global collocation to the optimization of a reactor-based flowsheet. Also, Liu et al. (1988) applied this global collocation approach for the optimization of a yeast fermentation system.

Using piecewise constant control profiles and orthogonal collocation on finite elements, this approach was further developed by Renfro (Renfro, 1986; Renfro et al., 1987) to deal with much larger problems. More recent simultaneous applications that involve SQP, orthogonal collocation, and piecewise constant control profiles have been presented by Patwardhan et al. (1988) for online control, and by Eaton and Rawlings (1988) for optimization of batch crystallizers. These studies have shown that simultaneous approaches can be applied successfully to small-scale applications with complex constraints.

However, all of these studies determine only approximate or parameterized optimal control profiles. Also, they do not consider the effect of approximation error in discretizing the ODEs to algebraic equations. In this section we therefore explore the potential of simultaneous methods for larger and more complex process optimization problems with ODE models. Given the characteristics of the simultaneous approach, it becomes important to consider the following topics:

- accurate algebraic representation of the differential equation model in (16),
- accuracy and stability for profile optimization problems, especially with respect to the control profile,
- generalization to large-scale problems.

Each of these questions will be explored in the next three subsections.

222   LORENZ T. BIEGLER

A. Accuracy of Representation of ODEs

A discretization of the ODEs given in (16) can easily be developed from families of ODE methods, since their approximation and stability properties are well studied. Since only self-starting methods can be used with a simultaneous, equation-based approach, we exclude linear multistep (LMS) methods and are forced to consider implicit Runge–Kutta (IRK) methods for the parameterization of these equations. Many stable and high-order methods from this class are collocation-based (Ascher and Bader, 1986). Coincidentally, they are also easy to apply to (16). By defining state and control profiles at $\tau_i \in [0, 1]$ (e.g., the K shifted roots of orthogonal [Legendre] polynomials [see Villadsen and Michelsen, 1978]), we can parameterize these profiles as Lagrange-form polynomials over $\tau \in [0, 1]$:

$$z_{K+1}(t) = \sum_{l=0}^{K} z_l \phi_l(t), \quad \phi_l(t) = \prod_{k=0, l}^{K} \frac{(t - t_k)}{(t_l - t_k)},$$

$$u_K(t) = \sum_{l=1}^{K} u_l \psi_l(t), \quad \psi_i(t) = \prod_{k=1, l}^{K} \frac{(t - t_k)}{(t_l - t_k)}. \quad (18)$$

Here K represents the number of collocation points and $z_{K+1}(t)$ and u_K are $(K + 1)$th-order (degree $< K + 1$) and Kth-order polynomials, respectively. (Here the notation $k = 0, l$ refers to the index k starting at zero but not equal to l.)

Note that state variable profiles are one order higher than the controls because they have explicit interpolation coefficients defined at the beginning of each element. With this representation of $Z(t)$ and $U(t)$, we can extend this approach to piecewise polynomials and apply orthogonal collocation on NE finite elements (of length $\Delta \alpha_i$). This leads to the following nonlinear algebraic equations:

$$z^i_{K+1}(t) = \sum_{l=0}^{K} z_{il} \phi_l(\tau), \quad \tau \in [0, 1], t \in [\alpha_i, \alpha_{i+1}], t = \alpha_i + \Delta \alpha_i \tau,$$

where

$\alpha_i =$ position of finite element i,
$\Delta \alpha_i =$ length of element $i(\alpha_{i+1} - \alpha_i)$, $i = 1$, NE,
$z^i_{K+1}(t) =$ piecewise polynomial approximation of $Z(t)$ over element i,
$u^i_K(t) =$ piecewise polynomial approximation of $U(t)$ over element i,
$z_{il} =$ coefficient of polynomial approximation, $z^i_{K+1}(t)$,
$u_{il} =$ coefficient of polynomial approximation, $u^i_K(t)$,

and

$$R(t_{il}, \Delta \alpha_i) = \sum_{j=0}^{K} z_{ij} \dot{\phi}_j(\tau_l) - \Delta \alpha_i f(x, u_{il}, z_{il}, \tau_l) = 0, \quad (19)$$

$$l = 1, \ldots K, i = 1, \ldots, \text{NE},$$

OPTIMIZATION STRATEGIES FOR COMPLEX PROCESS MODELS 223

with

$$z_{l,0} = Z_0 \quad \text{and} \quad z_{K+1}(\tau=1) = z_{i0}, \quad i = 2,\ldots,\text{NE},$$

where t_{il} represent (shifted) roots of Legendre polynomials over the ith element length. Note that state profiles satisfy continuity conditions across finite elements $\Delta\alpha_i$, while this property is not enforced for control profiles. Substituting this representation the ODE model into problem (16) leads to the following nonlinear program:

$$\begin{aligned}
\min_{x, u_{il}, z_{il}, \Delta\alpha_i} \quad & \Phi(x, u_{il}, z_{il}) \\
& c(x, u_{il}, z_{il}) = 0 \\
& g(x, u_{il}, z_{ik}) \leqslant 0 \\
& R(t_{il}, \Delta\alpha_i) = 0, \quad i = 1,\ldots,\text{NE} \\
& \quad\quad\quad\quad\quad\quad\quad\quad l = 1,\ldots,K \\
& z_{10} = Z_0 \\
& \Delta\alpha_i \geqslant 0, \quad i = 1,\ldots,\text{NE} \\
& z^{i-1}_{K+1}(\alpha_i) = z^i_{K+1}(\alpha_i), \quad i = 2,\ldots,\text{NE} \\
& \sum_{i=1}^{\text{NE}} \Delta\alpha_i - t_f = 0 \\
& x^L \leqslant x \leqslant x^U \\
& U^L \leqslant u_{il} \leqslant U^U \\
& Z^L \leqslant z_{il} \leqslant Z^U
\end{aligned} \quad (20)$$

By taking advantage of the orthogonal properties of the polynomial representation, Cuthrell and Biegler (1989) use a transformation suggested by Reddien (1979) in order to show that the Kuhn–Tucker conditions of (20) are exact parameterizations of the variational conditions of (16). Therefore, it is only required that we choose $\Delta\alpha_i$ appropriately, i.e., to be sufficiently small for accurate approximation of the differential equations and to be able to locate the breakpoints for the optimal control profile. More quantitative extensions of this property were later shown in Logsdon and Biegler (1989) and Vasantharajan and Biegler (1990). Selection of $\Delta\alpha_i$ to deal with an accurate approximation to the ODEs will be considered in this section. First, however, we consider *parameter* optimization problems where $U(t)$ is dropped from (16) and (20).

A detailed analysis of polynomial approximation using collocation at Legendre roots by de Boor (1978) shows that the *global error* $e(t) = Z(t) - z_{K+1}(t)$ satisfies the relation

$$\|e\|_i = \|z^i_{K+1}(t) - Z(t)\|_i = C^k \Delta\alpha_i^k \|T(t)\|_i + 0(\Delta\alpha^{k+1}) \quad \text{for } 1 \leqslant i \leqslant \text{NE}. \quad (21)$$

Here $\Delta\alpha = \max\{\Delta\alpha_i\}$, k is a positive integer less than $K + 1$, and $\| \ \|_i$ represents the max norm in t for element i. Also, the function $T(t)$ depends on the actual solution $Z(t)$; it is independent of the finite element partition on which the problem is solved. Here C^k is a computable constant. Assuming the state functions to be sufficiently smooth, the preceding error bound can be approximated by the following relation (Russell and Christiansen, 1978):

$$\|e\|_i = C_1^k \Delta\alpha_i \|Z^{(k)}(t)\|_i + 0(\Delta\alpha^{k+1}), \qquad (22)$$

which requires the kth time derivative $(Z^{(k)}(t))$, in element i, of the solution $Z(t)$. C_i^k depends only on k, K, and the order of the ODEs.

The leading term in the preceding error bound can be used to minimize the largest approximation error within a given number of elements. Generally, this can be formulated as the following minimax problem:

$$\underset{\Delta\alpha_i}{\text{Min}} \ \underset{j}{\text{Max}} \ \Delta a_j^k \|T(\bar{t})\|_j, \qquad i = 1, \text{NE}$$

$$j = 1, \text{NE}$$

$$\text{s.t.} \qquad \Delta\alpha_i \geqslant 0$$

$$\sum_{i=1}^{\text{NE}} \Delta\alpha_i = t_f \qquad (23)$$

where $\bar{t}$ is an intermediate point within element j. The objective of this optimization is to relocate the knots in order to minimize the maximum error estimate computed over all the finite elements. In this context, $T(t)$ is used to denote generically either the kth derivative of $z_{K+1}(t)$ or any other bound for the terms in the collocation equations, which can be directly estimated from the problem. Moreover, necessary and sufficient conditions for minimizing the approximation error for the above problem can be represented by

$$\Delta\alpha_i^k \|T(t)\|_i = \text{constant}. \qquad (24)$$

The equidistribution property in (24) was first proposed by Pereyra and Sewell (1975). DeBoor (1978) estimated the local error bound with a divided difference approximation of the nonzero derivative $z_{K+1}^K(t)$. By adding equality constraints based on this divided difference formulation for equidistribution, (18) was applied successfully by Cuthrell and Biegler (1987). However, nonconvexities and nondifferentiabilities in these equations (introduced by the presence of absolute-valued terms) require smoothing techniques for gradient calculations and careful initialization to guarantee the convergence of the NLP.

To overcome these limitations, Vasantharajan and Biegler (1990) propose a new formulation based on the residual function (see Russell and Christiansen, 1978). Since the ODE residual equations are satisfied only at the collocation points, a straightforward way to compute an error estimate would be to evaluate the ODE residual at a non-collocation point, t_{nc}. Thus, the equations governing the placement of elements now become

$$\|R_i(t_{nc})\| - \|R_{i+1}(t_{nc})\| = 0, \quad i = 1, \text{NE} - 1. \tag{25}$$

An advantage to a formulation using R_i is that it also permits direct quantitative local error estimates. Thus, one can relax the stringent requirement of equidistribution, and directly enforce the error restrictions as follows:

$$C_2^k \Delta \alpha_i^k \|R(t_{nc})\| \leqslant \varepsilon, \tag{26}$$

where C_2^k is also a computable constant (Ascher and Bader, 1986; Russell and Christiansen, 1978) and ε is any desired tolerance. Thus, the knot placement procedure can be tailored to satisfy tight error bounds.

The simultaneous formulation for parameter optimization problems with ODE models is given below.

$$\begin{aligned}
\min_{x, z_{il}, \Delta \alpha_i} \quad & \Phi(x, z_{il}) \\
\text{s.t.} \quad & c(x, z_{il}) = 0 \\
& g(x, z_{il}) \leqslant 0 \\
& R(t_{il}, \Delta \alpha_i) = 0, \quad i = 1, \ldots, \text{NE} \\
& \qquad\qquad\qquad\qquad l = 1, \ldots, K \\
& \begin{cases} \|R_i(t_{1/2})\| - \|R_{i+1}(t_{1/2})\| = 0, \quad i = 1, \ldots, \text{NE} - 1 \\ \text{or} \\ C_2 \Delta \alpha_i^k \|R(t_{1/2})\|_i \leqslant \varepsilon, \quad i = 1, \ldots, \text{NE} \end{cases} \\
& \Delta \alpha_i \geqslant 0, \quad i = 1, \ldots, \text{NE} \\
& \sum_{i=1}^{\text{NE}} \Delta \alpha_i - t_i = 0 \\
& X^L \leqslant x \leqslant X^U \\
& Z^L \leqslant z_{il} \leqslant Z^U
\end{aligned} \tag{27}$$

where ε is a small tolerance that satisfies a small error bound. Note that formulation (27) allows for either equidistribution or direct enforcement constraints. The former ensure that for a given number of elements, the error, as measured by the residual function, is minimized. However, this may not be adequate for problems with too few finite elements. The second set of constraints, on the other hand, requires that a sufficiently large number of finite elements be chosen. This leads to a more difficult problem, since the sufficient number of elements is rarely known *a priori*. Moreover, searching for a sufficient number of elements, by comparing converged solutions of (27), is likely to detract from the computational advantages of a simultaneous approach.

Instead, the simultaneous method can be extended to select adaptively a sufficient number of finite elements. Here, we note that even if we set any element length to zero, the collocation equations and the continuity equations are still satisfied. Thus, any number of zero length (or dummy) elements can be added without changing the control or state profiles, or the solution to the NLP. Vasantharajan and Biegler (1990) take advantage of this important property and propose an adaptive element addition approach embedded within the simultaneous solution strategy.

This approach operates in two phases. First, a sufficient number of elements is found in order to satisfy the linearization of all of the constraints at the initial point. In this way we guarantee that a feasible QP subproblem exists for (27). Second, to avoid convergence to a suboptimal solution with too few elements, we retain additional dummy elements in the formulation that are constrained to be *less* than or equal to a negligible element length. These elements can be placed at all nonzero element locations, but in practice they need only be associated with elements that have active error bounds at the QP solution. Now once the QP subproblem is solved, multipliers on the upper bounds of the dummy elements are checked for positive values. These indicate that the objective function can be further improved by relaxing the dummy element. After relaxation (which effectively adds another nonzero element to the problem), another dummy element is added in order to allow for any additional nonzero elements that may be needed.

The incorporation and relaxation of dummy elements for the simultaneous solution of *parameter* optimization problems can be embedded easily within the subproblems solved by the SQP algorithm. Again, both the model and optimization problem need only be solved once.

In order to illustrate this approach, we next consider the optimization of an ammonia synthesis reactor. Formulation of the reactor optimization problem includes the discretized modeling equations for a packed bed reactor, along with the set of knot placement constraints. The following case study illustrates how a differential-algebraic problem can be optimized efficiently using (27). In addition, suitable accuracy of the ODE model can be obtained at the optimum by directly enforcing error restrictions and adaptively adding elements. Finally, bounds on the continuous state profiles can be enforced directly in the optimization problem.

1. *Ammonia Synthesis Reactor Optimization*

To demonstrate the accurate approximation approach for parameter estimation, we summarize a case study presented in Vasantharajan and Biegler (1990). Ammonia synthesis performed in a Haber–Bosch reactor is operated at high pressures in an autothermal manner, and produces ammonia from the following catalyzed reaction: $N_2 + 3H_2 \Leftrightarrow 2NH_3$.

OPTIMIZATION STRATEGIES FOR COMPLEX PROCESS MODELS   227

In order for an autothermic process to be carried out, gas flow and heat exchange are arranged to control the increase in temperature associated with the exothermic reaction, and to suppress the need for an external source of heat once the reaction is started (Baddour et al., 1965). The TVA reactor, a particular design of the Haber–Bosch reactor, is illustrated in Fig. 4. Murase et al. (1970) considered the optimization of this reactor using variable heat transfer coefficients along the bed lengths. Here we briefly present their model and optimization problem, and apply our direct enforcement formulation. The following assumptions were made in deriving their model:

1. The Temkin and Pyzhev rate expression (Annable, 1952) is used, given by

$$R = F_c \left(k_1 \frac{P_{N_2} P_{H_2}^{3/2}}{P_{NH_3}} - k_2 \frac{P_{NH_3}}{P_{H_2}^{3/2}} \right), \qquad (28)$$

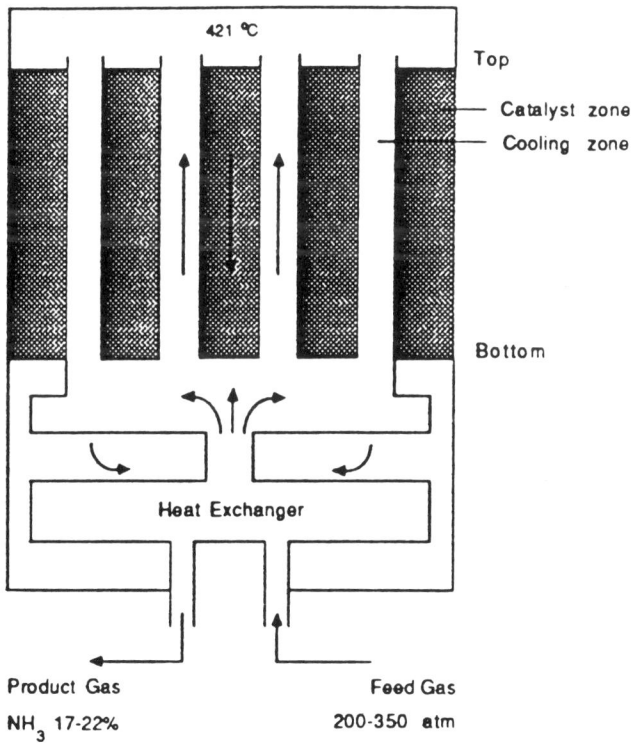

FIG. 4. Schematic of ammonia synthesis reactor. Reprinted with permission from *Comp. & Chem. Eng.*, **14**, No. 10, 1083–1100, S. Vasantharajan and L. T. Biegler, "Simultaneous Optimization of Differential/Algebraic Systems with Error Criterion Adjustment," Copyright 1990, Pergamon Press PLC.

where

F_c = catalyst activity,
P_{H_2} = partial pressure of hydrogen,
P_{N_2} = partial pressure of nitrogen,
P_{NH_3} partial pressure of ammonia,
k_1, k_2 = rate constants,
R = reaction rate, kg mol of $N_2/(h\,m^2)$.

TABLE II
NOTATION AND DATA FOR AMMONIA SYNTHESIS REACTOR*

Independent and Dependent Variables

x	Reactor length, m
N_{N^2}	Mole flow of N_2 per unit area catalyst, kg mol/(h m^2)
T_f	Temperature of feed gas, K
T_g	Temperature of reacting gas, K

Parameters

C_{fg}	Heat capacity of feed gas = 0.707 kcal/(kg)(K)
C_{rg}	Heat capacity of reacting gas = 0.719 kcal/(kg)(K)
R	Ideal gas constant. 1.987 kcal/(kg mol)(K)
k_1	Rate constant 1 = $1.78954 \times 10^4 \exp(-20.800/(RT))$
k_2	Rate constant 2 = $2.5714 \times 10^{10} \exp(-47.400/(RT))$
F_C	Catalyst activity = 1.0
ΔH	Heat of reaction = -26.600 kcal/kg mol N_2
N_i	Mass flow of component i through catalyst zone
T_H	Hours of operation per year = 8,330
P_i	Partial pressure of component i
P	Reactor pressure = 286 atm
S_t	Surface area of catalyst tubes per unit reactor length = 52 m
S_2	Cross-sectional area of catalyst zone = 0.78 m^2
T_{ref}	Reference temperature = 127°C = 400.15 K
L	Reference length = 10 m
U	Overall heat transfer coefficient = 500.58 kcal/(m^2)(h)(K)
W	Mass flow rate = 26.400 kg/h
ρ_{fg}	Density of feed gas = 10.5 kg/kg mol
G	Molar flow at top of reactor = 3.223.44 kg mol/(h m^2)
F	Economic factor for heat exchanger cost = 1.01
C_H	Cost of heat = 4.453×10^{-3} $/kcal
C_A	Cost of ammonia = 1.3147 $/kg mol
Φ	Profit objective function, $/year

*Reprinted with permission from *Comp. & Chem. Eng.*, **14**, No. 10, 1083–1100, S. Vasantharajan and L. T. Biegler, "Simultaneous Optimization of Differential/Algebraic Systems with Error Criterion Adjustment," Copyright 1990, Pergamon Press PLC.

OPTIMIZATION STRATEGIES FOR COMPLEX PROCESS MODELS 229

2. The rate constants are assumed to obey the Arrhenius equation,

$$k_i = a_i \exp(k_i/RT). \tag{29}$$

3. Heat and mass diffusion in the longitudinal direction are negligible.
4. The gas temperature in the catalytic zone is also the temperature of the catalytic particles.
5. The heat capacities of the reacting gas and feed gas are independent of temperature.
6. The catalyst activity, F_c, is uniform along the reactor and equal to unity.
7. Pressure drop across the reactor is negligible compared to the total pressure of the system.
8. Notation and data are listed in Table II, and the feed composition is given in Table III.

Using the parameters defined in Table II, the ODE system characterizing the ammonia converter can be reduced to dimensionless form with the following dimensionless groups:

$$\begin{aligned} t &= x/L\varepsilon[0,1], \\ X_1 &= T_f/T_{ref}\varepsilon[1,2], \\ X_2 &= T_g/T_{ref} \geq 1, \\ X_3 &= N_{N_2}/G\varepsilon[0,1], \end{aligned} \tag{30}$$

with the equations written as

$$\begin{aligned} \frac{dX_1}{dt} &= -K_1 U(X_2 - X_1), \\ \frac{dX_2}{dt} &= -K_2 U(X_2 - X_1) + K_3 R, \\ \frac{dX_3}{dt} &= K_4 R, \end{aligned} \tag{31}$$

TABLE III
FEED GAS COMPOSITION FOR AMMONIA REACTOR*

	N_2	H_2	NH_3	CH_4	Ar
Mol %		0.2175	0.6525	0.05	0.04
Molar Rate N_i		701.10	2,103.30	161.17	128.94

*Reprinted with permission from Comp. & Chem. Eng., 14, No. 10, 1083-1100, S. Vasantharajan and L. T. Biegler, "Simultaneous Optimization of Differential/Algebraic Systems with Error Criterion Adjustment," Copyright 1990, Pergamon Press PLC.

where

$$K_1 = (S_1 L)/(W C_{pp}),$$
$$K_2 = (S_1 L)/(W C_{pg}),$$
$$K_3 = (-\Delta H S_2)/(W C_{pg}) L/T_{ref}),$$
$$K_4 = L/G.$$
(32)

The reactor design problem considered by Murase et al. (1970) is to maximize the following profit function:

$$F(\times 10^6) = -0.27981 X_1 + 0.28174 X_2 \\ + 55.07 X_3 - 10^6 (BD^2 + CD^2 Lt)^{1/2} + 13.3543$$
(33)

subject to the dimensionless reactor model equations and variable bounds. The objective function is based on the difference between the value of the product gas (heating value and ammonia value) and the value of the feed gas (as a source of heat) minus the amortization of reactor capital costs (Murase et al., 1970). Derivation of this objective function can be found in Vasantharajan (1989).

The reactor was optimized using (27) with the direct enforcement error criterion and the reduced SQP algorithm. Here the approximation error tolerance, ε, was set to 10^{-3}, and the dummy elements were added only at elements with active error constraints. In addition, four different choices of initial number of elements (NE = 2, 3, 4, and 5) were considered in initializing the element partition. The initial and final element partitions are shown in Table IV. The number of SQP iterations and the error norms, for each of these four cases, are also presented there. Initial and final optimal values for the state variables, measured at exit conditions, and the objective function are given in Table V. In addition, the calculated values of exit ammonia

TABLE IV
OPTIMAL ELEMENT PARTIONS CHARACTERISTICS AND ERROR NORMS*

Initial NE	Final NE	# of Interations	$\|T_f^{err}(x)\|$	$\|T_g^{err}(x)\|$	$\|N_{N_2}^{err}(x)\|$
2	6	24	0.657E-3	0.619E-2	0.664E-3
3	5	22	0.655E-3	0.619E-2	0.662E-3
4	5	14	0.655E-3	0.619E-2	0.664E-3
5	5	14	0.659E-3	0.619E-2	0.664E-3

*Reprinted with permission from Comp. & Chem. Eng., 14, No. 10, 1083–1100, S. Vasantharajan and L. T. Biegler, "Simultaneous Optimization of Differential/Algebraic Systems with Error Criterion Adjustment," Copyright 1990, Pergamon Press PLC.

TABLE V
INITIAL AND OPTIMAL REACTOR EXIT CONDITIONS*

Variable	Initial Values	Optimal Values
$T_f(K)$	497.403	400.15
$T_g(K)$	714.724	634.699
N_N^2(kg mol/(h m^2))	505.284	491.007
x(m)	5.18	6.582
NH_3(mol %)	19.52	20.74
$\Phi(\times 10^6)$	4.7758	5.0185

*Reprinted with permission from *Comp. & Chem. Eng.*, **14**, No. 10, 1083–1100, S. Vasantharajan and L. T. Biegler, "Simultaneous Optimization of Differential/Algebraic Systems with Error Criterion Adjustment," Copyright 1990, Pergamon Press PLC.

concentration are also given. The final nitrogen molar flow profile approximation, and the final temperature profiles, are presented in Figs. 5 and 6, respectively. Also, a solution obtained with LSODE at the optimal parameter values has been plotted along with the collocation solution.

From the optimal solution, it can be seen that a long reactor is preferred. The feed gas temperature is at its lower bound. These conditions result in an

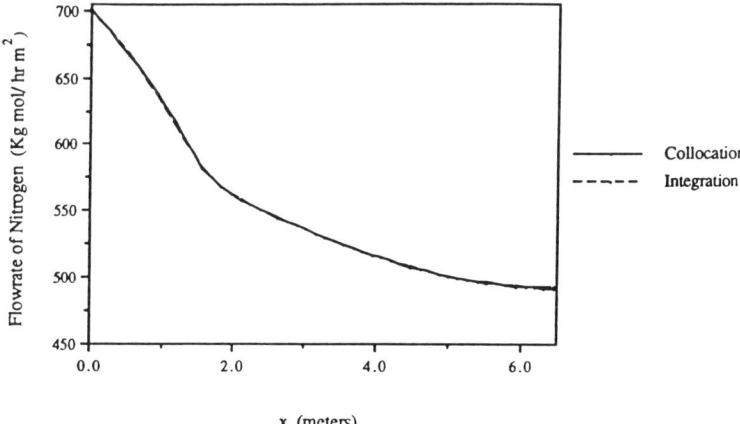

FIG. 5. Constant U: optimal state profile. Reprinted with permission from *Comp. & Chem. Eng.*, **14**, No. 10, 1083–1100, S. Vasantharajan and L. T. Biegler, "Simultaneous Optimization of Differential/Algebraic Systems with Error Criterion Adjustment," Copyright 1990, Pergamon Press PLC.

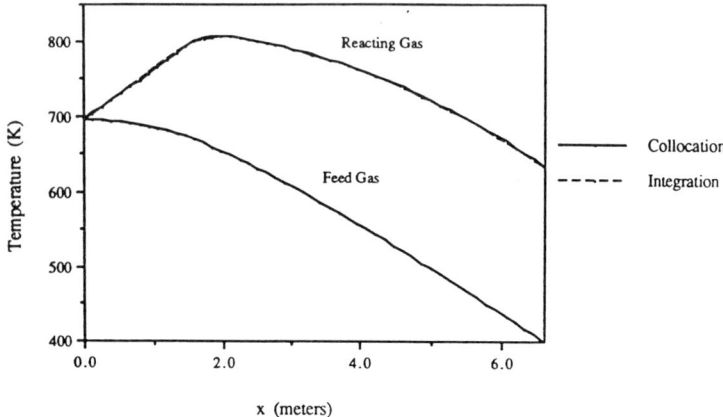

FIG. 6. Constant U: optimal temperature profiles. Reprinted with permission from *Comp. & Chem. Eng.*, **14**, No. 10, 1083–1100, S. Vasantharajan and L. T. Biegler, "Simultaneous Optimization of Differential/Algebraic Systems with Error Criterion Adjustment," Copyright 1990, Pergamon Press PLC.

increase of about 6.2% in the exit ammonia concentration (over the initial parameter set), and the corresponding increase in the objective function is about 5.1%. From Figs. 5 and 6 it can be seen that accurate approximations are obtained for the continuous profiles, with the maximum errors less than 0.005.

This design was performed assuming a constant value of the heat transfer coefficient, $U = 500$, throughout the reactor. However, better designs exist if the heat transfer coefficient can be varied along the feed gas tube. If such a design is possible, the heat transfer between the feed and reacting gases at the bottom of the reactor can be enhanced, resulting in increased conversion and an improved objective function value. Murase et al. (1970) determined $U(t)$ with the following general design policy to maximize reactor yield:

For $0 \leqslant t < x_1$, set $U(t) = U^{\max}$;

For $x_1 \leqslant t \leqslant L$, choose $U(t)$ variable (so that $\partial R/\partial x_2 = 0$).

With this result, the reactor is treated as if composed of two sections. In addition, from the form of the profile in Murase et al., $U(t)$ for the latter section can be approximated by a quadratic profile:

$$U(t) = at^2 + bt + c, \qquad (34)$$

where a, b, and c are now parameters determined by the optimization algorithm. The lengths of the two reactor sections, x_1 and x_2, are also parameters in the optimization problem and determine the optimal switching point between the two reactors. Note that this formulation is employed because of the discontinuous nature the heat transfer coefficient profile between the two reactor sections. However, state profile continuity across the reactors is still enforced.

The optimization problem is again solved with formulation (27), using residual based inequalities and the adaptive strategy to add elements. Also, we label the number of elements in the two reactors as NE1 and NE2, respectively. The optimization problem was solved repeatedly for several initial choices of NE1 and NE2. In addition, for each of these cases, different initial reactor segments, characterized by x_1 and x_2, were considered to confirm our results. The initial and final element partition characteristics and the computational details for four different cases of initial NE1 and NE2 choices are given in Table VI. The optimal heat transfer coefficient profile is presented in Fig. 7. The nitrogen molar flow rate and temperature profile approximations, along with their integrated counterparts, are shown in Figs. 8 and 9. Initial and optimal values of the variables, as well as computed values of the ammonia concentration and objective function, are given in Table VII.

The results show clearly that a much better design can be obtained with a variable heat transfer coefficient. The length of the original reactor is now at its upper bound, which translates into increased conversion of nitrogen to ammonia. This is reflected in the exit ammonia concentration of 23.93 mol %, a 15.2% increase over the initial value. Further, this also represents a 15.4% increase over the corresponding result with a constant U case. The final objective function value of 5.6359 denotes a 12.25% rise over the initial value. In addition, the accuracy of the approximation profiles and adequacy of the final element partitions are also evident.

Finally, Murase et al. (1970) recommend, as an extension of their work, that an optimization case be considered where the reaction temperature is bounded in order to prevent catalyst deactivation. This can be accomplished

TABLE VI
OPTIMAL PARTITIONS WITH RESULTS WITH VARIABLE U*

Initial NE1	Final NE1	Initial NE2	Final NE2	# of Iterations	KKT Tolerance
1	4	1	2	32	9E-07
1	3	2	2	29	5E-08
2	3	2	2	25	9E-07
3	3	3	3	28	1E-08

*Reprinted with permission from Comp. & Chem. Eng., 14, No. 10, 1083–1100, S. Vasantharajan and L. T. Biegler, "Simultaneous Optimization of Differential/Algebraic Systems with Error Criterion Adjustment," Copyright 1990, Pergamon Press PLC.

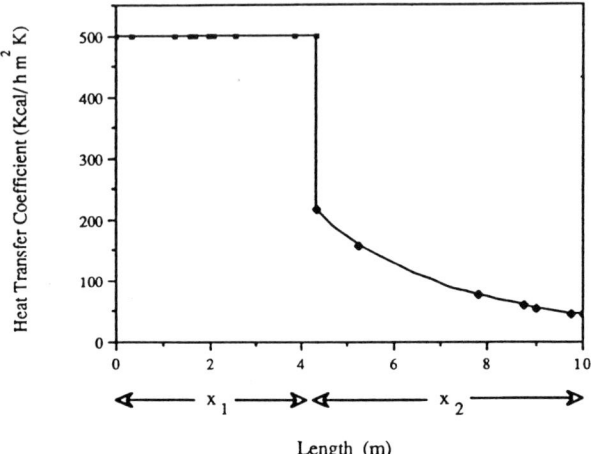

FIG. 7. Optimal profile for $U(t)$. Reprinted with permission from *Comp. & Chem. Eng.*, **14**, No. 10, 1083–1100, S. Vasantharajan and L. T. Biegler, "Simultaneous Optimization of Differential/Algebraic Systems with Error Criterion Adjustment," Copyright 1990, Pergamon Press PLC.

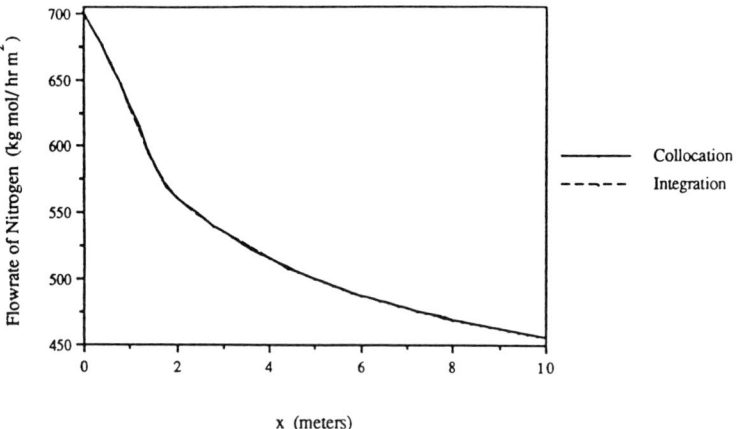

FIG. 8. Optimal nitrogen flow rate for variable U. Reprinted with permission from *Comp. & Chem. Eng.*, **14**, No. 10, 1083–1100, S. Vasantharajan and L. T. Biegler, "Simultaneous Optimization of Differential/Algebraic Systems with Error Criterion Adjustment," Copyright 1990, Pergamon Press PLC.

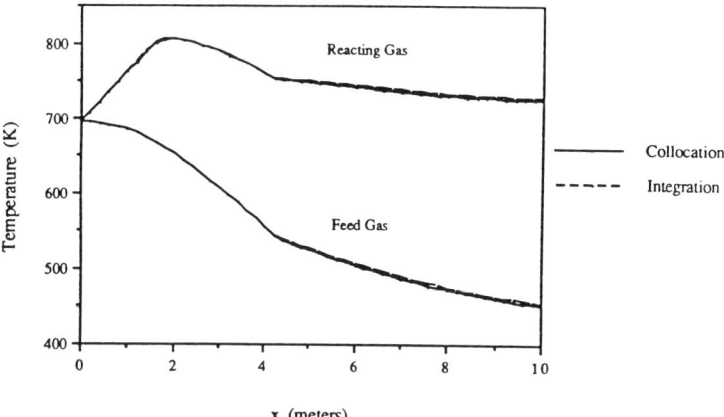

FIG. 9. Optimal temperature profiles for variable U. Reprinted with permission from *Comp. & Chem. Eng.*, **14**, No. 10, 1083–1100, S. Vasantharajan and L. T. Biegler, "Simultaneous Optimization of Differential/Algebraic Systems with Error Criterion Adjustment," Copyright 1990, Pergamon Press PLC.

TABLE VII
INITIAL AND OPTIMAL EXIT CONDITIONS WITH VARIABLE U*

Variable	Initial Values	Optimal Values
$T_f(K)$	487.0	450.974
$T_g(K)$	714.126	724.122
N_{N^2} (kg mol/(h m^2))	496.482	454.919
NH$_3$ (mol %)	20.269	23.929
x_1 (meters)	2.59	4.314
x_2 (meters)	2.59	5.686
$\Phi(\times 10^6)$	4.9330	5.5359
$\|T_f^{crr}\|$	0.481E-3	
$\|T_g^{crr}\|$	0.627E-2	
$\|N_{N_2}^{crr}\|$	0.677E-3	

*Reprinted with permission from *Comp. & Chem. Eng.*, **14**, No. 10, 1083–1100, S. Vasantharajan and L. T. Biegler, "Simultaneous Optimization of Differential/Algebraic Systems with Error Criterion Adjustment," Copyright 1990, Pergamon Press PLC.

directly in our approach by bounding the appropriate polynomial coefficients. Here we impose an upper bound of 1.935 on the normalized temperature profile (equivalent to a bound of 774 K on T_g). Also, we increased the upper bound for U to 1,000. The two-section formulation was initialized with NE1 = 3 and NE2 = 2 using the results obtained for the previous case. The final element partitions were characterized by NE1 = 8 and NE2 = 3. The optimization with the reduced SQP algorithm required 30 QP iterations. The optimal profiles obtained for molar flow of nitrogen and temperature are shown in Figs. 10 and 11 and are compared with profiles solved with LSODE at the optimal point. The optimal profile for U is given in Figure 12. Initial and optimal values of the variables, the computed values of the ammonia concentration and objective function are given in Table VIII. The same trends in the optimal solution are displayed as in the unconstrained case. In addition, we see how easily bounds or constraints can be imposed on the continuous profiles with the simultaneous approach. Additional cases have also been considered by Vasantharajan (1989).

In summary, formulation of (27) with appropriate placement of finite elements works well for parameter optimization problems. In the next subsection, however, we consider additional difficulties when control profiles are introduced. Stated briefly, the reason for these difficulties lies in the nature of the discretized variational conditions of (16). As shown in Logsdon and Biegler (1989), optimality conditions for parameter optimization problems take the form of two point boundary value problems. For optimal control

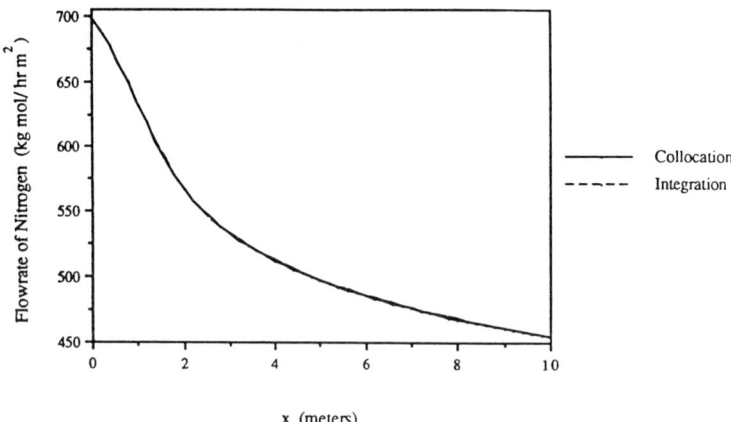

FIG. 10. Optimal nitrogen flow rate for T_g constrained. Reprinted with permission from *Comp. & Chem. Eng.*, **14**, No. 10, 1083–1100, S. Vasantharajan and L. T. Biegler, "Simultaneous Optimization of Differential/Algebraic Systems with Error Criterion Adjustment," Copyright 1990, Pergamon Press PLC.

OPTIMIZATION STRATEGIES FOR COMPLEX PROCESS MODELS 237

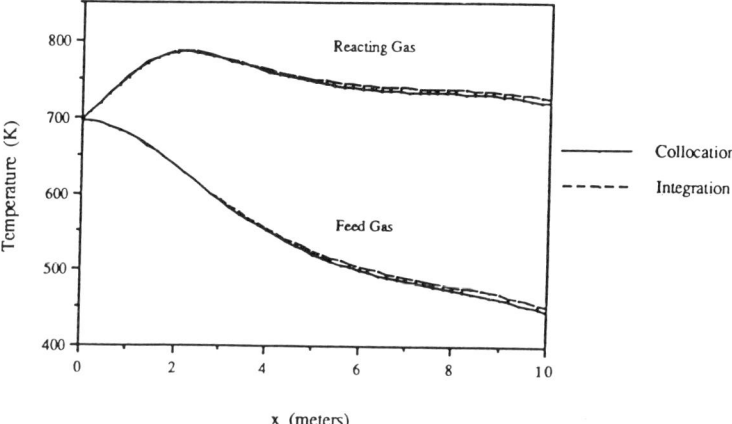

FIG. 11. Optimal temperature profiles for T_g constrained. Reprinted with permission from *Comp. & Chem. Eng.*, **14**, No. 10, 1083–1100, S. Vasantharajan and L. T. Biegler, "Simultaneous Optimization of Differential/Algebraic Systems with Error Criterion Adjustment," Copyright 1990, Pergamon Press PLC.

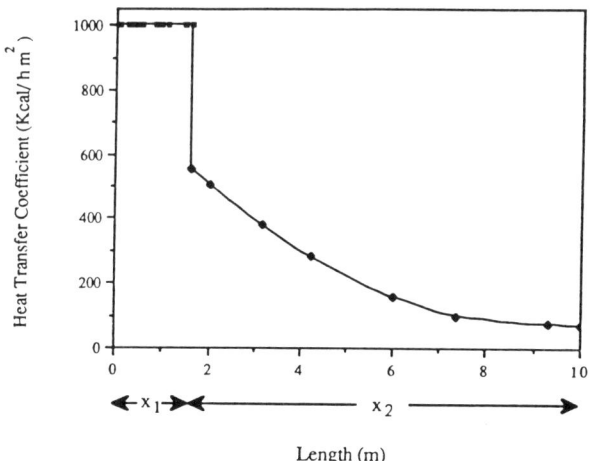

FIG. 12. Optimal U profile for T_g constrained. Reprinted with permission from *Comp. & Chem. Eng.*, **14**, No. 10, 1083–1100, S. Vasantharajan and L. T. Biegler, "Simultaneous Optimization of Differential/Algebraic Systems with Error Criterion Adjustment," Copyright 1990, Pergamon Press PLC.

TABLE VIII
Initial and Optimal Conditions for Constrained T_g Case*

Variable	Initial Values	Optimal Values
$T_f(K)$	487.001	473.403
$T_g(K)$	716.308	739.279
N_{N_2} (kg mol/(h m^2)	496.482	461/230
NH_3 (mol %)	20.269	23.359
x_1 (meters)	2.59	4.686
x_2 (meters)	2.59	5.314
$\Phi(\times 10^6)$	4.9346	5.52308
$\|T_f^{\text{err}}\|$		0.922E-3
$\|T_g^{\text{err}}\|$		0.683E-2
$\|N_{N_2}^{\text{err}}\|$		0.879E-2

*Reprinted with permission from *Comp. & Chem. Eng.*, **14**, No. 10, 1083–1100, S. Vasantharajan and L. T. Biegler, "Simultaneous Optimization of Differential/Algebraic Systems with Error Criterion Adjustment," Copyright 1990, Pergamon Press PLC.

problems, on the other hand, the optimality conditions are represented by systems of differential algebraic equations (DAEs) with boundary conditions.

B. Simultaneous Formulations for Profile Optimization

As mentioned in Section IV. A, a straightforward way to deal with optimal control problems is to parameterize them as piecewise polynomial functions on a predefined set of time zones. This suboptimal representation has a number of advantages. First, the approaches developed in the previous subsection can be applied directly. Secondly, for many process control applications, control moves are actually implemented as piecewise constants on fixed time intervals, so the parameterization is adequate for this application.

However the accurate treatment of state variable inequality constraints presents a few problems. Parameter optimization problems obtained by discretizing the control profile generally allow inequality constraints to be active *only at a finite set of points*, simply because a finite set of decisions cannot influence an infinite number of values (i.e., keeping the state fixed at every point in a finite time period).

However, if we are interested in an accurate representation of the *exact* optimal control profile, the problem becomes more complicated. First, from (16) we recognize the possibility that the optimal solution can cause time-dependent inequality constraints to become active for a finite period of time.

OPTIMIZATION STRATEGIES FOR COMPLEX PROCESS MODELS 239

For these time periods, the ODEs and active algebraic constraints influence the state and control variables. For these active sets, we therefore need to be able to analyze and implicitly solve the DAE system. To represent the control profiles at the *same level of approximation* as for the state profiles, approximation and stability properties for DAE (rather than ODE) solvers must be considered. Moreover, the variational conditions for problem (16), with different active constraint sets over time, lead to a multizone set of DAE systems. Consequently, the analogous Kuhn–Tucker conditions from (27) must have stability and approximation properties capable of handling all of these DAE systems.

Finally, finite elements are added as decision variables in (27) not just to ensure accurate approximation (of the state *and* control profiles), *but also* to provide optimal points of discontinuity for the control profile. This dual purpose led Cuthrell and Biegler (1987) to distinguish some elements as finite- and super-elements. These roles can be combined, however, if one considers the NLP formulation of the optimal control problem given below:

$$\min_{x, u_{il}, z_{il}, \Delta\alpha_i} F(x, u_{il}, z_{il})$$

$$\text{s.t.} \quad c(x, u_{il}, z_{il}) = 0$$

$$g(x, u_{il}, z_{il}) \leq 0$$

$$R(t_{il}, \Delta\alpha_i) = 0, \quad i = 1, \ldots, NE$$
$$\quad l = 1, \ldots, K$$

$$z_{10} = Z_0$$

$$\|R(t_{nc})\|_i \leq \varepsilon, \quad i = 1, \ldots, NE$$

$$\Delta\alpha_i \geq 0, \quad i = 1, \ldots, NE \quad (35)$$

$$z_{K+1}^{i-1}(\alpha_i) = z_{K+1}^{i}(\alpha_i), \quad i = 2, \ldots, NE$$

$$\sum_{i=1}^{NE} \Delta\alpha_i - t_f = 0$$

$$x^L \leq x \leq X^U$$

$$U^L \leq u_{il} \leq U^U$$

$$Z^L \leq z_{il} \leq Z^U$$

Note that potential control profile discontinuities are allowed at each element location with error restrictions directly enforced for each element. For a sufficient number of elements (which can be determined by the algorithm in the previous section), the element can be as large as allowed by an active error constraint, or it can act as a degree of freedom for the control profile discontinuity, with its corresponding error constraint inactive. Otherwise, (35) is based on the implicit Runge–Kutta (IRK) or collocation strategy discussed in the previous section. Before demonstrating this formulation, we first consider how it is influenced by the properties of DAE systems. Petzold (1982) systematically describes and classifies difficulties with DAEs according to the index of the system. For the common semi-implicit form of DAEs,

$$c(U(t), Z(t)) = 0,$$
$$\dot{Z}(t) = f(U(t), Z(t)), \qquad (36)$$

the index of this system is defined as the number of times $c(U, Z)$ must be differentiated (and corresponding ODEs substituted) in order to yield a system of ODEs.

The solution of this problem in its higher index form was analyzed for linear multistep methods such as the backward difference (BDF) formulas. These are currently used in codes such as DASSL and LSODI. Convergence proofs have been established for fixed step-size BDF methods for index 2 and index 3 problems by Petzold (1986). Theoretical properties for variable step-size BDF for index 2 systems were established by Gear et al. (1985). Moreover, for process systems Pantelides (1988a) developed a structural approach to detect higher index systems, as well as sufficient conditions to derive a consistent set of initial conditions prior to solution of the DAE system. These concepts have also been incorporated into the SPEEDUP dynamic simulator (Pantelides, 1988b; Pantelides et al., 1988). An efficient approach for solving index 2 problems is discussed by Gritsis et al. (1988). Here, properties of index 2 systems are analyzed and error criteria in DASSL are modified to allow for more efficient step-size algorithms. To handle index 1 problems that arise in process problems, Holl et al. (1988) develop the DIVA dynamic simulator. In addition, Bachmann et al. (1989) develop a reformulation technique for DAEs and discuss numerical difficulties that arise if DAE systems are reformulated incorrectly. Finally, Chung and Westerberg (1990) develop an approach similar to that of Bachmann et al. and present some useful insights for consistent initialization and reformulation of DAEs.

For simultaneous solution of (16), however, the equivalent set of DAEs (and the problem index) changes over the time domain as different constraints are active. Therefore, reformulation strategies cannot be applied since the active sets are unknown *a priori*. Instead, we need to determine a maximum index for (16) and apply a suitable discretization, if it exists. Moreover, BDF and other linear multistep methods are also not appropriate for (16), since they are not self-starting. Therefore, implicit Runge–Kutta (IRK) methods, including orthogonal collocation, need to be considered.

Properties of Runge–Kutta methods for DAEs have also been analyzed in a number of recent studies. Petzold (1986) showed that some Runge–Kutta methods can suffer order reduction for index one problems. Brenan and Petzold (1987) extended this approach to consider the order, stability, and convergence of implicit Runge–Kutta (IRK) methods applied to higher index DAE systems. Burrage and Petzold (1988) further refined the convergence and stability properties of index 1 systems solved by IRK methods. Along with Brenan and Petzold (1987), they showed that orthogonal collocation on finite elements, an A-stable IRK method, is sufficiently stable and accurate for index 1 problems. For higher index problems, however, L-stable (i.e., strongly A-stable) IRK methods must be applied in order to guarantee stable solutions to the error propagation difference equations. Methods with this property can be constructed from orthogonal collocation, simply by adding an additional collocation point at the end of each element. Compared to *orthogonal* collocation, however, this is a lower-order discretization and may require more finite elements. Finally, Ascher and Petzold (1991) considered the solution of boundary value DAEs with IRK methods. In addition to verifying the L-stable properties, they showed how A-stable methods can be reformulated and applied to index 2 DAE systems.

In addition to stability considerations, the order of the approximation error is also a function of the system index. From the results of Brenan and Petzold (1987), systems of equations of higher index can be considered simply by choosing the appropriate IRK method with the appropriate integration error constraints. Based on error and stability considerations, Logsdon and Biegler (1989) concluded that minimum order requirements for collocation methods are the following:

Index 1 problems: two-point orthogonal collocation (A-stable);
Index 2 problems: three-point collocation (L-stable if many high index elements);
Index 3 problems: four-point collocation (L-stable if many high index elements).

From the variational conditions of (16), Logsdon and Biegler (1989) showed that all profile optimization problems are at least of index 1. For inequality constraints that become active over a finite period of time, the additional algebraic equation increases the index over that time period by at least one if it is entirely a function of the state variables. Finally, if (16) exhibits singular arc solutions, i.e., it is linear in the control variable and $\partial H/\partial u$ becomes identically zero over a finite time period, then the index of the singular portion is at least three (Lewis, 1980; Logsdon and Biegler, 1989). Finally, arbitrarily high index systems can be constructed simply by making the DAE model or the structure of the state variable constraints more complex.

In summary, we see that problem (16) cannot be reformulated to a lower index system of DAEs (or ODEs), since the location of high index active constraints and singular profiles is unknown *a priori*. Instead, one is forced to analyze for the maximum potential index of (16), and to choose a stable and accurate IRK discretization for this index. If the analysis then leads to an index that is too high for any IRK method, the only recourse may be to parameterize the control profile and solve a suboptimal problem using the methods in Section IV or V.A. To illustrate further the concepts of DAE systems applied to optimal control, we consider next the optimal mixing of catalyst in a tubular reactor as presented in Logsdon and Biegler (1989).

1. Catalyst Mixing Problem

To illustrate the concept of a maximum index for problem (16), we consider the singular catalyst mixing problem originally due to Gunn and Thomas (1965), with an analytic solution by Jackson (1968). The optimization problem is given by

$$\text{Max } (1 - z_1(t_f) - z_2(t_f))$$

$$\frac{dz_1}{dt} = f(k_2 z_2 - k_1 z_1)$$

$$\frac{dz_2}{dt} = -f(k_2 z_2 - k_1 z_1) - (1 - f)k_3 z_2$$

$$z_1(0) = 1$$

$$z_2(0) = 0 \tag{37}$$

The kinetic sequence is given by A ↔ B → C; catalyst I deals with the first, reversible reaction, while catalyst II deals with the second. The control profile is the fraction of catalyst I ($f(t)$) over the packed bed reactor. Here $z_1(t)$ is the mole fraction profile of reactant A and $z_2(t)$ is the mole fraction profile of B. Since the conversion to C (the objective function) is limited by the first reaction's equilibrium, it can be shown that for a sufficiently long reactor length, the system admits a singular arc segment. For this segment, a mixture of the two catalysts is optimal. The problem description is given in Fig. 13, and the analytical solution is given in Fig. 14. Note that the solution trajectory consists of mixed index portions, with the first and last (index 1) control sections at their upper and lower bounds, respectively. The middle section, however, is a singular arc (index three), and this is a severe test for the simultaneous approach.

To deal with this system, a high-order collocation method should be applied, and if a large number of elements are required to describe the singular segment, the method must be *L*-stable in order to limit growth of the propagation error. To test these conditions, we attempted to solve this problem using only two-point collocation and carefully controlling the integration error for each element. As seen in Logsdon and Biegler (1989), we

OPTIMIZATION STRATEGIES FOR COMPLEX PROCESS MODELS   243

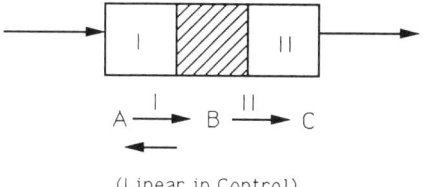

(Linear in Control)

FIG. 13. Brief description of singular control problem (see Gunn and Thomas, 1965, and Jackson, 1968). Reprinted with permission from *Ind. Eng. Chem. Res.* **28**, 1628–1645, J. S. Logsdon and L. T. Biegler, "On the Accurate Solution of D-A Optimization Problems," Copyright 1989 American Chemical Society.

achieve a solution only by specifying error tolerances in each element by trial and error so that the solution matches the analytical profile. We note that especially tight error tolerances must be specified for the singular segment. Otherwise, the optimization fails or converges to suboptimal solutions.

With four-point collocation using gaussian roots (the A-stable case), we fail to find an optimal solution. Because of this, we include an additional constraint on the *control profile error* (third time derivative of the control profile less than a tolerance, see Russell and Christiansen, 1978). The resulting

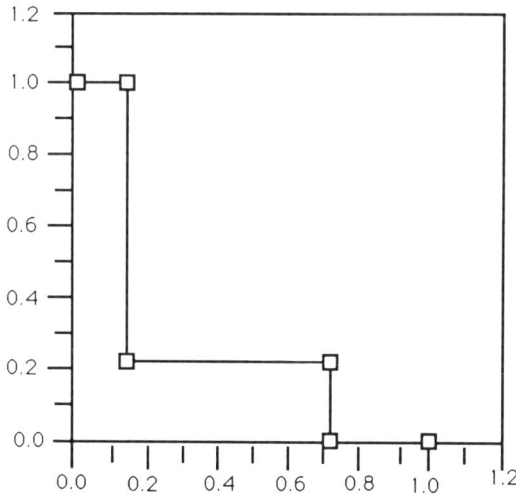

FIG. 14. Optimal catalysis mixing profile: analytical solution. Note the singular arc in the reactor midsection. Reprinted with permission from *Ind. Eng. Chem. Res.* **28**, 1628–1645, J. S. Logsdon and L. T. Biegler, "On the Accurate Solution of D-A Optimization Problems," Copyright 1989 American Chemical Society.

formulation yields a numerical solution that matches the analytical profile in Fig. 14. Note that controlling the integration error on the control variable is similar to error control strategies for higher index variables used by DAE solvers.

With this example we are fortunate that the singular portion could be approximated by a single element. The A-stable four-point collocation method therefore produces a solution because there is no unstable propagation of error in the index three portion. To handle this propagation we also apply an L-stable collocation method to this problem. Here four-point collocation (using the three Legendre roots and the element endpoint for the collocation points) is used to illustrate the L-stable method. Although we apply a lower order method, convergence to the analytical solution in Fig. 14 is obtained.

Finally, the catalyst mixing problem can be converted from an index three problem to an index zero problem by parameterizing the control profile using variable length piecewise constant functions. (This approach is acceptable because of the known form of the optimal control profile.) The solution using this approach also matches the analytical solution within numerical tolerances.

This example illustrates some of the theoretical difficulties with profile optimization problems. Again, we see that care must be applied to the discretization of the DAE system. Also, if the discretization cannot handle high index problems that are encountered in the optimization, we may be forced to use a suboptimal parameter optimization approach.

C. GENERALIZATION TO LARGE-SCALE PROBLEMS

In Section II, the SQP strategy for solving NLPs was extended to large-scale problems that have a relatively small number of degrees of freedom. Here we further discuss how these approaches can be tailored to the solution of differential/algebraic optimization problems. In the previous examples, these problems were solved using the reduced SQP approach derived ealier. Note that this approach solves quadratic programming (QP) subproblems only in the degrees of freedom; equations and dependent variables that describe the DAE model itself are eliminated in the decomposition step. Here Vasantharajan (1989) shows that this approach provides a significant saving for large-scale optimization.

The reduced SQP algorithm assumes that the $Z^T BY$ term in (9) is unimportant, and thus need not be constructed. To help insure this, Y is chosen so that the range step direction, Yd_y, is a least-squares step to recover feasibility of the linearized constraints. Note also that if we include the $Z^T BY$

term, the choice of Y may be arbitrary as long as the $[YZ]$ matrix is nonsingular. However, estimating $Z^T BY$ (an $(n - m) \times m$ matrix) can be expensive for general-purpose problems. An exception to this arises with parameter estimation problems discussed in Section II, where with good model/measurement agreement, the B matrix can be given analytically by the Hessian of the objective function (Tjoa and Biegler, 1991).

While the reduced SQP algorithm is often suitable for parameter optimization problems, it can become inefficient for optimal control problems with many degrees of freedom (the control variables). Logsdon et al. (1990) noted this property in determining optimal reflux policies for batch distillation columns. Here, the reduced SQP method was quite successful in dealing with DAOP problems with state and control profile constraints. However, the degrees of freedom (for control variables) increase linearly with the number of elements. Consequently, if many elements are required, the effectiveness of the reduced SQP algorithm is reduced. This is due to three effects:

- computation of the least squares projection step (Ydy) requires a Householder factorization (see Vasantharajan and Biegler, 1988a) and a dense matrix decomposition, where the matrix order is equal to the degrees of freedom;
- the projected Hessian approximation, B, becomes large and dense;
- the number of state and control variable inequalities remains large despite the decomposition.

To deal with the first point, we note that a related decomposition approach by Locke et al. (1983) is analogous to setting $Y = [0\ I]^T$ in Eq. (9) and neglecting $Z^T BY$. Computing the range space step Yd_y becomes much easier since a least squares step is not required. However, Yd_y may be large relative to the range space move, Zd_z, and neglecting $Z^T BYd_y$ may lead to slow convergence. Nevertheless, Logsdon and Biegler (1990) adapted this approach to the solution of optimal control problems. Here the finite element calculations can be organized in a modular form, as shown in Fig. 15. In

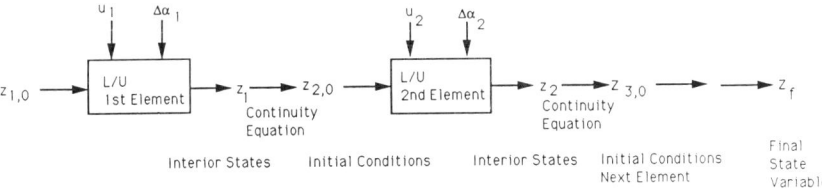

FIG. 15. ODE solver for state differential equations using collocation on finite elements with information processed from element to element.

particular, construction of the terms in the reduced QP require a forward chainruling of the elements. Note that the collocation equations are given by

$$\sum_{j=0}^{K} z_{ij}\dot{\phi}_j - \Delta\alpha_i f(x, u_{il}, z_{il}) = 0, \qquad l=1,\ldots K, \; i=1,\ldots NE, \qquad (38)$$

where

$$v = z_{il},$$
$$u = [u_{il}, \Delta\alpha_i].$$

Now denoting the state variables as the dependent variables v, we form Z by

$$Z = [I \; -\nabla_u h(\nabla_v h)^{-1}]^T, \qquad (39)$$

where

$$\nabla_u h(\nabla_v h)^{-1} = \left.\frac{dv}{du}\right|_{h=0}.$$

Now for each element i, dz_{ij}/du_{il} and $dz_{ij}/d\Delta\alpha_i$ are obtained by the linearized collocation equations (38). Also, since $z_{i,f} = \Sigma_{i=0}^{K} z_{ij}\phi_j(\tau = 1.0)$ and $z_{if} = z_{i+1,0}$, $dz_{i,f}/du$ is obtained easily. By chainruling, reduced gradients for $F(z_f)$ and $g(z_f)$ are obtained by

$$Z^T F = \frac{dF}{du_{ij}} = \left(\frac{dz_{if}}{du_{ij}}\right) = \left(\frac{dz_{if}}{du_{ij}}\right)\left(\frac{dz_{i+if}}{dz_{i+1,0}}\right)\cdots\left(\frac{dz_{NE,f}}{dz_{NE,0}}\right)\left(\frac{\partial F}{\partial z_{NE,f}}\right).$$

Note that this operation can be done for each element simply by fixing u_{ij} and $\Delta\alpha_i$ as inputs to the finite elements, and recovering the state variables in each element through (state) linearization of the collocation equations. This information is then chainruled to the next element. The process continues until reduced gradients $Z^T \nabla F$, $Z^T \nabla g$ are calculated for the objective function and constraints after the last element. From (16) it can be observed that this approach takes advantage of an implicit lower triangular structure of the collocation equations; this is manifested in the forward chainruling scheme shown in Fig. 15. As expected, calculation of the Lagrange multipliers for these equations can also be recovered by backward chainruling (derived from an implicit upper triangular structure, the transpose of the collocation equations).

Moreover, this algorithm is greatly simplified if the differential equations are linear in the state variables. Here the linearization with u_{ij} and $\Delta\alpha_i$ fixed, and construction of the reduced gradients, is always done in a feasible space. Consequently, the Lagrange multiplier calculation is not required for SQP, and only forward chaining needs to implemented. Finally, Logsdon and Biegler (1990) showed that by solving nonlinear collocation equations in each element, one obtains a tailored formulation of the generalized reduced gradient method. In this study, the forward chaining approach was applied to several optimal control problems; solutions were obtained much faster here than with the general-purpose reduced SQP algorithm. Moreover, the largest problem they solved contained over 2,500 variables and 285 degrees of freedom.

However, even this approach becomes inefficient if the degrees of freedom and the reduced Hessian matrix become large. Consider a problem with NU control variables per element and NE finite elements. Note that even after decomposition, a Newton method applied to a dense, reduced system with NU × NE degrees of freedom requires computational effort on the order of $(NU \times NE)^3$.

For unconstrained optimal control problems, methods exist, however, where the computational effort grows only linearly with NE. To deal with more complex problems such as (16), it therefore is necessary to reconsider these methods and exploit the structure of the problem. Consider the simpler discrete time optimal control problem given as follows:

$$\text{Min } F(z^{NE+1})$$
$$\text{s.t. } z^1 = a \qquad (40)$$
$$z^{i+1} = f_i(z^i, u^i), \quad i = 1, NE$$

Note that the equations are given in explicit algebraic form. Comparing to problem (35), note that all additional constraints have disappeared and the differential equations have had a simple, low-order Euler discretization applied to them. However, as with (35) we note that the Lagrange function for this problem,

$$L(z, u) = F(z^{NE+1}) + \sum_{i=1}^{NE} \lambda_{i+1}^T (f(z^i, u^i) - z^{i+1}), \qquad (41)$$

has nonlinear terms that are localized within each element i; the only interaction between elements in (16) is due to linear continuity equations. Consequently, the Hessian matrix of the Lagrange function can be written independently for each element. The optimality conditions for (41) can be written easily as follows:

$$\frac{\partial L}{\partial z^{NE+1}} = \frac{\partial F}{\partial z^{NE+1}} + \lambda_{NE+1} = 0,$$

$$\frac{\partial L}{\partial z^i} = \lambda_i + \frac{\partial f^i}{\partial z^i} \lambda_{i-1} = 0,$$

$$\frac{\partial L}{\partial u^i} = \frac{\partial f^i}{\partial u^i} \lambda_{i-1} = 0,$$

$$z^{i+1} = f(z^i, u^i), \qquad (42)$$

for $i = 1, \ldots, NE$. Linearizing these equations to apply a Newton method and rearranging them leads to the following banded form (Wright, 1989):

$$A = \begin{bmatrix} 0 & -I \\ -I & T^1 & V^1 & A^{1T} \\ & V^{1T} & S^1 & C^{1T} \\ & A^1 & C^1 & 0 & -I \\ & & & -I & T^2 & V^2 & A^{2T} \\ & & & & V^{2T} & S^2 & C^{2T} \\ & & & & A^2 & C^2 & 0 \\ & & & & & & -I & -I \\ & & & & & & & T^3 & V^3 & A^{3T} \\ & & & & & & & V^{3T} & S^3 & C^{3T} \\ & & & & & & & \vdots & \vdots & \vdots \end{bmatrix} \quad (43)$$

and $A\xi = b$ where

$\xi^T = [\Delta\lambda^1, \Delta z^1, \Delta u^1, \Delta\lambda^2, \Delta z^2, \Delta u^2 \ldots];$
$b = [0, -\partial L/\partial z^1, -\partial L/\partial u^1, -s^1, -\partial L/\partial z^2, -\partial L/\partial u^2, -s^2 \ldots];$
$A^i = \partial f^i/\partial z^i;$
$C^i = \partial f^i/\partial u^i;$
$V^i = \partial^2 L/\partial z^i \partial u^i;$
$T^i = \partial^2 L/(\partial z^i)^2;$
$S^i = \partial^2 L/(\partial u^i)^2;$
$s^i = z^{i+1} - f(z^i, u^i)$, at current point.

Solution of this banded form can be handled in two ways. First, a straightforward transformation due to Dunn and Bertsekas (1989) can be applied directly to the Newton step as follows:

$$\Delta\lambda_i = P^i \Delta z + p^i,$$
$$\Delta u^i = Q^i \Delta z + q^i, \quad (44)$$

where the vectors p^i and q^i are functions of the states evaluated at the current point. Here the linearized equations can be decoupled so that p^i, q^i, P^i, and Q^i are solved in the reverse direction, and state and control variable moves are recovered in the forward direction. This approach is analogous to the well-known Riccati transformation for linear/quadratic optimal control problems. The result is an algorithm whose work increases only linearly with NE.

Wright (1989) addressed this problem by noting that the Newton step (43) can also be solved directly by a banded matrix solver implemented on a parallel processing machine. Here the block matrices are rearranged, and pivoting and factorization of the submatrices is done in parallel. Compared to the Dunn and Bertsekas approach, which is inherently sequential, the parallel band solver is much more efficient and offers speedups of two to eight on an Alliant FX/8 (eight-processor) computer. Both Dunn (1988) and Wright (1989) consider the extension of problem (40) to include constraints of the form $a \leqslant u \leqslant b$. Here a projected gradient approach is applied, and the parallel band solver algorithm can be readily adapted simply by fixing the bounded control variables and altering a few elements in the Newton matrix. More general constraints than these, however, tend to destroy the banded structure in the Newton step and thus lead to higher computational expense.

Therefore, for large optimal control problems, the efficient exploitation of the structure (to obtain $O(NE)$ algorithms) still remains an unsolved problem. As seen above, the structure of the problem can be complicated greatly by general inequality constraints. Moreover, the number of these constraints will also grow linearly with the number of elements. One can, in fact, formulate an infinite number of constraints for these problems to keep the profiles bounded. Of course, only a small number will be active at the optimal solution; thus, adaptive constraint addition algorithms can be constructed for selecting active constraints.

To address this question, Hettich and Gramlich (1990) discuss an adaptive scheme based on defining constraints on a coarse grid in the time domain. When they solve the resulting problem (in this case, a quadratic program), determine an active constraint set, and further refine the grid, only a small set of inequality constraints are actually considered. To date this approach has been applied to approximation and data smoothing problems, although similar strategies can be tailored to optimal control problems with profile constraints.

VI. Summary and Conclusions

This survey paper provides a brief review of simultaneous optimization strategies for process engineering. Over the past decade the recognition of the effectiveness of sophisticated nonlinear programming algorithms, such as SQP, has led to the formulation of larger and more difficult optimization problems. The key to this advance lies in more flexible formulations of the optimization problems. Here time-consuming procedures that had to be accessed repeatedly by the optimizer could be incorporated and solved simultaneously along with the optimization problem. Inequality constraints that could not be handled directly within procedures are also straightforward to handle by the NLP algorithm. Moreover, inefficient convergence algorithms that were incorporated within a calculation procedure can now be replaced with a simultaneous Newton-type algorithm. Finally, simultaneous solution and optimization strategies have been extended and demonstrated on large optimization problems. In particular, the extension of successive quadratic programming to large-scale systems is discussed and illustrated.

We first considered applications of this approach within process engineering. Steady-state flowsheeting or simulation tools are the workhorse for most process design studies; the application of simultaneous optimization strategies has allowed optimization of these designs to be performed within an order of magnitude of the effort required for the simulation problem. An application of this strategy to an ammonia synthesis process was presented. Currently, flowsheet optimization is widely available commercially and has also been installed on the FLOWTRAN simulator for academic use.

The optimization of models described by differential/algebraic equations (DAEs) also shares many characteristics of flowsheet optimization. Again, the simultaneous approach allows the direct enforcement of profile constraints for state and control variables. Also, the SQP algorithms can be tailored to the DAE system to allow for moving finite elements and the accurate determination of state and optimal control profiles. Here we considered two problem classes. Parameter optimization is frequently encountered in process design and analysis. As with flowsheet optimization, this problem is characterized by few degrees of freedom, despite the complexity of the DAE model. As a result these problems can be solved quite successfully by the SQP decomposition strategies described in Section II. As an illustration of these problems, we considered the optimal design of an ammonia reactor.

Optimal control problems have more interesting features in that control profiles are literally infinite-dimensional and attention must be paid to approximating them accurately. Here the optimality conditions can be represented implicitly by high-index DAE systems, and consequently a stable and accurate discretization is required. To demonstrate these features, the classical catalyst mixing problem of Jackson (1968) was solved with the simultaneous approach. In addition to theoretical properties of the discretization, the structure of the optimal control problem was also exploited through a chainruling strategy.

Unlike parameter optimization, the optimal control problem has degrees of freedom that increase linearly with the number of finite elements. Here, for problems with many finite elements, the decomposition strategy for SQP becomes less efficient. As an alternative, we discussed the application of Newton-type algorithms for unconstrained optimal control problems. Through the application of Riccati-like transformations, as well as parallel solvers for banded matrices, these problems can be solved very efficiently. However, the efficient solution of large optimal control problems with general inequality constraints remains a challenging research problem. Although a number of approaches appear promising, they still need to be evaluated and extended to more general features of problem (16).

Finally, the solution of large-scale optimization problems has been made much easier by the availability of powerful equation-based modeling tools such as GAMS (Brooke et al., 1988) and ASCEND (Piela et al., 1989). These greatly simplify the formulation of process optimization problems. Also, great strides have also been made in the efficient and automatic calculation of accurate derivatives (Griewank, 1989). Here, derivative calculations parallel the procedural calculations for functions and apply to "black box" models and other applications where symbolic expressions for derivatives are difficult to use. Griewank (1989) describes several preprocessing programs that construct derivative "subroutines" automatically from the function routines. As a result, many of the labor-intensive objections toward applying sophisticated optimization methods are disappearing. Instead, sophisticated and efficient optimization tools will be available to the practicing engineer on a routine and frequent basis.

Lastly, this study has concentrated on nonlinear programming methods and formulations, with little mention of initialization and proper problem formulation. While the SQP method can be made globally convergent, through line search and trust region strategies, performance greatly depends on the user's knowledge of the problem, as well as his or her understanding of the optimization method. Briefly stated, SQP has characteristics similar to those of Newton's method; nonsmooth functions and singularities introduce difficulties, and performance depends on the degree of nonlinearity of the objective and constraint functions, as well as proximity to the solution. However, these characteristics vary from problem to problem and require careful attention from the user. Nevertheless, proper formulations can be guided by heuristics that can even be automated. One such approach for NLP and MINLP modeling is provided by Amarger et al. (1991). Consequently, with efficient NLP algorithms and modeling tools, as well as proper education about their strengths and weaknesses, more powerful analyses can be performed and better designs will result in process engineering.

References

Achenie, L. K. E., and Biegler, L. T., "A superstructure based approach to chemical reactor network synthesis," *Comp. and Chem. Engr.* **14**(1), 23 (1990).

Amarger, R., Grossmann, I. E., and Biegler, L. T., "An intelligent modeling interface for design optimization," *EDRC Report*, Carnegie Mellon University (1991).

Annable, D., "Application of the Temkin kinetic equation to ammonia synthesis in large-scale reactors," *Chem. Eng. Sci.* **I**(4), 145–153 (1952).

Ascher, U and Bader, G., "Stability of collocation at Gaussian points," *SIAM J. Numer. Anal.* **23**(2), 412–422 (1986).

Ascher, U. M., and Petzold, L. R., "Projected implicit Runge–Kutta methods for differential algebraic equations," *SIAM J. Num. Anal.* **28**(4), 1097 (1991).

Bachmann, R., Bruell L., Pallaske, U., and Mrzglod, T., "A contribution to the numerical treatment of differential equations arising in chemical engineering," *Dechema Monographs* **116**, 343 (1989).

Baddour, R. F., Brian, P. L. T., Logeais, B. A., and Eymery, J. P., "Steady state simulations of an ammonia synthesis converter," *Chem. Eng. Sci.* **20**, 81 (1965).

Berna, T. J., Locke, M. H., and Westerberg, A. W., "A new approach to optimization of chemical processes," *AIChE J.* **26**(1), 37–43 (1980).

Biegler, L. T., "Solution of dynamic optimization problems by successive quadratic programming and orthogonal collocation," *Comp. and Chem. Eng.* **8**(3/4), 243–248 (1984).

Biegler, L. T., "Chemical process simulation: a concise survey," *Chemical Engineering Progress* **85**(10), 50 (1989).

Biegler, L. T., and Cuthrell, J. E., "Improved infeasible path optimization for sequential modular simulators—II: The optimization algorithm," *Comp. and Chem. Eng.* **9**(3), 257–267 (1985).

Biegler, L. T., and Hughes, R. R., "Infeasible path optimization of sequential modular simulators," *AIChE J.* **28**(6), 994 (1982).

Biegler, L. T., Damiano, J. J., and Blau, G. E., "Nonlinear parameter estimation: a case study comparison," *AIChE J.* **32**(1), 29 (1986).

Brenan, K. E., and Petzold, L. R., "The numerical solution of higher index differential/algebraic equations by implicit Runge–Kutta methods," UCRL-95905, preprint, Lawrence Livermore National Laboratories, Livermore, California (1987).

Brenan and Petzold, *SIAM J. Num. Anal.* **26**, 976 (1987).

Britt, H. I., Smith, J. A., Wareck, J. S., "A computer aided process synthesis and analysis environment," in "Foundations of Computer Aided Process Design 3," (Siirola, J. J., Grossmann, I. E., and Stephanopoulos, G. N., eds.), Cache Corporation, Snowmass, Colorado (1990).

Brooke, A., Kendrick, D., and Meeraus, A., "GAMS, A User's Guide." Scientific Press, Redwood City, California, (1988).

Bryson, A. E., and Denham, W. F., *J. Appl. Mech.* **29**, 247 (1962).

Bryson, A. E., and Ho, Y.-C. "Applied Optimal Control." Hemisphere Publishing, Washington, D.C., 1975.

Bunch, J. R., and Parlett, B. N., "Direct methods for solving symmetric indefinite systems of linear equations," *SIAM J. Num. Anal.* **8**, 639 (1971).

Burrage, K., and Petzold, L. R., "On order reduction for Runge–Kutta methods applied to differential/algebraic systems and to stiff systems of ODEs," Lawrence Livermore National Laboratory, UCR-98046 preprint (1988).

Canon, M. D., Cullum, J., and Polak, E., "Theory of Optimal Control and Math. Programming." McGraw-Hill, New York, 1970.

Caracotsios, M., and Petrellis, N. C., "ACCOPT—An integrated software package for the optimization of discrete and continuous processes," presented at Annual AIChE Meeting, San Francisco (1989).

Caracotsios, M., and Stewart, W. E., "Sensitivity analysis of initial value problems with mixed ODEs and algebraic equations," *Comp. and Chem. Eng.* **9**, 359–365 (1985).

Chen, H.-S., and Stadtherr, M. A., "A simultaneous–modular approach to process flowsheeting and optimization," Parts I, II and III, *AIChE J.* **31**(11), 1843 (1985).

Chung, Y., and Westerberg, A. W., "A proposed numerical algorithm for solving non-linear index problems," *I&EC Res.* **29**, 1234 (1990).

Clark, P. A., and Westerberg, A. W., "Optimization for design problems having more than one objective," *Computers and Chemical Engineering* **7**, 259–278 (1983).

Coleman, T. F., and Conn, A. R., "Nonlinear programming via an exact penalty function: asymptotic analysis," *Math. Prog.* **24**, 123 (1982).

Cuthrell, J. E., and Biegler, L. T., "Simultaneous solution and optimization of process flowsheets with differential equation models," *Chem. Eng. Res. Des.* **64**, 341 (1986).

Cuthrell, J. E., and Biegler, L. T., "On the optimization of differential-algebraic process systems," *AIChE J.* **33**(8), 282 (1987).

Cuthrell, J. E., and Biegler, L. T., "Simultaneous optimization and solution methods for batch reactor control profiles," *Comp. and Chem. Eng.* **13**(1/2), 49–62 (1989).

de Boor, C., "A Practical Guide to Splines." Springer-Verlag, Berlin, 1978.

Douglas, J.(M., "Conceptual Design of Chemical Processes." McGraw-Hill, New York, 1988.

Dunn, J. C., "A projected Newton method for minimization problems with nonlinear inequality constraints," *Numer. Math.* **53**, 377 (1988).

Dunn, J. C., and Bertsekas, D., "Efficient dynamic programming implementations of Newton's method for unconstrained optimal control problems," *J. Opt. Theo. Applics.* **63**(1), 23 (1989).

Duran, M. A., and Grossmann, I. E., "Simultaneous optimization and heat integration of chemical processes," *AIChE J.* **32**, 123 (1986).

Eaton, J. W., and Rawlings, J. B., "Feedback control of chemical processes using on-line optimization techniques," presented at Annual AIChE Meeting, Washington, D.C. (1988).

Edgar, T. F., and Himmelblau, D. M., "Optimization of Chemical Processes." McGraw-Hill, New York, 1988.

Fletcher, R., and Xu, C., "Hybrid methods for nonlinear least squares," *IMA J. Num. Anal.* **7**, 371 (1987).

Floudas, C. A., Aggarwal, A., and Ciric, A. R., "Global optimum search for nonconvex NLP and MINLP problems," *Comp. and Chem Engr.* **13**(10), 1172 (1989).

Gabay, D. "Reduced quasi-Newton methods with feasibility improvement for nonlinearly constrained optimization," *Math. Prog. Study* **16** 18 (1982).

Gaines, L. D., and Gaddy, J. L., "Process optimization by flowsheet simulation," *I&EC Proc. Des. Dev.* **16**, 337 (1977).

Gear, C. W., Leimkuhler, B., and Gupta, G. K., "Automatic integration Euler–Lagrange equations with constraints," *Journal of Computational and Applied Mathematics* **12** and **13**, 77–90 (1985).

Gill, P. E., Murray, W., and Wright, M. H., "Practical Optimization," 6th Ed. Academic Press, New York, 1981.

Glasser, D., Hildebrandt, D., and Crowe, C., "A geometric approach to steady flow reactors: The attainable region and optimization in concentration space," *I&EC Research* **26**, 1803 (1987).

Goh, C. J., and Teo, K. L., "MISER: A FORTRAN Program for solving optimal control problems," *Adv. Eng. Software* **10**(2), 90 (1988).

Griewank, A., "On automatic differentiation," in "Mathematical Programming: Recent Developments and Applications." (M. Iri and K. Tanebe, eds.). Kluwer Academic Publishers, 1989, 83–108.

Gritsis, D., Pantelides, C. C., and Sargent, R. W. H., "The dynamic simulation of transient systems described by index two differential-algebraic equations," *Proc. Third International PSE Symposium*, Sydney, Australia, p. 132 (1988).

Gunn, D. J., and Thomas, W. J., "Mass transport and chemical reaction in multifunctional catalyst systems," *Chemical Engineering Science* **20**, 89–100 (1965).

Gurwitz, C. B., and Overton, M., "SQP methods based on approximating a projected Hessian matrix," *SIAM J. Sci. Stat. Comput.* **10**(4) 631 (1989).

Han, S.-P., "A globally convergent method for nonlinear programming," *J. Opt. Theory and Applications* **22**(3), 297 (1977).

Harsh, M. G., Saderne, P., and Biegler, L. T., "A mixed integer flowsheet optimization strategy for process retrofits: The debottlenecking problem," *Comp. and Chem. Eng.* **13**(8), 947 (1989).

Hettich, R., and Gramlich, G., "A note on an implementation of a method for quadratic semi-infinite programming," *Math. Prog.* **46**, 249 (1990).

Hindmarsh, A. C., "LSODE and LSODI, two new initial value ordinary differential equation solvers," *ACM-SIGNUM Newsletter* **15**(4), 10–11 (1980).

Hock, W., and Schittkowski, K., "Test Examples for Nonlinear Programming Codes," Lecture Notes in Economics #187. Springer, Berlin, 1981.

Holl, P., Marquardt, W., and Gilles, E. D., "DIVA—A powerful tool for dynamic process simulation," *Comp. Chem. Eng.* **12**(5), 421 (1988).

Hutchison, H. P., Jackson, D. J., and Morton, W., "The development of an equation-oriented flowsheet simulation and optimization package—II. Examples and results," *Comp. and Chem Eng.* **10**(1), 31 (1986).

Jackson, R., "Optimal use of mixed catalysts for two successive chemical reactions," *Journal of Optimization Theory and Applications* **2**(1), 27 (1968).

Jang, S. S., Joseph, B., and Mukai, H., "On-line optimization of constrained multivariable chemical processes," *AIChE J.* **33**(1), 26 (1987).

Jones, D. I., and Finch, J. W., "Comparison of optimization algorithms," *Int. J. Cont.* **40**, 747 (1984).

Kaijaluoto, S., "Process optimization by flowsheet simulation," Technical Research Center of Finland, Publication #20 (1984).

Kelley, C. T., and Sachs, E. W., "Quasi-Newton method and unconstrained optimal control problem," *SIAM J. Control & Optimiz.* **25**(6), 1503 (1987).

Kisala, T. P., Boston, J. F., Britt, H. I., and Evans, L. B., "SQP in sequential modular process optimization," *Comp. Chem. Engr.* **11**(6), 566 (1987).

Kocis, G. R., and Grossmann, E. E., "A modeling and decomposition strategy for the MINLP optimization of process flowsheets," *Comp. and Chem. Engr.* **13**, 797 (1988).

Kokossis, A. C., and Floudas, C. A., "Optimization of complex reactor networks," *Chem. Engr. Sci.* **45**(3), 595 (1990).

Kravanja, Z., and Grossmann, I. E., "PROSYN—An MINLP process synthesizer," presented at Annual AIChE Meeting, San Francisco (1989).

Kumar, A., and Lucia, A., "Separation process optimization calculations," *IMACS J. Appl. Num. Math.* **3**, 409–425 (1987).

Lang, Y.-D., and Biegler, L. T. "A unified algorithm for flowsheet optimization," *Comp and Chem. Eng.* **11**, 143 (1987).

Lang, Y.-D., Biegler, L. T., and Grossmann, I. E., "Simulataneous optimization and heat integration with process simulators," *Comp. Chem. Engr.* **12**(4), 311 (1988).

Lasdon, L. S., *IEEE Trans. Aut. Control* **AC-15**, 268 (1970).

Lasdon, L. S., Mitter, S. K., and Warren, A. D., *IEEE Trans. Aut. Control* **AC-12**, 132 (1967).

Lewis, R. M., "Definitions of order and junction conditions in singular optimal control problems," *SIAM. J. Control And Optimization* **18**(1) p .21 (1980).

Linnhoff, B., Townsend, D. W., Boland, D., Hewitt, G. F., Thomas, B. E., Guy, A. R., and Marsland, R. H., "User guide on process integration for the efficient use of energy," *Inst. Chem. Engrs.*, Rugby, U.K. (1982).

Liu, D. C., and Nocedal, J., "On the limited memory BFGS method for large scale optimization," Tech. Rep. NAM 03, Northwestern University (1988).

Liu, L.-C., Prokopakis, G. J., and Asenjo, J. A., "Optimization of enzymatic lysis of yeast," *Bioetech Bioeng.* **32**(9), 1113 (1988).

Locke, M. H., Edahl, R., and Westerberg, A. W., "An improved successive quadratic programm-

OPTIMIZATION STRATEGIES FOR COMPLEX PROCESS MODELS   255

ing optimization algorithm for engineering design problems," *AIChE J.* **29**(5), 871 (1983).

Logsdon, J. S., and Biegler, L. T., "On the accurate solution of differential-algebraic optimization problems," *I&EC Research* **28**, 1628 (1989).

Logsdon, J. S., and Biegler, L. T., "Decomposition strategies for large scale dynamic optimization problems, in press, *Chem. Eng. Sci.* (1991).

Logsdon, J. S., Diwekar, U. M., and Biegler, L. T., "On jhe simultaneous optimal design and operation of batch distillation columns," *Trans. IChE* **68A**, 1134 (1990).

Lucia, A., and Kumar, A., "Distillation optimization," *Comp. and Chem. Eng.* **12**(12), 1263–1266 (1988).

Lucia, A., and Xu, J., "Chemical process optimization using Newton-like methods," *Comp. and Chem. Engr.* **14**(2), 119 (1990).

Luus, R., and Lapidus, L., "Optimal Control of Engineering Processes." Blaisdell, Waltham, Massachusetts (1967).

Lynn, L. L., Parkin, E. S., and Zahradnik, R. L., "Near-optimal control by trajectory approximation," *I&EC Fundamentals* **9**(1), 58–63 (1970).

Mahidhara, D., and Lasdon, L., "Theory and implementation of a large-scale SQP algorithm," presented at TIMS/ORSA Meeting, New York, October, 1989.

Morison, K., "Optimal control of processes described by systems of differential-algebraic equations," Ph.D. thesis, University of London (1984).

Morshedi, A. M., "Universal dynamic matrix control," "Chemical Process Control Conf. III" (Morari and McAvoy, eds.). CACHE Corp., 1986, p. 547.

Morshedi, A. M., Lin, H. Y., and Luecke, R. H., "Rapid computation of the Jacobian matrix for optimization of nonlinear dynamic processes," *Comp. and Chem. Eng.* **10**(4), 367 (1986).

Mujtaba, I. M., and Macchietto, S., "Optimal control of batch distillation," presented at 12th IMACS World Congress, Paris, July (1988).

Murase, Akira, Roberts, Howard L., and Converse, Alvin O., "Optimal thermal design of an autothermal ammonia synthesis reactor," *I&EC Proc. Des. Dev.* **9**(4), 503 (1970).

Murray, W., and Wright, M., "Projected Lagrangian methods based on trajectories of barrier and penalty functions," SOL Report 78–23, Stanford University, Stanford, California (1978).

Murtagh, B. A., and Saunders, M. A., "Minos/Augmented Supplementary User's Manual," Report OR/78/6, The University of New South Wales, Australia (1978).

Neuman, C. P., and Sen, A., "A suboptimal control algorithm for constrained problems using cubic splines," *Automatica* **9**, 601–613 (1973).

Nickel, R. H., and Tolle, J. W., "A sparse sequential quadratic programming algorithm," *J. Opt. Theo. Applics.* **60**, 452 (1989).

Nocedal, Jorge, and Overton, Michael L., "Projected Hessian updating algorithms for nonlinearly constrained optimization," *SIAM J. Numer. Anal.* **22**(5), 821 (1985).

Oh, S. H., and Luus, R., "Use of orthogonal collocation method in optimal control problems," *Int. J. Control* **26**(5), 657 (1977).

Pantelides, C. C., "The consistent initialisation algorithm," *SIAM J. Sci. Stat. Comput.* **9**, 213 (1988a).

Pantelides, C. C., "SPEEDUP—Recent Advances in Process Simulation," *Comp. and Chem. Eng.* **13**(6), 745 (1988b).

Pantelides, C. C., Gritsis, D., Morison, K. R., and Sargent, R. W. H., "The Mathematical Modeling of Transient Systems Using Differential-Algebraic Equations," *Comp. and Chem. Eng.* **12**(5), 449 (1988).

Patwardhan, A. A., Rawlings, J. B., and Edgar, T. F., "Model predictive control of nonlinear processes in the presence of constraints," presented at annual AIChE Meeting, Washington, D.C. (1988).

Pereyra, V., and Sewell, E. G., "Mesh selection for discrete solution of boundary problems in

ordinary differential equations," *Numer. Math.* **23**, 261–268 (1975).
Petzold, L. R., "Differential/algebraic equations are not ODEs," *SIAM Journal on Scientific and Statistical Computing*, No. 3, pp. 367–385 (1982).
Petzold, L. R., "Order results for implicit Runge–Kutta methods applied to differential/algebraic," *SIAM Journal on Numerical Analysis*, No. 4, pp. 837–852 (1986).
Piela, P. C., Epperly, T. G., Westerberg, K. M., and Westerberg, A. W., "ASCEND: An object oriented computer environment for modeling and analysis, *Comp. and Chem. Eng.* **15**(1), 53 (1991).
Powell, M. J. D., "A fast algorithm for nonlinearly constrained optimization calculations," paper presented at the Dundee Conf. on Numerical Analysis, Dundee, Scotland (1977).
Powell, M. J. D., "The convergence of variable metric methods for nonlinearly constrained optimization calculations," *in* Nonlinear Programming 3" (Mangasarian, O. L., Meyer, R., Robinson, S., eds.). Academic Press, New York, 1978.
Reddien, G. W., "Collocation at Gauss points as a discretization in optimal control," *SIAM J. Cont. Opt.* **17**(2), 298 (1979).
Renfro, J. G., Ph.D. thesis, University of Houston, Houston (1986).
Renfro, J. G., Morshedi, A. M., and Asbjornsen, O. A., "Simultaneous optimization and solution of systems described by differential/algebraic equations," *Comp. and Chem. Eng.* **11**(5), 503–517 (1987).
Rosenbrock, H. H., and Storey, C., "Computational Techniques for Chemical Engineers." Pergamon, New York, 1966.
Russell, R. D., and Christiansen, J., "Adaptive mesh selection strategies for solving boundary value problems," *SIAM J. Numer. Anal.* **15**(1), 59–80 (1978).
Sargent, R. W. H., and Sullivan, G. R., "The development of an efficient optimal control package," Proceedings of the 8th IFIP Conference on Optimization Techniques, Pt. 2 (1977).
Sargent, R. W. H., and Sullivan, G. R., "Development of feed changeover policies for refinery distillation units," *I&EC Proc. Des. Dev.* **18**(1), 113 (1979).
Seader, J. D., Seider, W. D., and Pauls, A. C., "FLOWTRAN Simulation—An Introduction," 3rd Ed. CACHE Corp., 1987.
Tjoa, I.-B., and Biegler, L. T., "Simultaneous solution and optimization strategies for parameter estimation of differential-algebraic equations systems," *I&EC Research*, **30**, 376 (1991).
Tsang, T. H., Himmelblau, D. M., and Edgar, T. F., "Optimal control via collocation and nonlinear programming," *Int. J. Control* **21**(5), 763–768 (1975).
Vasantharajan, S., Ph.D. thesis, Carnegie Mellon University, Pittsburgh (1989).
Vasantharajan, S., and Biegler, L. T., "Large-scale decomposition strategies for successive quadratic programming," *Comp. and Chem. Eng.* **12**(11), 1087–1101 (1988a).
Vasantharajan, S., and Biegler, L. T., "Simultaneous solution of reactor models within flowsheet optimization," *Chem. Eng. Res. Des.* **66**, 396 (1988b).
Vasantharajan, S., and Biegler, L. T., "Simultaneous parameter optimization of differential-algebraic systems," *Computers and Chemical Engineering* **14**(10), 1083–1100 (1990).
Vasantharajan, S., Viswanathan, J., and Biegler, L. T., "Reduced SQP implementation for large scale optimization problems," *Comp. and Chem. Engr.* **14**(8), 907–917 (1990).
Villadsen, J., and Michelsen, M. L., "Solution of Differential Equation Models by Polynomial Approximation." Prentice-Hall Inc., New York, 1978.
Westerberg, A. W., "The synthesis of distillation based separation systems," *Comp. Chem. Engrs.* **9**, 421 (1985).
Wright, M. H., Ph.D. thesis, Stanford University, Stanford, California (1976).
Wright, S., "Solution of discrete-time optimal control problems on parallel computers," preprint MCS-P89-0789, Argonne National Lab, Argonne, Illinois (1989).

INDEX

A

Absorption spectroscopy, 25
Accomodation coefficient, 65
Activity coefficient, measurement, 68
Adjoint equations, 219
Adsorption, 173
Aerocolloidal particles, 1
Aerodynamic particle size, 18
Aerosol
  chemical reaction, 12
  forces on
    aerodynamic drag, 16, 17, 18
    diffusiophoresis, 23
    photophoresis, 23, 24, 25
    radiation pressure, 23, 28
    thermophoresis, 23
  measurements
    charge, 12, 14, 19
    mass, 12, 13
    size, 18, 19
Ammonia process flowsheet, 210
Ammonia synthesis reactor, 226
AM1, 110
Approximation error estimates, 223
Aqueous solution droplets, 54
ASCEND, 208, 251
ASPEN, 213
Association reactions, 149
Atomic radii, 128
Automatic differentiation, 251
Automatic reformulation, 251

B

β-scission rule, 140
Basis sets, 108
Bimolecular reactions, 144
Block diagram, 209

Bond additivity, 113
Bond dissociation energies, 124
Bond-energy bond-order (BEBO) method, 147

C

Catalyst mixing problem, 242
Chainruling, 246
Charge loss, particles, 21
Chemical reaction processes, 132
Coated droplets, 67
Complex unimolecular reactions, 139
Condensation, 55
Constraints, 199
Continuous variables, 199
Control profiles, 218
Cyclization reactions, 143

D

DCKM (detailed chemical kinetic modeling), 96
Decomposition of CH_3Cl, 176
Desorption, 174
Differential-algebraic equations (DAE), 216
  index of DAE system, 240
Diffusion coefficient, measurement, 57, 61
Diffusion-controlled evaporation, 55
Droplet
  charge-to-mass ratio, 12
  explosion, 21

E

Electrodynamic balance
  balance constant, 11

257

Electrodynamic balance (*Continued*)
  bihyperboloidal electrode configuration, 6, 7
  double ring electrodes, 27
  stability, 7
Electrodynamic thermogravimetric analyzer, 75, 79
Electron stepping, 14
Electromagnetic energy transfer, 45, 75
Electromagnetic heating, 75
Electrostatic balance, 3, 4
Energy transfer limited reactions, 160
Equation oriented simulation mode, 207
Equidistribution, 224
Evaporation coefficient, 65
Evaporation, droplet
  binary system, 64, 65
  immiscible system, 65, 66, 67
  multicomponent, 63
  pure component, 55

F

Fall-off parameters, 165, 166
Feasible path, 217
Finite elements, 222, 245
Flowsheet optimization, 207
FLOWTRAN, 210

G

GAMESS, 110
GAMS, 251
Gas/particle chemical reaction, 81
GAUSSIAN, 110
Gaussian type orbitals (GTO), 108
Group additivity, 115

H

Hartree–Fock equation, 107
Heat integration and flowsheet optimization, 213
Heat source function, 46, 47
Heat transfer, 55, 72
Heterogeneous reactions, 172
Hirschfelder's rules, 147
Hydrogenlike atom wave functions, 103

I

Implicit Runge–Kutta (IRK) methods, 239
Integer variables, 199

K

Kuhn–Tucker conditions, 200

L

Lagrange polynomials, 222
Large-scale optimization, 203, 244
Legendre polynomials, 222
Lennard–Jones potential, 58
Light scattering
  cross-sections, 39
    absorption, 40, 45
    extinction, 39
    scattering, 40
  elastic, 32
    Mie theory, 33
    morphological resonance, 41
    phase function, 41, 43
    Rayleigh scattering, 33
  inelastic, 47
    fluorescence, 48
    phosphorescence, 47
    photoluminescence, 47
    Raman
Lindemann–Hinshelwood analysis, 162
LSODE, 231

M

Marginal stability state, 10
Mass transfer, droplet, 55, 72
Metathesis reactions, 145
Methods of weighted residuals (MWR), 220
Methyl radical addition, 159
Microparticle spectroscopy, 81
  fluorescence, 48
  optical resonance, 81
  photophoretic, 25
  Raman, 51, 84
Mie theory, 33
Millikan condenser, 3
MINOS, 205

INDEX

Mixed integer linear programming, (MILP), 199
Mixed integer nonlinear programming (MINLP), 199, 213
MNDO, 110
Modular simulation mode, 208
MO-LCAO method, 106
Molecular orbital theory, 103, 106

N

Negative activation energies, 151
Neglect of differential overlap (NDO), 109
Newton-type method, 201
Non-Arrhenius behavior, 146
Nonlinear programming (NLP), 198
Null space, 203

O

Objective function, 199
Optical resonance spectroscopy, 81
Optical gradient trap, 29, 30
Optimal control, 238
Optimal reactor profiles, 230
Optimization, 198
Orthogonal collocation, 214, 222

P

Parallel band solver, 249
Parallel computing, 249
Parameter estimation, 206
Parameter optimization, 225
Parameter vector, 218
Particle traps, 3
Photochemical reaction, 49
Photophoretic spectrum, 25–27
Pinch technology, 198
Polanyi relationships, 147
Process optimization, 197
Projected Hessian, 204

Q

QRRK analysis, 167
Quadratic programming (QP), 201
Quadrupole trap, 5, 79
Quantum chemistry, 101

QUASILIN, 208
Quasi-Newton updates, 201

R

Radiation pressure, 29, 39
Radicals, 122
Radioactive aerosol, 15
Raman spectroscopy, 51, 84
Range space, 203
Rate constants for prototype reactions, 136
Rayleigh limit of charge, 19, 21
Rayleigh scattering, 32
Reactor optimization, 214
Refractive index, measurement, 79
Resonance stabilization, 123
Retrofit optimization, 213
Riccati transformation, 248
RRKM analysis, 164–165

S

Scattering coefficients, 36
Scattering functions, 38
Schrödinger's equation, 102
SCOPT, 211
Sensitivity analysis, 98
Sensitivity equations, 219
Sequential modular, 208
Silane decomposition, 154
Simple fission reactions, 136
Simultaneous modular, 208
Singular control arc, 242
Slater-type orbitals (STO), 108
Solution thermodynamics, 71
SPEEDUP, 208, 220
Spring point voltage, 11
State constrained reactor profiles, 236
State profiles, 218
Statistical mechanics, 115, 118
Sticking coefficient, 173
Stiff differential equations, 97
Strain energy, 123
Structural resonance spectrum, 37
Successive quadratic programming (SQP), 201
  decomposition, 203
  sparse SQP methods, 203
Symmetry numbers, 114

T

Thermochemical sources, 97
Thermochemistry, 111
Troe's formalism, 165

U

Unimolecular reactions, 134

V

Vapor pressure, measurement, 57, 60, 61
Variational methods, 217

Z

Z-matrix, 130

Contents of Volumes in This Serial

Volume 1

J. W. Westwater, *Boiling of Liquids*
A. B. Metzner, *Non-Newtonian Technology: Fluid Mechanics, Mixing, and Heat Transfer*
R. Byron Bird, *Theory of Diffusion*
J. B. Opfell and B. H. Sage, *Turbulence in Thermal and Material Transport*
Robert E. Treybal, *Mechanically Aided Liquid Extraction*
Robert W. Schrage, *The Automatic Computer in the Control and Planning of Manufacturing Operations*
Ernest J. Henley and Nathaniel F. Barr, *Ionizing Radiation Applied to Chemical Processes and to Food and Drug Processing*

Volume 2

J. W. Westwater, *Boiling of Liquids*
Ernest F. Johnson, *Automatic Process Control*
Bernard Manowitz, *Treatment and Disposal of Wastes in Nuclear Chemical Technology*
George A. Sofer and Harold C. Weingartner, *High Vacuum Technology*
Theodore Vermeulen, *Separation by Adsorption Methods*
Sherman S. Weidenbaum, *Mixing of Solids*

Volume 3

C. S. Grove, Jr., Robert V. Jelinek, and Herbert M. Schoen, *Crystallization from Solution*
F. Alan Ferguson and Russell C. Phillips, *High Temperature Technology*
Daniel Hyman, *Mixing and Agitation*
John Beek, *Design of Packed Catalytic Reactors*
Douglass J. Wilde, *Optimization Methods*

Volume 4

J. T. Davies, *Mass-Transfer and Interfacial Phenomena*
R. C. Kintner, *Drop Phenomena Affecting Liquid Extraction*
Octave Levenspiel and Kenneth B. Bischoff, *Patterns of Flow in Chemical Process Vessels*
Donald S. Scott, *Properties of Cocurrent Gas–Liquid Flow*
D. N. Hanson and G. F. Somerville, *A General Program for Computing Multistage Vapor–Liquid Processes*

Volume 5

J. F. Wehner, *Flame Processes–Theoretical and Experimental*
J. H. Sinfelt, *Bifunctional Catalysts*
S. G. Bankoff, *Heat Conduction or Diffusion with Change of Phase*
George D. Fulford, *The Flow of Liquids in Thin Films*
K. Rietema, *Segregation in Liquid–Liquid Dispersions and Its Effect on Chemical Reactions*

Volume 6

S. G. Bankoff, *Diffusion-Controlled Bubble Growth*
John C. Berg, Andreas Acrivos, and Michel Boudart, *Evaporation Convection*
H. M. Tsuchiya, A. G. Fredrickson, and R. Aris, *Dynamics of Microbial Cell Populations*
Samuel Sideman, *Direct Contact Heat Transfer between Immiscible Liquids*
Howard Brenner, *Hydrodynamic Resistance of Particles at Small Reynolds Numbers*

Volume 7

Robert S. Brown, Ralph Anderson, and Larry J. Shannon, *Ignition and Combustion of Solid Rocket Propellants*
Knud Østergaard, *Gas–Liquid–Particle Operations in Chemical Reaction Engineering*
J. M. Prausnitz, *Thermodynamics of Fluid–Phase Equilibria at High Pressures*
Robert V. Macbeth, *The Burn-Out Phenomenon in Forced-Convection Boiling*
William Resnick and Benjamin Gal-Or, *Gas–Liquid Dispersions*

Volume 8

C. E. Lapple, *Electrostatic Phenomena with Particulates*
J. R. Kittrell, *Mathematical Modeling of Chemical Reactions*
W. P. Ledet and D. M. Himmelblau, *Decomposition Procedures for the Solving of Large Scale Systems*
R. Kumar and N. R. Kuloor, *The Formation of Bubbles and Drops*

Volume 9

Renato G. Bautista, *Hydrometallurgy*
Kishan B. Mathur and Norman Epstein, *Dynamics of Spouted Beds*
W. C. Reynolds, *Recent Advances in the Computation of Turbulent Flows*
R. E. Peck and D. T. Wasan, *Drying of Solid Particles and Sheets*

Volume 10

G. E. O'Connor and T. W. F. Russell, *Heat Transfer in Tubular Fluid–Fluid Systems*
P. C. Kapur, *Balling and Granulation*
Richard S. H. Mah and Mordechai Shacham, *Pipeline Network Design and Synthesis*
J. Robert Selman and Charles W. Tobias, *Mass-Transfer Measurements by the Limiting-Current Technique*

Volume 11

Jean-Claude Charpentier, *Mass-Transfer Rates in Gas–Liquid Absorbers and Reactors*
Dee H. Barker and C. R. Mitra, *The Indian Chemical Industry–Its Development and Needs*
Lawrence L. Tavlarides and Michael Stamatoudis, *The Analysis of Interphase Reactions and Mass Transfer in Liquid–Liquid Dispersions*
Terukatsu Miyauchi, Shintaro Furusaki, Shigeharu Morooka, and Yoneichi Ikeda, *Transport Phenomena and Reaction in Fluidized Catalyst Beds*

Volume 12

C. D. Prater, J. Wei, V. W. Weekman, Jr., and B. Gross, *A Reaction Engineering Case History: Coke Burning in Thermofor Catalytic Cracking Regenerators*
Costel D. Denson, *Stripping Operations in Polymer Processing*
Robert C. Reid, *Rapid Phase Transitions from Liquid to Vapor*
John H. Seinfeld, *Atmospheric Diffusion Theory*

Volume 13

Edward G. Jefferson, *Future Opportunities in Chemical Engineering*
Eli Ruckenstein, *Analysis of Transport Phenomena Using Scaling and Physical Models*
Rohit Khanna and John H. Seinfeld, *Mathematical Modeling of Packed Bed Reactors: Numerical Solutions and Control Model Development*
Michael P. Ramage, Kenneth R. Graziani, Paul H. Schipper, Frederick J. Krambeck, and Byung C. Choi, *KINPTR (Mobil's Kinetic Reforming Model): A Review of Mobil's Industrial Process Modeling Philosophy*

Volume 14

Richard D. Colberg and Manfred Morarı, *Analysis and Synthesis of Resilient Heat Exchanger Networks*
Richard J. Quann, Robert A. Ware, Chi-Wen Hung, and James Wei, *Catalytic Hydrodemetallation of Petroleum*
Kent Davis, *The Safety Matrix: People Applying Technology to Yield Safe Chemical Plants and Products*

Volume 15

Pierre M. Adler, Ali Nadim, and Howard Brenner, *Rheological Models of Suspensions*
Stanley M. Englund, *Opportunities in the Design of Inherently Safer Chemical Plants*
H. J. Ploehn and W. B. Russel, *Interactions between Colloidal Particles and Soluble Polymers*

Volume 16

Perspectives in Chemical Engineering: Research and Education

Clark K. Colton, *Editor*

Historical Perspective and Overview

L. E. Scriven, *On the Emergence and Evolution of Chemical Engineering*
Ralph Landau, *Academic–Industrial Interaction in the Early Development of Chemical Engineering*
James Wei, *Future Directions of Chemical Engineering*

Fluid Mechanics and Transport

L. G. Leal, *Challenges and Opportunities in Fluid Mechanics and Transport Phenomena*
William B. Russel, *Fluid Mechanics and Transport Research in Chemical Engineering*
J. R. A. Pearson, *Fluid Mechanics and Transport Phenomena*

Thermodynamics

Keith E. Gubbins, *Thermodynamics*
J. M. Prausnitz, *Chemical Engineering Thermodynamics: Continuity and Expanding Frontiers*
H. Ted Davis, *Future Opportunities in Thermodynamics*

Kinetics, Catalysis, and Reactor Engineering

Alexis T. Bell, *Reflections on the Current Status and Future Directions of Chemical Reaction Engineering*
James R. Katzer and S. S. Wong, *Frontiers in Chemical Reaction Engineering*
L. Louis Hegedus, *Catalyst Design*

Environmental Protection and Energy

John H. Seinfeld, *Environmental Chemical Engineering*
T. W. F. Russell, *Energy and Environmental Concerns*
Janos M. Beer, Jack B. Howard, John P. Longwell, and Adel F. Sarofim, *The Role of Chemical Engineering in Fuel Manufacture and Use of Fuels*

Polymers

Matthew Tirrell, *Polymer Science in Chemical Engineering*
Richard A. Register and Stuart L. Cooper, *Chemical Engineers in Polymer Science: The Need for an Interdisciplinary Approach*

Microelectronic and Optical Materials

Larry F. Thompson, *Chemical Engineering Research Opportunities in Electronic and Optical Materials Research*
Klavs F. Jensen, *Chemical Engineering in the Processing of Electronic and Optical Materials: A Discussion*

Bioengineering

James E. Bailey, *Bioprocess Engineering*
Arthur E. Humphrey, *Some Unsolved Problems of Biotechnology*
Channing Robertson, *Chemical Engineering: Its Role in the Medical and Health Sciences*

Process Engineering

Arthur W. Westerberg, *Process Engineering*
Manfred Morari, *Process Control Theory: Reflections on the Past Decade and Goals for the Next*
James M. Douglas, *The Paradigm After Next*
George Stephanopoulos, *Symbolic Computing and Artificial Intelligence in Chemical Engineering: A New Challenge*

The Identity of Our Profession

Morton M. Denn, *The Identity of Our Profession*

Volume 17

Y. T. Shah, *Design Parameters for Mechanically Agitated Reactors*
Mooson Kwauk, *Particulate Fluidization: An Overview*

Volume 18

E. James Davis, *Microchemical Engineering: The Physics and Chemistry of the Microparticle*
Selim M. Senkan, *Detailed Chemical Kinetic Modeling: Chemical Reaction Engineering of the Future*
Lorenz T. Biegler, *Optimization Strategies for Complex Process Models*